BIOLOGICAL PHYSICS SERIES

_____BIOLOGICAL PHYSICS SERIES

Continued after index

George B. Benedek
Felix M.H. Villars

Physics With Illustrative Examples From Medicine and Biology

MECHANICS   *Second Edition*

Foreword by Irving M. London

With 227 Illustrations

George B. Benedek
Department of Physics
Massachusetts Institute of Technology
Cambridge, MA 02139-4037
USA

Felix M.H. Villars
Department of Physics
Massachusetts Institute of Technology
Cambridge, MA 02139-4037
USA

Cover illustration: (Top right) forces acting at the hip joint; (bottom right) the centrifugal pendulum: feedback and control systems; (bottom left) virus particle subject to forces in a centrifuge.

Library of Congress Cataloging-in-Publication Data
Benedek, George Bernard, 1928–
    Physics with illustrative examples from medicine and biology/George B. Benedek, Felix
M.H. Villars. — 2nd ed.
        p. cm.
    Contents: v. 1. Mechanics — v. 2. Statistical physics — v. 3. Electricity and magnetism.
    ISBN 978-1-4612-7051-5   ISBN 978-1-4612-1228-7 (eBook)
    DOI 10.1007/978-1-4612-1228-7
    1. Physics. 2. Medicine. 3. Biology. I. Villars, Felix, 1921– . II. Title.
QC23 .B424 2000
530—dc21   99-033040

Printed on acid-free paper.

Production managed by Frank McGuckin; manufacturing supervised by Joe Quatela.
Typeset by Integre Technical Publishing Co., Inc., Albuquerque, NM.

9 8 7 6 5 4 3 2 1

ISBN 978-1-4612-7051-5   SPIN 10708650

Foreword to the First Edition

During the past several decades the biological and medical sciences have flourished with the growth of knowledge of the chemistry of living matter. Ours is a biochemical era. Increasingly, however, biology and medicine are dependent on the physical sciences for a deeper and more quantitative understanding of important biomedical problems and for the development of successful solutions to these problems. Accordingly, students of the life sciences, and specifically of human biology and medicine, have a growing need to acquire an effective working knowledge of the physical sciences.

To help meet this need, Professors George Benedek and Felix Villars have developed a rigorous and, indeed, challenging introductory course in physics with illustrative content derived from biology and medicine. The course not only provides an excellent means for the study of physics, but it also demonstrates the value and utility of the quantitative approach of the physicist in affording illuminating insights into a wide range of important biological problems. This book and the course on which it is based have been sponsored by the Harvard–MIT Program in Health Sciences and Technology. One of the objectives of the Program is the advancement of the health sciences by promoting their productive interaction with the physical sciences and engineering. This course of study is designed to help meet this objective. The enthusiastic reception of the course at MIT prompted the recommendation that Professors Benedek and Villars make it available to a wider audience. They have responded, fortunately, by the writing of this book.

Gassin, France   IRVING M. LONDON
July 1973

v

Series Preface

The field of biological physics is a broad, multidisciplinary, and dynamic one, touching on many areas of research in physics, biology, chemistry and medicine. New findings are published in a large number of publications within these disciplines, making it difficult for students and scientists working in biological physics to keep up with advances occurring in disciplines other than their own. The Biological Physics Series is intended therefore to be a comprehensive one covering a broad range of topics important to the study of biological physics. Its goal is to provide scientists and engineers with text books, monographs and reference books to address the growing need for information.

Books in the Biological Physics Series will emphasize frontier areas of science including molecular, membrane, and mathematical biophysics; photosynthetic energy harvesting and conversion; information processing; physical principles of genetics; sensory communications; automata networks, neural networks, and cellular automata. Equally important will be coverage of current and potential applied aspects of biological physics such as biomolecular electronic components and devices, biosensors, medicine, imaging, physical principles of renewable energy production, and environmental control and engineering.

We are fortunate to have a distinguished roster of consulting editors on the Editorial Board, reflecting the breadth of biological physics. We believe that the Biological Physics Series can help advance the knowledge in the field by providing a home for publications in the field and that scientists and practitioners from many disciplines will find much to learn from the upcoming volumes.

Oak Ridge, Tennessee   ELIAS GREENBAUM
Series Editor-in-Chief

vii

Preface to the Second Edition

Already in the early 1970's it was becoming evident that the biological and medical sciences were on the threshold of major new advances. It was also clear that the promotion of these advances required interdisciplinary interactions between the physical sciences and engineering on the one hand and the biological and medical sciences on the other. To facilitate such interactions, the Harvard–MIT Division of Health Sciences and Technology was created. Under the supportive auspices of this unique inter-university structure, we created, in the period 1973 to 1979, the first edition of this three-volume text. Its purpose was to teach students the principles of physics using examples and applications taken from biology and medicine. By this means we sought to help educate scientists who could work effectively at the fruitful interface between the physical and biological disciplines.

The first edition of these texts, which were soft cover photocopies of a type-written manuscript, illustrated by the authors, ultimately was allowed to go out of print in January of 1990. However, the extraordinary advances at the interface between the physical and biological sciences have continued to accelerate. Increasingly, science and engineering departments are recognizing the need to train students and faculty to be capable of research and teaching at this active interface. As a result, growing numbers of our colleagues, who had found these texts to be uniquely useful, have urgently requested that they be made available once again.

This increasing need was clearly understood by our editor, Maria Taylor. At her initiative, the American Institute of Physics and Springer-Verlag have undertaken to publish this second edition. We reviewed our original text and, even though our colleagues still accorded it high value, we decided to update various sections. We also have included supplementary references so as to help make connections with the current literature.

We hope that this beautifully printed, bound, and illustrated second edition will attract the careful attention and interest of yet a new generation of intellectually adventurous students and teachers.

Cambridge, Massachusetts   GEORGE B. BENEDEK
January 2000   FELIX M.H. VILLARS

Acknowledgments for the Second Edition

A renewed publication more than twenty years after the original issue can only take place with the aid of far-sighted supporters. We are grateful to Maria Taylor, our editor, whose vision and advocacy was essential to initiate the entire project. All the effort associated with the preparation of an updated, accurate second edition required reallocation of the authors' time from their normal academic teaching responsibilities. The authors wish to express their appreciation to Professors Martha Gray, and Joseph Bonventre, the Co-Directors of the Harvard–MIT Division of Health Sciences and Technology for their continued support of our teaching and writing efforts. Finally, we wish to give thanks to those many colleagues, both in the United States and abroad who have closely studied these books, understood their value and requested its renewed availability. Their urgent requests led to the reappearance of our work in the form of this Second Edition.

Cambridge, Massachusetts   GEORGE B. BENEDEK
January 2000   FELIX M.H. VILLARS

Preface to the First Edition

This is a unique book. It is an introductory textbook of physics in which the development of the principles of physics is interwoven with the quantitative analysis of a wide range of biological and medical phenomena. Conversely, the biological and medical examples serve to vitalize and motivate the learning of physics. By its very nature, this book not only teaches physics, but also exposes the student to topics in fields such as anatomy, orthopedic medicine, physiology, and the principles of homeostatic control.

This book, and its follow-up, Volumes II and III, grew out of an introductory physics course which we have offered to freshman and sophomores at MIT since 1970. The stimulus for this course came from Professor Irving M. London, MD, Director of the Harvard–MIT Program in Health Sciences and Technology. He convinced us that continued advances in the biological and medical sciences demand that students, researchers, and physicians should be capable of applying the quantitative methods of the physical sciences to problems in the life sciences. We have written this book in the hope that students of the life sciences will come to appreciate the value of training in physics in helping them to formulate, analyze, and solve problems in their own fields.

Also, our own experience as teachers of physics convinces us that the continued attractiveness of physics as an undergraduate course requires that the principles of physics be applied at the earliest stages of study to an understanding of the directly observable phenomena of our daily lives. We very much want more students to know physics, to use physics, and also to have a better understanding of how physical principles were discovered. In this connection, we present a few of these discoveries by quoting directly the words of the scientists, in order to illustrate their mode of thought and their deep, human commitment to science.

We believe that this book can be used in many different ways. For students who are concurrently taking a course in elementary differential and integral calculus, or who have had a high school course in calculus, this book can serve as a textbook in freshman physics. At MIT we have taught this course with considerable success to freshmen with career objectives in the life sciences. Many of these students initially had little interest in physics and mathematics, but later came to enjoy these subjects when they saw how effectively they clarified biological and medical problems.

There are, of course, places in this book at which the mathematical matter used may not be in phase with the concurrent calculus course. We believe that extra help from the instructor can guide students over or around these occasional steep places and lead them to more comfortable and gently paced slopes. It is perhaps worth mentioning that, from a strictly technical point of view, there are no calculations used, either in the text or in the problems, that require more than the ability to differentiate and integrate the functions $\sin x$, $\cos x$, e^x, and x^n.

This text could be used in conjunction with a traditional college physics text-book.

Medical school faculties who wish to infuse the physical sciences into their courses in the basic medical sciences may wish to use this book as part of the MD curriculum in the preclinical years.

Students with career interests in bioengineering will find this book valuable in learning physics and quantiative physiology. Their attention is drawn particularly to Chapter 6 on Feedback and Control theory.

Finally, this book can also serve as a source book, or a reader, for the teacher of physics, biology, or medicine for it constitutes a storehouse of many medical and biological examples ready for use in the enrichment of the educational diet of his students.

The experienced teacher will recognize that this book contains more material than can normally be covered in a single semester. We encourage him to make selections (and perhaps even rearrange the order of presentation) so as to best serve his own purposes.

This book is the first volume of a three-volume series which represents, in part, the content of the introductory sequence in physics we offer students at MIT. Volume II is entitled *Statistical Physics: With Illustrative Examples from Medicine and Biology*. Volume III is entitled *Electricity and Magnetism: With Illustrative Examples from Medicine and Biology*.

Cambridge, Massachusetts   GEORGE B. BENEDEK
August 1973   FELIX M.H. VILLARS

Acknowledgments for the First Edition

The authors wish to express their deep gratitude to Irving M. London, MD, Director of the Harvard–MIT Program in Health Sciences and Technology, who provided the initial stimulus for this effort, and gave us his constant support, encouragement, and penetrating understanding. Dr. Phillip Aisen, a gifted clinician with great knowledge of physics and mathematics, played a vital role in educating G.B.B., and in identifying useful medical examples at the outset of this project. Dr. Jerome Wiesner, then Provost of MIT, at a pivotal meeting, gave the crucial support needed to obtain committee approval to go ahead with this course. Professor J. D. Litster, who also taught this course, made a number of important contributions to Chapter 6 and has supplied many excellent homework problems.

Finally, we wish to express our profound appreciation to Miss Mary Patton Fitzgerald, our exceptional secretary. Despite very heavy commitment to other duties, she typed this manuscript with great speed and accuracy, and is responsible for its very lucid format.

Cambridge, Massachusetts George B. Benedek
August 1973 Felix M.H. Villars

Acknowledgements for the First Edition

Contents

CHAPTER 1

Kinematics

1.1 Introduction

Kinematics is the quantitative description of motion. In the case of a particle which moves along some trajectory in space, kinematics enables us to describe precisely how the position, the velocity, and the acceleration of the particle changes in time. In the case of a solid body, kinematics gives us a mathematical description of the location and the orientation of the body as it moves, tumbles, and spins through space. Kinematics *describes* motion accurately without concerning itself with the sources of that motion. Kinematics is the anatomy of mechanics.

Once we know quantitatively the motion of a body, be it a planet or a paramecium, we can begin to discover the effect of the forces which actuate that motion. Similarly, the physiologist needs quantitative anatomy (morphometry) in order to learn the basis of organ function.

Today many of us have, as part of our education, the powerful ideas of differential and integral calculus. Within this framework we can easily understand and use kinematics. It can be difficult for us to appreciate that the quantitative description of motion was a profound mystery to man until the extraordinary discoveries of Galileo Galilei (1564–1642). It is a great testimony to the unity of knowledge that Galileo's discoveries in kinematics and astronomy shook the very foundations of the order in his society. He became a threat to the church and was made a prisoner of the Inquisition. To those who feel that science today is a challenge to our society, it is well to remember that others felt so in the seventeenth century.

The mystery of motion was not just a symptom of the Dark Ages of Medieval Europe. The brilliant Greek philosophers and mathematicians also had grave difficulties with the quantitative description of motion. The paradoxes of Zeno of Elea (495–435 B.C.) are particularly telling examples of their difficulties. One of the best known of the paradoxes of Zeno is the racetrack paradox (which is essentially the same as the familiar paradox of Achilles and the tortoise). It is useful to re-count and solve the paradox here because it illustrates how one can go wrong by jumping to a conclusion even in an apparently quantitative analysis. The solution of the paradox illustrates how important it is to be as quantitative as possible to the very end of your reasoning. The paradox can be expressed as follows. A runner can never reach the end of the racetrack because he must first cover one-half of the track ($\frac{1}{2}$). He then must cover half of what remains ($\frac{1}{4}$), and then half of that part still remaining, *and so ad infinitum*. The final phrase contains the essential point in the paradox for it implicitly expresses the view that it must take an infinite amount of time to pass an infinite number of line segments. The quantitative essence of the difficulty can be exposed and the paradox resolved quite easily by expressing it quantitatively, with the aid of Figure 1.1.

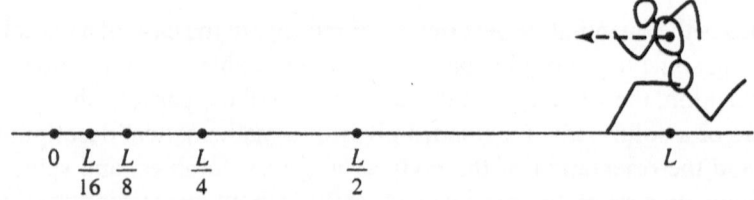

Figure 1.1. Runner reaching his goal in a finite time.

The runner runs with speed V. The length of the track is L. The time, T, needed to reach the end of the track is the sum of the times needed to pass each of the successive halfway marks. If the speed of the runner is constant, since speed is the distance traveled in each unit of time, we have that

$$T = \frac{(L/2)}{V} + \frac{(L/4)}{V} + \frac{(L/8)}{V} + \cdots,$$

$$T = \frac{L}{V}\left(\frac{1}{2} + \frac{1}{4} + \frac{1}{8} + \cdots\right).$$

Here we see clearly exposed the essence of the difficulty. The Greeks felt that the sum of the infinite number of finite numbers had to be infinite. While this appears to be reasonable, any high school student knows that this is not generally true. For example, the sum $S = 0.3 + 0.03 + 0.003 + 0.0003 + 0.00003 + \cdots$ is not only finite but is equal precisely to one-third. In the equation for the total elapsed time for the runner the sum is quite finite and is in fact equal exactly to unity. You can convince yourself by computing the sum for any finite number of terms, i.e.,

$$S_n = \left(\frac{1}{2} + \frac{1}{4} + \frac{1}{8} + \cdots + \frac{1}{2^n} \right).$$

Do this for $n = 1$, $S_1 = \frac{1}{2}$; $n = 2$, $S_2 = (\frac{1}{2} + \frac{1}{4}) = \frac{3}{4}$; $n = 3$, $S_3 = (\frac{1}{2} + \frac{1}{4} + \frac{1}{8}) = \frac{7}{8}$; and you can quickly convince yourself that the sum after n terms is equal exactly to

$$S_n = \left(1 - \frac{1}{2^n} \right).$$

Thus in the case of an infinite number of terms $S_\infty = 1$ and $T = (L/V)$. The paradox is resolved.

For us the resolution was simple because we know how to deal with infinite series. For the Greeks this was not so, and the mathematical description of motion presented very grave obstacles. Of course, with the aid of calculus, the solution to this problem is so simple that it is hard to imagine that it could cause any difficulty.

1.2   Motion in One Dimension: Velocity and Acceleration

1.2.A.   Velocity

In the case that a particle moves along a straight line the instantaneous velocity of the particle is the time rate of change of its position. In the notation of calculus

$$V = \left(\frac{dr}{dt} \right). \tag{1-1}$$

Here r is the position of the particle, t is the time, and V is the instantaneous velocity.

The calculus enables us to solve the racetrack paradox with trivial ease. The time required for the particle to move between r and $r + \Delta r$ is

$$\Delta t = \frac{\Delta r}{V}.$$

The time T required to move the entire racetrack distance from 0 to L is the sum of the times to cover all the small distances Δr, i.e.,

$$T = \sum \Delta t = \sum \left(\frac{\Delta r}{V} \right).$$

Going now to the limit where Δt and Δr become differentials the finite sums go over into integrals and we have

$$T = \int_{r=0}^{r=L} \frac{dr}{V}.$$

In the case that the velocity is a constant independent of the position of the particle the integral is trivial and we have at once

$$T = \frac{L}{V}.$$

The integral calculus in fact gives us the sum of the racetrack infinite series immediately. In general, the time required for a particle to move a certain distance is given by

$$T = \int_{t=0}^{t=T} dt = \int_{r_i}^{r_f} \frac{dr}{V}. \tag{1-2}$$

Here T is the elapsed time, r_i and r_f are the initial and final positions, and V is the velocity, whose variation along the path must be known in order to calculate the integral. Alternatively, if one desires to know the distance traveled when the elapsed time and velocity are known, we use the definition of the velocity to obtain

$$(r_f - r_i) = \int_{r_i}^{r_f} dr = \int_{t=0}^{t=T} V \, dt. \tag{1-3}$$

Here, of course, one needs to know the velocity as a function of time to compute the integral.

In the simple case of constant velocity, of course we have the result

$$(r_f - r_i) = Vt. \tag{1-4}$$

Examples.   A few elementary examples of motion will be helpful in clarifying the use and role of velocity.

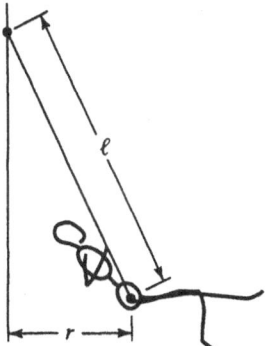

Figure 1.2. Child on a playground swing.

(a) Let us imagine a child swinging gently back and forth on a playground swing, as indicated in Figure 1.2. Let us imagine that the position of the swing, as measured from the vertical line, is r, and that r varies sinusoidally with time, i.e.,

$$r(t) = r_0 \sin \frac{2\pi t}{T}.$$

At $T = 0$ the position r is zero; at $t = T/4$ the swing is farthest out to the right; at $t = T/2$ the swing is back at the origin; at $t = 3T/4$ the swing is furthest to the left; at $t = T$ the swing returns to $r = 0$. The entire cycle is complete after a period of time T, and so T is called the period of the motion.

The velocity of the swing can be computed from the definition $V = dr/dt$ which gives

$$V = \left(\frac{2\pi r_0}{T}\right) \cos\left(\frac{2\pi t}{T}\right).$$

We see that when the displacement of the swing is a maximum either to the left or to the right ($t = T/4$ or $t = 3T/4$) the velocity is zero. The speed is greatest when the swing passes through the vertical line. The maximum speed of the swing is proportional to the displacement r_0. That is why children, who want the sensation of speed, "pump" the swing; they thereby increase the amplitude r_0. The maximum speed is also inversely proportional to the period T. We shall see later in the course that the period T is directly proportional to the square root of the length l of the support. Thus, for fixed amplitude (r_0) of the oscillation, one gets a greater maximum speed the shorter the length of the supporting chain or rope.

(b) In the example above the position of the object was known as a function of time and we computed the velocity. Often the velocity is the known quantity and the position of the object is desired. It is natural to inquire how the velocity is measured without first knowing the position of the body as a function of time. One such means of measuring speed is the so-called Doppler effect. It enables one to measure the speed of a body, be it a rocket, or a red blood cell, by measuring the "tone" or the frequency of a sound wave (or light wave) scattered from or emitted by the moving body. The simplest and most familiar situation is that of the sound of a whistle on a train. If the frequency of the whistle is ν_0 when the train is stationary, then the frequency ν as detected by an observer alongside the track is different from ν_0 by an amount which depends on the speed (V) of the train. The fractional change in the apparent frequency of the whistle, i.e., $(\nu - \nu_0)/\nu_0$ is related to the velocity of the train by the formula

$$\frac{\nu - \nu_0}{\nu_0} = \pm \left(\frac{V}{C_s}\right).$$

Here C_s is the speed of sound. The $+$ sign applies when the train approaches the observer. The $-$ sign applies when the train moves away from the trackside observer.

This method of determining velocity is in fact used by so-called Doppler radar systems in tracking aircraft and rockets. In this case microwaves are used. The Doppler shift of sound waves scattered from moving erythrocytes (red blood cells) is used to measure the speed of blood flow (see [4]–[7]). Also the Doppler shift of light waves scattered by moving blood in the retina can be used to study blood flow there [8].

Suppose then we imagine that, by a Doppler shift measurement, we determine the velocity of a train as it builds up speed leaving the station. Suppose the train's maximum speed is V_0 and that it builds up to that speed in accordance with the formula

$$V(t) = V_0 \left(1 - e^{-t/\tau_0}\right).$$

The velocity is zero when $t = 0$. As $t \to \infty$ the velocity approaches V_0. The parameter τ_0 is a measure of the amount of time needed for the train to get near to top speed. For example, when $t = \tau_0$ the train has reached 0.635 of its final velocity. At $t = 2\tau_0$ the velocity is equal to 0.865 of its final value. We show graphically in Figure 1.3 the velocity as a function of time.

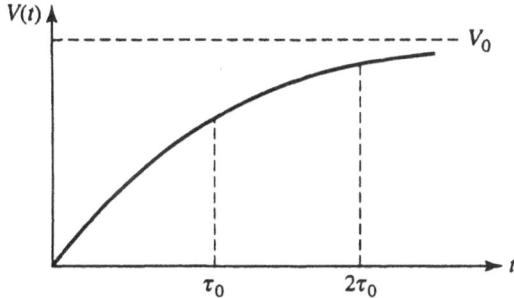

Figure 1.3. Exponential approach of a train's speed toward its maximum value.

We may compute the position of the train as a function of time knowing the velocity as follows. Since

$$\left(\frac{dr}{dt}\right) = V_0 \left(1 - e^{-t/\tau_0}\right),$$

we may integrate this equation to find that after the elapse of time t' the position r of the train is

$$\int_0^r dr = \int_0^{t'} dt \, V_0 \left(1 - e^{-t/\tau_0}\right).$$

On carrying out the integration and putting in the upper and lower limits we find

$$r(t') = V_0 t' - V_0 \tau_0 + V_0 \tau_0 e^{-t'/\tau_0}.$$

The mathematics is all here. Let us now see what it means. Let us look first at the position versus time at short times, i.e., for times $t' \ll \tau_0$. When this is the case, e^{-t'/τ_0} can be approximated by

$$e^{-t'/\tau_0} = 1 - \frac{t'}{\tau_0} + \frac{1}{2}\frac{t'^2}{\tau_0^2} + \cdots.$$

If we use this in the expression above for $r(t')$ we find that all terms linear in t and in τ_0 cancel, and the position of the train increases as the square of the time when

$t \ll \tau_0$. In fact one obtains

$$r(t') \simeq \left(\frac{V_0\tau_0}{2}\right)\left(\frac{t'}{\tau_0}\right)^2,$$

$$t' < \tau_0.$$

This is illustrated in Figure 1.4.

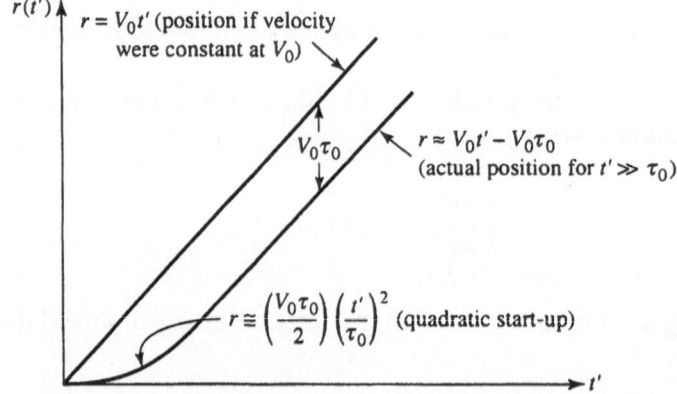

Figure 1.4. Position $r(t')$ of a train, showing both the quadratic startup phase and the subsequent constant speed trajectory.

For times $t' \gg \tau_0$ the position of the train obeys the equation

$$r(t') \simeq V_0t' - V_0\tau_0.$$

This has the interpretation that at the beginning of the motion of the train, it did not move much, and in effect lost a distance $V_0\tau_0$ that it would have traveled had it had maximum speed V_0 right from the start. We show this below in the graph $r(t)$ for the entire motion.

1.2.B.   Acceleration

Just as velocity is defined as the rate of change of position, one may define a quantity called the acceleration a, which is the rate of change of velocity, viz.:

$$\text{acceleration} = a = \left(\frac{dV}{dt}\right). \tag{1-5}$$

Why do we need the idea of acceleration in kinematics? Part of the answer to this question was provided by Galileo in his study of the kinematics of falling bodies. Quoting from Galileo's *Two New Sciences* (see [1] and [11]):

> When therefore, I observe a stone initially at rest falling from an elevated position and continually acquiring new increments of speed, why should I not believe that such increases take place in a manner which is exceedingly simple and fairly easily apprehended by everybody? If now we examine the matter carefully, we find no addition or increment more simple than that which repeats itself always in the same manner."... "just as uniform motion is defined by and thought of in terms of equal time intervals and equal distances (thus we call a motion uniform when equal distances are traversed in equal time intervals) so also we may in similar manner, by thinking in terms of equal time intervals, conceive additions of velocity as taking place without complication.
>
> Hence the definition of the motion we are about to discuss may be stated as follows:
>
> A body is said to be uniformly accelerated when, starting from rest, it acquires equal increments of velocity during equal time intervals.

Galileo then goes on to demonstrate experimentally that the motion of falling bodies is that of uniform acceleration. It is an interesting side-light that as a means of measuring time Galileo used the pulse beats of his heart!

It remained for Isaac Newton (1642–1727) to explain why falling bodies have constant acceleration regardless of their weight. More fundamentally, however, Newton showed that the forces acting on a body determine its acceleration. Thus if one can determine the total *force* on a body this fixes its *acceleration*. From the acceleration then the motion can be computed by integrating the equations of kinematics twice. By integrating the expression for acceleration one finds the velocity and by integrating the velocity one finds the position. Each integration requires one constant of integration. Thus, in general, we see that when the acceleration is known as a function of time one can obtain the position, provided that the position and velocity are known at some time. Generally these are known at the beginning of the motion at $t = 0$.

1.2.C. Examples of Accelerated Motion

The simplest example of accelerated motion is that considered by Galileo: that is, motion with constant acceleration. We consider this case first, and assume that the body starts its motion with some initial velocity V_0 and position r_0. We continue to restrict ourselves to motion in one dimension. The expression for the acceleration and velocity can be integrated very simply to give the entire motion as follows:

$$a = \text{constant} = \left(\frac{dV}{dt} \right). \tag{1-6}$$

Integrating once to get the velocity as a function of time we find

$$V(t) = V_0 + at. \tag{1-7}$$

In using this expression it is important to take into account the sign of the acceleration and the initial velocity. If the body is accelerated, say, to the right and its initial velocity is to the left, then the quantities a and V_0 enter in (1-7) with opposite signs.

We can obtain the position as a function of time by relating the velocity to r through $V = dr/dt$, i.e.,

$$\left(\frac{dr}{dt} \right) = V_0 + at. \tag{1-8}$$

Integrating this gives

$$r = r_0 + V_0 t + \tfrac{1}{2} a t^2. \tag{1-9}$$

We may put this result in another, often useful, form by completing the square in the quadratic expression involving the time. Writing (1-9) as

$$r = r_0 + \tfrac{1}{2} a \left[t^2 + \tfrac{2V_0}{a} t \right],$$

and completing the square by adding and subtracting $(V_0/a)^2 (\tfrac{1}{2} a)$ we have

$$r = r_0 + \frac{1}{2} a \left[t^2 + \frac{2V_0 t}{a} + \left(\frac{V_0}{a} \right)^2 \right] - \frac{V_0^2}{2a},$$

or

$$r = r_0 - \frac{V_0^2}{2a} + \tfrac{1}{2}a \left[t + \frac{V_0}{a} \right]^2 . \tag{1-10}$$

The physical meaning of this result can be readily understood by considering a familiar situation which it describes: that is, the motion of a ball thrown vertically into the air. The position of the ball at each moment is r, the initial position is r_0. The initial velocity is V_0 and since it is directed upward we associate with it a positive sign. The acceleration on the other hand is downward, and so must have a negative sign. To denote the fact that the acceleration is that due to gravity, we shall denote it by g. Thus we place $a = -g$, where g is the same for all falling bodies, as Galileo first discovered. Under these conditions the height r of the ball at each instant of time follows the equation

$$(r - r_0') = -\frac{g}{2} \left(t - \frac{V_0}{g} \right)^2 , \tag{1-11}$$

where

$$r_0' = r_0 + \frac{V_0^2}{2g}.$$

If we make a plot of r versus time with r plotted vertically and t horizontally, we will obtain the parabola shown in Figure 1.5.

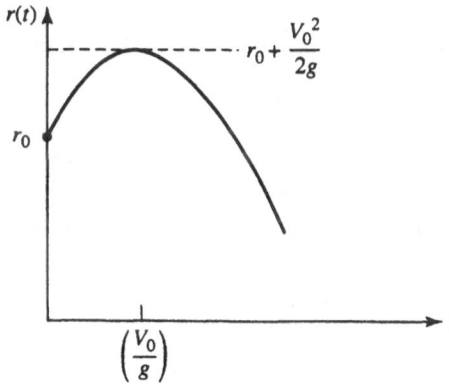

Figure 1.5. Parabolic trajectory $r(t)$ of a ball initially thrown upward.

The maximum of the parabola occurs at $t_{max} = (V_0/g)$ and $r_{max} = r_0' = (r_0 + V_0^2/2g)$. Thus the ball rises at first, slowing down all the time until it stops when it reaches a height $(V_0^2/2g)$ higher than its starting position. The velocity is zero when $t = V_0/g$. After this the ball falls ever faster. The total trajectory in the r–t plane is a parabola, and this is true regardless of the initial velocity or initial position. These quantities only determine the location of the parabola's maximum or minimum in the plane. Clearly, if the initial velocity were zero the maximum of the parabola would be at $t = 0$, $r = r_0$, and the maximum height is just at the position where the ball is dropped. If, for example, the ball were thrown initially downward, the parabola in the r–t plane would have its maximum at the negative time $t = (-V_0/g)$ and the ball would follow a path predicted by that part of the parabola in the half-plane corresponding to $t > 0$.

Motion with unconstant acceleration: In general when the acceleration is known as a function of the time, we can integrate twice to find the motion. In the example of the moving train whose velocity built up gradually from $V = 0$ to a final value V_0 according to the formula $V = V_0(1 - e^{-t/\tau_0})$, the acceleration is

$$\left(\frac{dV}{dt} \right) = \frac{V_0}{\tau_0} e^{-t/\tau_0}$$

and this starts at a large value and falls off to zero exponentially with time. The motion of the train has already been computed in our earlier discussion. It is interesting to observe that in this example the position versus time depends not on the velocity at $t = 0$, but on the velocity at infinite time.

1.3   Motion in Two and Three Dimensions: Velocity and Acceleration

In two and three dimensions the position of a point is specified by the length and orientation of a vector $(\vec{r})$ from some arbitrary origin to the point in question. In a short appendix at the end of this chapter (pp. 23 ff), we present the basic mathematical properties of vectors. This appendix provides all the elements of vector algebra that will be needed in this course.

1.3.A.   Velocity

The velocity is the instantaneous rate of change of the vector $\vec{r}$. The velocity itself is a vector whose magnitude is generally called the speed. Mathematically the velocity is defined by the vector equation

$$\vec{V} = \left(\frac{d\vec{r}}{dt}\right).$$
(1-12)

To obtain the magnitude and direction of the velocity vector at some point in the motion, one proceeds in principle as follows. First, one obtains the position vector at say time t; then one obtains the position vector $(\vec{r} + \Delta\vec{r})$ at time $t + \Delta t$. The velocity vector is found by taking the difference between $\vec{r} + \Delta\vec{r}$ and $\vec{r}$ and then going to the limit at $\Delta t \to 0$. This process is indicated diagramatically in Figure 1.6:

$$\vec{V} = \lim_{\Delta t \to 0} \left(\frac{\Delta\vec{r}}{\Delta t}\right).$$
(1-13)

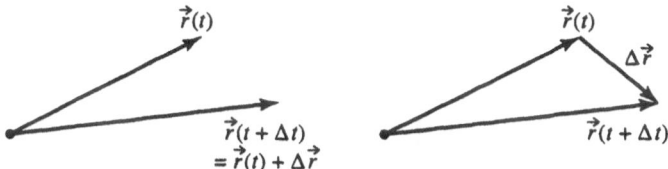

Figure 1.6. Position vectors $\vec{r}(t)$ and $\vec{r}(t + \Delta t)$ at times t and $t + \Delta t$, respectively.

1.3.B.   Acceleration

The acceleration in two and three dimensions is, as in the one-dimensional case, the rate of change of the velocity. However, in the multidimensional case, one must compute the rate of change of the vector velocity. The acceleration itself is again a vector having a magnitude and direction

$$\vec{a} = \frac{d\vec{V}}{dt} = \lim_{\Delta t \to 0} \frac{(\vec{V} + \Delta\vec{V}) - \vec{V}}{\Delta t}.$$
(1-14)

We can understand more clearly the meaning of the velocity and acceleration in two and three dimensions by considering some specific examples. Consider first the case of uniform circular motion:

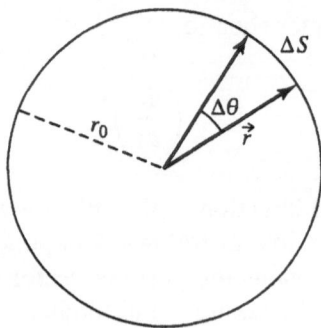

Figure 1.7. Position vector $\vec{r}(t)$ of an object in circular motion.

1.3.C.   Uniform Circular Motion

In this case a point travels in a circle in such a way that its *speed* along the periphery is constant. By saying that the speed along the periphery is constant we mean that the rate of change of the peripheral distance is constant in time, i.e.,

$$\lim_{\Delta t \to 0} \left(\frac{\Delta S}{\Delta t} \right) = \text{constant}.$$

Since the peripheral distance ΔS traveled in time Δt is related to the small angle $\Delta \theta$ turned by the radius r_0 in the same time, we see that the constancy of peripheral speed is the same as saying that the rate of change of the angular orientation of the particle is constant in time, i.e.,

$$\left(\frac{dS}{dt} \right) = \lim_{t \to 0} \left(\frac{\Delta S}{\Delta t} \right) = r_0 \lim_{t \to 0} \left(\frac{\Delta \theta}{\Delta t} \right) = r_0 \left(\frac{d\theta}{dt} \right). \qquad (1\text{-}15)$$

The quantity $(d\theta/dt)$ is called the "angular velocity" and is usually denoted by the symbol ω. This is the rate of change of the angular coordinate of the particle. In our example ω is a constant at all times in the motion. We now inquire as to the instantaneous vector velocity of the particle. We can find this from Figure 1.8, which shows the change in the vector position $\vec{r}$ in a small time Δt. The velocity is

$$\vec{V} = \frac{d\vec{r}}{dt} = \lim_{\Delta t \to 0} \left(\frac{\Delta \vec{r}}{\Delta t} \right) = \lim_{\Delta t \to 0} \left(\frac{r_0 \Delta \theta}{\Delta t} \right), \qquad (1\text{-}16)$$

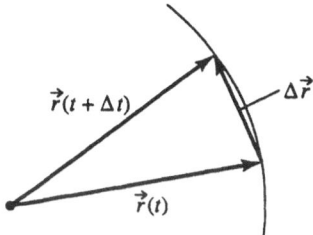

Figure 1.8. Displacement $\Delta\vec{r}$ occurring in time Δt for an object in circular motion.

since the length of the vector r is r_0. We see that as $\Delta t \to 0$ the chord $\Delta\vec{r}$ becomes tangent to the circle. Thus the velocity vector at each point in the motion is tangent to the circle. The magnitude of the velocity vector is

$$|V| = r_0 \left(\frac{d\theta}{dt} \right) = r_0\omega, \tag{1-17}$$

where ω is the angular velocity as defined above

$$\omega \equiv \left(\frac{d\theta}{dt} \right). \tag{1-18}$$

Thus at each instant the velocity vector can be expressed graphically in Figure 1.9 as a vector tangential to the circle with length $r_0\omega$.

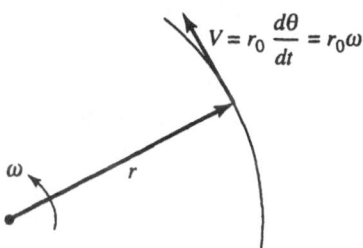

Figure 1.9. Angular velocity ω and tangential velocity $\vec{v}$ in circular motion.

The acceleration in this uniform circular motion can now be computed. According to the definition of acceleration we must compute $[\vec{V}(t + \Delta t) - \vec{V}(t)]/\Delta t$

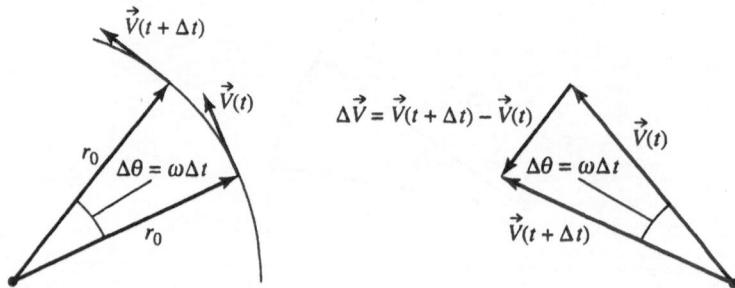

Figure 1.10. Relation between the angular velocity ω and the rate of change of the tangential velocity in circular motion.

and find the magnitude and direction of this vector in the limit $\Delta t \rightarrow 0$. This is done with the aid of Figure 1.10: This shows that the velocity vector has the same length at time t and $t + \Delta t$, but that its direction changes as shown. To obtain the difference between $\vec{V}(t + \Delta t)$ and $\vec{V}(t)$ we translate the two vectors until their origins are in contact. The angle between the two vectors is $\Delta\theta$:

$$\left| \left(\frac{\Delta \vec{V}}{\Delta t} \right) \right| = V \left(\frac{\Delta\theta}{\Delta t} \right). \qquad (1\text{-}19)$$

In the limit, of course, as $\Delta t \rightarrow 0$, $(\Delta\theta/\Delta t) = \omega$, the angular velocity. The magnitude of the acceleration therefore is

$$|a| = V\omega. \qquad (1\text{-}20)$$

By considering the direction of $\Delta \vec{V}$ we see that the acceleration points inward along the radius vector at each instant of time. Thus we have the striking result that while the velocity vector is always tangential to the circle, the acceleration on the other hand is "centripetal" or central seeking—it points at each instant toward the center of the circular motion. The result is shown in Figure 1.11. Since, in uniform circular motion, the magnitude of the velocity is $r_0\omega$ one can also express the magnitude of the acceleration as

$$|a| = V\omega = r_0\omega^2 = \frac{V^2}{r_0}. \qquad (1\text{-}21)$$

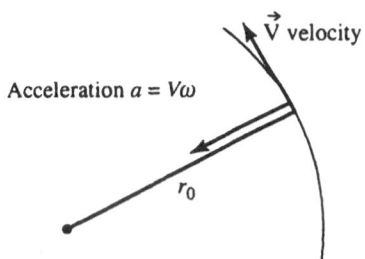

Figure 1.11. Radial acceleration of a mass in circular motion.

The result that the acceleration, in uniform circular motion, is directed toward the center of the circle is a consequence of the vector nature of the velocity and the acceleration. Even though the speed of the particle remains constant, the direction of the velocity is constantly turning inward toward the center of the circle. This realization, that in circular motion the acceleration, or the change in velocity, is directed toward the center is at the very heart of Newton's explanation of the origin of planetary motion. For if the planet's acceleration is central, so there must be some force pulling it toward the center of its orbit. This force in the planetary case we now understand as the force of gravitational attraction between the Sun, at the center, and each of the planets. And so we see how essential it is that we understand first the kinematics of motion, so that we can discover the forces that produce it.

We shall not consider here more complex motions, such as motion in a circle in which the angular velocity is not a constant. For a discussion of the form of the velocity and acceleration vectors in the general case of motion described in polar coordinates, see Kleppner and Kolenkow's book [9]. Here the acceleration is computed in the general case that there is angular velocity ω, angular acceleration $(d\omega/dt)$, radial velocity (dr_0/dt), and radial acceleration d^2r_0/dt^2.

1.3.D.   Projectile Motion

One of Galileo's great achievements was his correct mathematical analysis of the motion of projectiles. The flight of a ball fired from a cannon, a stone dropped from the mast of a moving ship, or in more modern terms the flight of a baseball hit by a batter or thrown by a fielder are all examples of projectile motion.

The essential mathematical ingredients of projectile motion are:

(1) motion with constant acceleration; and

(2) an initial velocity which points in a direction different from the acceleration.

Analysis of projectile motion permits us to understand the necessity for a vectorial description of motion. More importantly, however, it leads to powerful general ideas used again and again in physics; namely, the notion of "superposition" and the idea of "Galilean Equivalence." As we proceed we shall explain what these terms mean and see how they emerge from the analysis of projectile motion. The essential features of projectile motion can be seen by considering the motion of a projectile whose initial velocity V_0 is horizontal, and whose acceleration (g) is vertical. To formulate the equations governing the motion we need to know how the vertical position of the projectile changes in time, and how the horizontal position changes in time.

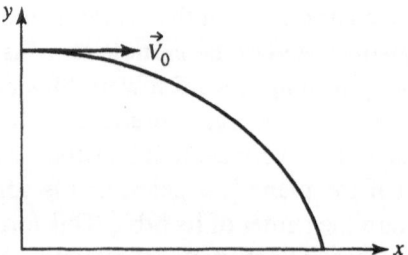

Figure 1.12. Parabolic motion—under gravity—of a mass with horizontal initial velocity $\vec{V}$.

If we measure, experimentally, the time for the body to hit the ground ($y = 0$) we will find that this is the *same* regardless of the magnitude of V_0. If the particle drops downward with $V_0 = 0$ it hits the ground at the same time as does a body moving with a high horizontal speed V_0. Galileo put it this way [1] and [10]:

If a cannon were leveled on the top of a tower, and fired point-blank, that is, horizontally, and whether the charge were small or large with the ball falling sometimes a thousand yards distant, sometimes four thousand, sometimes ten, etc., all these shots shall come to the ground in times equal to each other. And every one equal to the time that the ball would take to pass from the mouth of the piece to the ground, if, without other impulse, it falls simply downward in a perpendicular line. Now, it seems a very admirable thing that, in the same short time of its falling perpendicularly down to the ground from the height, say, of a hundred yards, equal balls, fired violently out of the piece, should be able to pass four hundred, a thousand, even ten thousand yards. All the balls in all the shots made horizontally shall remain in the air an equal time.

Thus Galileo saw, and we see, that the vertical component of the motion is that of uniform acceleration downward. Also, a vertical motion is clearly "independent" or unaffected by the horizontal motion.

We may now inquire as to whether the horizontal motion in turn is, or is not, affected by the vertical motion. If we examine the horizontal speed of the body we will find, experimentally, that it remains constant throughout the motion and equal to the initial horizontal velocity. Thus, experimentally, one concludes that the horizontal motion is in fact independent of the vertical motion.

The total motion is simply the sum or the "superposition" of the two independent motions. This "superposition principle" appears again and again in physics. It affords very great simplification of complex situations, particularly in wave motions, for it permits the complex motion to be regarded as a superposition of independent noninteracting simple motions.

In the case at hand, the sum of the vertical and horizontal motions is a vector sum, and this is in fact the reason that we use vectors to characterize the position, velocity, and acceleration. The components of these physical quantities combine in addition and subtraction exactly in accordance with the laws of vector algebra. These physical facts are all we need to write out the exact form of the equations of projectile motion. It is then quite straightforward to compute the precise form of the trajectory. While it is tempting to carry out this program directly, let us hold back for a little longer to see that all the physical facts needed can, in fact, be deduced from a more general principle: that is, the Principle of Galilean Equivalence.

Let us imagine the motion of a coin dropped by a passenger flying in an airplane. It is natural to believe that the time required for the coin to land on the cabin floor is the same whether the plane is at rest or moving at six hundred miles per hour. It is also natural for us to believe that the coin will land at the same spot on the floor regardless of the speed of the plane. This latter fact means that the horizontal speed of the coin (as observed from outside the airplane) is maintained constant as the coin drops. Now the statements made above are essentially identical with the experimental facts mentioned above which are needed to establish the form of the equations of projectile motion. These beliefs will in fact be correct if the Principle of Galilean Equivalence is correct. We can state this principle as follows: Any mechanical experiment conducted on a body within some coordinate system will give the same result if conducted in a second coordinate system provided that the two systems move with constant velocity relative to one another.

Considerations as general as this are at the very basis of kinematics and dynamics. Such principles of equivalence and the laws of transformation of the equations of motion between moving coordinate systems led Einstein to his Theory of Relativity, and continually serve as powerful modes of thought in the synthesis and simplification of experimental findings, which is at the heart of physics.

We now can write the equations of motion of projectile motion. Since the vertical (y) motion is one of constant downward acceleration $(-g)$, we see that the velocity component (V_y) in the y-direction obeys the equation

$$\frac{dV_y}{dt} = -g. \tag{1-22a}$$

The condition that the horizontal motion takes place with no change in horizontal velocity gives the equation

$$\frac{dV_x}{dt} = 0. \tag{1-22b}$$

To integrate these two equations twice we need four initial conditions: The initial x- and y-coordinates, which we take as $x = 0$, $y = y_0$, and the initial x- and y-components of the velocity which we take as V_{0x}, V_{0y} respectively. Notice that by assuming that the y-component of the velocity is not zero, we will be able to have a more general final result than in the special case that the initial velocity is horizontal. The initial conditions are shown in Figure 1.13. We may now solve (1-22a) and (1-22b) subject to the initial conditions separately for the position (x, y) as a function of time. We can also eliminate time between the two equations and obtain the shape of the trajectory, i.e., $y(x)$.

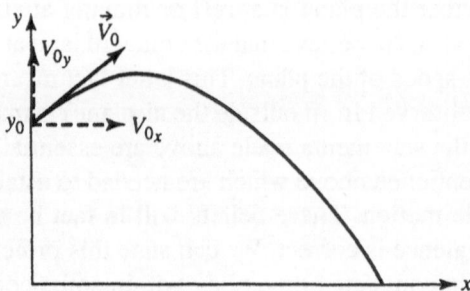

Figure 1.13. Parabolic motion of a mass—under gravity—and an initial velocity with both a horizontal and vertical component.

We have, in fact, already integrated an equation identical to (1-22a), for this is the same as that of constant acceleration in one dimension. We did this in (1-8) and (1-9). With the present symbols the integration of (1-22a) gives the following results for $V_y(t)$ and $y(t)$:

$$V_y(t) = V_{0y} - gt \qquad (1\text{-}23)$$

and

$$y = y_0 + V_{0y} - \tfrac{1}{2}gt^2 \qquad (1\text{-}24)$$

or, completing the square in this quadratic expression in t gives

$$y(t) = y_0 + \frac{V_{0y}^2}{2g} - \frac{g}{2}\left(t - \frac{V_{0y}}{g}\right)^2. \qquad (1\text{-}25)$$

The x motion is

$$V_x(t) = V_{0x} \qquad (1\text{-}26)$$

and

$$x(t) = V_{0x}t. \qquad (1\text{-}27)$$

Equations (1-25) and (1-27) give us x and y, the coordinates of the projectile as a function of the time. In many cases it is more interesting to know the actual path of the projectile without reference to the time as the path is traversed. To do this, we need only combine (1-27) and (1-25) in such a way as to eliminate the time between them. This is done by noting that $t = x/V_{0x}$ from (1-27), and on substituting this into (1-25) we find

$$y = \left(y_0 + \frac{V_{0y}^2}{2g}\right) - \frac{g}{2V_{0x}^2}\left(x - \left(\frac{V_{0x}V_{0y}}{g}\right)\right)^2. \qquad (1\text{-}28)$$

From this equation we see that the path of the projectile is a parabola. The maximum of the parabola occurs at a value of

$$(x)_{\text{max}} = \left(\frac{V_{0x}V_{0y}}{g}\right), \qquad (1\text{-}29)$$

$$(y)_{\text{max}} = y_0 + \left(\frac{V_{0y}^2}{2g}\right). \tag{1-30}$$

The total distance traveled by the projectile is the value of x when $y = 0$. If, for convenience, we consider the case that the projectile starts from ground level ($y_0 = 0$), then the total distance traveled is easily found from (1-28) to be

$$x_{\text{total}} \equiv \text{range} = 2\frac{V_{0x}V_{0y}}{g}. \tag{1-31}$$

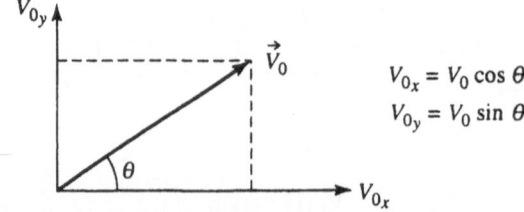

$$V_{0x} = V_0 \cos\theta$$
$$V_{0y} = V_0 \sin\theta$$

Figure 1.14. Horizontal and vertical components of a general initial velocity $\vec{V}$.

We can use this result to compute the angle at which the projectile should be fired to get a maximum range. If we consider the direction of the initial velocity $\vec{V}_0$, and denote the angle θ as the angle between $\vec{V}_0$ and the horizontal, then

$$V_{0x}V_{0y} = V_0 \sin\theta \cos\theta.$$

But from trigonometry $2\sin\theta \cos\theta = \sin 2\theta$:

$$\text{range} = \frac{V_0^2 \sin 2\theta}{g}. \tag{1-32}$$

Clearly, since the sine is maximum at $90°$, the angle θ for maximum range is $2\theta = 90°$ or $\theta = 45°$. Furthermore, since the sine is symmetric about $90°$ the range will be equal amounts smaller than the maximum range for angles which differ equally from $45°$. Thus, for example, the range will be the same for $\theta = 45° + 15°$ and $\theta = 45° - 15°$. Galileo knew these results. He obtained them of course without the use of calculus. It is interesting to hear his own reaction to his discovery. (See [1] and [10].)

The force of rigorous demonstrations such as occur only by the use of mathematics fills me with wonder and delight. From accounts given by gunners, I was already aware of the fact that in the use of cannon and morters, the maximum range, that is, the one in which the shot goes farthest, is obtained when the elevation is 45°...; but to understand *why* this happens far outweighs the mere information obtained by the testimony of others or even by repeated experiment.... The knowledge of a single effect apprehended through its cause opens the mind to understand and ascertain other facts without recourse to experiment, precisely as in the present case, where, having won by demonstration the certainty that the maximum range occurs wherever the elevation (θ) is 45°, the Author (Galileo) demonstrates what has perhaps never been observed in practise, namely, that for elevations which exceed or fall short of 45° by equal amounts, the ranges are equal....

1.4   Units and Conversion Factors

In this course we shall generally use as units of length the meter or the centimeter. The unit of time will be the second. When the meter is used as a unit of length, the units of velocity are meter/second, and the units of acceleration are meters/second^2 or m/s^2. The acceleration of gravity in these units is $\sim$ 9.8 m/s^2. When the centimeter is used as a unit of length, the velocity units are cm/s, and the acceleration units are cm/s^2. The acceleration of gravity in cm/s^2 is $\sim$ 980 cm/s^2.

Sometimes it is convenient to use the foot as a unit of distance. One foot is equal to 30.48 cm. The acceleration of gravity in ft/s^2 is 32 ft/s^2. The mile, also sometimes convenient, is 5280 ft, or 1.61×10^3 m.

Appendix to Chapter 1: Vectors

Every student of physics must acquire an understanding of the concept of vectors, and become familiar with the simplest vector operations: decomposition of a vector into components, addition of vectors, and the use and meaning of the scalar and cross products of two vectors.

This appendix offers a very concise but sufficient presentation of all the essentials needed in this volume.

(1) *Definition.* A vector is a quantity specified by *magnitude* and a *direction*. Examples of vectors occurring in physics are force, velocity of a moving body, or the position of a body relative to some reference point 0.

Nonvectorial quantities, such as mass, temperature, pressure, or pure numbers, are called *scalars*.

The example of the velocity illustrates a crucial point: The motion of a mass in space is described completely only if we know how *fast* it moves (that is, the speed, or *magnitude* of the velocity vector), and also in what *direction*. The statement that the velocity of an object is known *always* implies that both speed and direction are known. Similarly, a force is in part specified by its strength (magnitude), but a complete description requires a statement of its direction.

Now, "direction" is a meaningless statement without some *frame of reference*. Such a frame can be established by means of any three mutually perpendicular axes pointing, for example, north, east, and upward. More generally, the physicist will choose any three such axes, and label them as x-, y-, and z-axes, respectively. These axes define a Cartesian coordinate system.

It is current practice to represent a vector *algebraically* by a letter symbol topped by an arrow; for instance: $\vec{v}$ for velocity. The symbol v or $|\vec{v}|$ is then used to indicate the magnitude of the vector; in our example, v (or $|\vec{v}|$) stands for the speed of a moving object. *Geometrically*, a vector will be represented by a properly oriented arrow, whose length is a measure of its magnitude.

(2) *Cartesian components.* A vector $\vec{v}$ and a coordinate system are drawn in Figure A1.1.

It is seen that $\vec{v}$ may be viewed as the main diagonal of a parallelepiped, whose edges intercept the three axes at points v_x, v_y, v_z. Assuming some scale of velocities has been introduced along the axes, the algebraic values of these intercepts are called the Cartesian *components* of the vector $\vec{v}$. In Figure A1.1, v_y and v_z

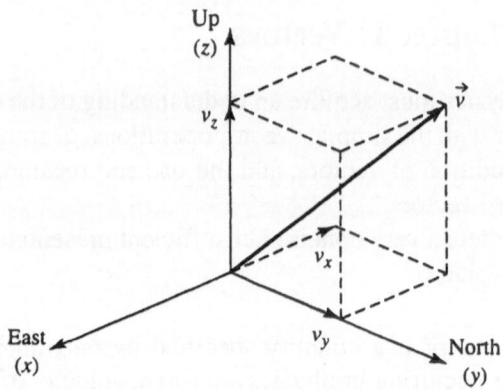

Figure A1.1. A vector $\vec{v}$ and a Cartesian coordinate system.

are positive, v_x is negative. Clearly, if the values of v_x, v_y, and v_z are given, the vector $\vec{v}$ can be constructed; conversely, if the vector $\vec{v}$ is given, v_x, v_y, and v_z can be found. Figure A1.1 also shows that the magnitude v of the vector $\vec{v}$ may be expressed in terms of the components v_x, v_y, v_z as

$$v = \sqrt{v_x^2 + v_y^2 + v_z^2}. \tag{A1-1}$$

(3) *Equality of two vectors.* The statement that two vectors $\vec{a}$ and $\vec{b}$ are equal is expressed by means of the *vector equation*

$$\vec{a} = \vec{b}. \tag{A1-2a}$$

Since a vector is specified completely only by giving the values of its three components in a coordinate system, (A1-2a) represents, in fact, a set of three algebraic equations

$$\begin{aligned} a_x &= b_x, \\ a_y &= b_y, \\ a_z &= b_z. \end{aligned} \tag{A1-2b}$$

So, on the one hand, we may view (A1-2a) simply as a convenient "shorthand" notation for the set of three equations, (A1-2b). There is, however, a deeper point expressed in (A1-2a). Equation (A1-2b) requires a coordinate system for expression, but (A1-2a) does not. This should remind the user of the fact that the statement $\vec{a}$ and $\vec{b}$ implies the equality of the *corresponding* vector components of $\vec{a}$ and $\vec{b}$ in *any* coordinate system.

(4) *Linear vector operations.* (a) Let μ be a pure number, or else a dimensional scalar, such as the mass of an object, etc. Starting with any vector $\vec{a}$, we can then define a vector

$$\vec{b} = \mu\vec{a} \tag{A1-3}$$

as a multiple of $\vec{a}$. $\vec{b}$ has the same direction as $\vec{a}$, but a different magnitude; Cartesian components are scaled by a factor μ:

$$b_x = \mu a_x,$$

$$b_y = \mu a_y, \qquad\qquad\qquad\text{(A1-4)}$$

$$b_z = \mu a_z.$$

and its magnitude by a factor $|\mu|$: $b = |\mu|\, a$. The special case $\mu = -1$ defines the vector $-\vec{a}$, whose components have their sign reversed. An example where the multiplicative factor has a dimension is the velocity momentum relation $\vec{p} = m\vec{v}$, m being the mass of an object, $\vec{p}$ the momentum of a moving mass, and $\vec{v}$ its velocity.

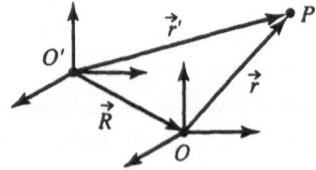

O and O' are the origins of two coordinate systems:
$\vec{r}$ = position of P relative to O.
$\vec{R}$ = position of O relative to O'.
$\vec{r}' = \vec{R} + \vec{r}$ = position of P
relative to O'.

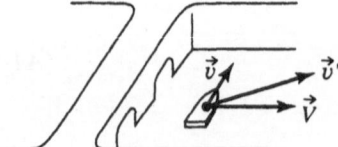

$\vec{v}$ = velocity of boat relative to water.
$\vec{V}$ = velocity of water relative to ground.
$\vec{v}' = \vec{v} + \vec{V}$ = velocity of boat relative
to ground.

Figure A1.2. Addition of vectors.

(b) Vector addition: The need for this operation emerges naturally, as the two examples in Figure A1.2 show. The statement that a vector $\vec{c}$ is the sum of two vectors $\vec{a}$ and $\vec{b}$:

$$\vec{c} = \vec{a} + \vec{b}$$

is algebraically defined by the statement that any Cartesian component of $\vec{c}$ is equal to the sum of the corresponding components of $\vec{a}$ and $\vec{b}$:

$$c_x = a_x + b_x,$$

$$c_y = a_y + b_y, \qquad\qquad\qquad\text{(A1-5)}$$

$$c_z = a_z + b_z.$$

Appendix to Chapter 1: Vectors

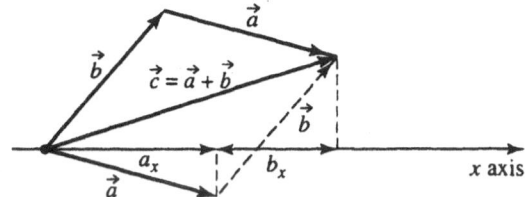

Figure A1.3. Addition of two vectors.

There is an *equivalent* geometric representation of this rule, illustrated in Figure A1.3. It is clear that this construction is equivalent to the algebraic statement, (A1-5). Notice also that, for the purpose of this geometrical construction of the sum, we may parallel displace either $\vec{a}$ and $\vec{b}$, so as to place the end of one arrow against the tip of the other.

(c) The two operations described in (a) and (b) may be combined so as to define a vector

$$\vec{c} = \lambda\vec{a} + \mu\vec{b},$$

λ, μ being two scalars. Combining (A1-4) and (A1-5), it follows that the components of $\vec{c}$ are

$$c_x = \lambda a_x + \mu b_x,$$
$$c_y = \lambda a_y + \mu b_y,$$
$$c_z = \lambda a_z + \mu b_z. \tag{A1-6}$$

Such an operation occurs, for example, in the definition of the center of mass $\vec{R}$ of two bodies of mass m_1 and m_2, located at positions $\vec{r}_1$ and $\vec{r}_2$ (see Chapter 3):

$$\vec{R} = \left(\frac{m_1}{m_1 + m_2}\right)\vec{r}_1 + \left(\frac{m_2}{m_1 + m_2}\right)\vec{r}_2.$$

It may be observed that from the rules of composition, (A1-6), it follows that

$$\lambda(\vec{a} + \vec{b}) = \lambda\vec{a} + \lambda\vec{b}. \tag{A1-7}$$

Also there is an obvious generalization of (A1-6) and (A1-7) to the case of addition of three or more vectors.

(5) *The unit vectors* $\hat{\imath}$, $\hat{\jmath}$, $\hat{k}$. *Decomposition of a vector.*   Any Cartesian coordinate system may be defined by means of three mutually perpendicular vectors of magnitude 1 ("unit vectors"), whose directions define the direction of the x-, y-, and z-axes, respectively. Such a triplet of unit vectors is customarily written as $\hat{\imath}$, $\hat{\jmath}$, $\hat{k}$, the caret being used *conventionally* instead of the arrow to identify a vector of unit length. The x-, y-, z-components of $\hat{\imath}$, $\hat{\jmath}$, $\hat{k}$ are, by definition, $(1, 0, 0)$, $(0, 1, 0)$, and $(0, 0, 1)$, respectively. It is then seen that, by the generalized rule, (A1-7), the vector

$$\vec{a} = \lambda\hat{\imath} + \mu\hat{\jmath} + \nu\hat{k}$$

has the components

$$a_x = \lambda$$

and, similarly, $a_y = \mu$, $a_z = \nu$, so that we may write

$$\vec{a} = a_x\hat{\imath} + a_y\hat{\jmath} + a_z\hat{k}. \qquad\qquad \text{(A1-8)}$$

This decomposition rule for a vector $\vec{a}$ also has a simple geometric interpretation, as seen in Figure A1.4:

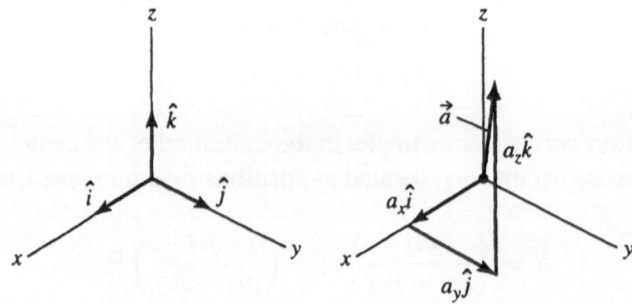

Figure A1.4. Decomposition of vector $\vec{a}$.

(6) *The scalar product of two vectors.*   Two nonparallel vectors $\vec{a}$ and $\vec{b}$ define a triangle. Define a vector $\vec{c}$ by $\vec{c} = \vec{b} - \vec{a}$. The length of this third side c is given by

$c = |\vec{b} - \vec{a}|$. We know from elementary trigonometry that

$$c^2 = |\vec{b} - \vec{a}|^2 = b^2 + a^2 - 2ab\cos\theta. \tag{A1-9}$$

We see that this expression for $|\vec{b} - \vec{a}|^2$, the square magnitude of the difference of two vectors, has a form (A1-9) very similar to that obtained by "multiplying out" the square of the difference of two numbers:

$$(b - a)^2 = b^2 + a^2 - 2ab.$$

This suggests that we define a multiplication of two vectors $\vec{a}$ and $\vec{b}$, the scalar product, which we shall write as $(\vec{a} \cdot \vec{b})$, and whose value is the number (scalar) $ab\cos\theta$:

$$(\vec{a} \cdot \vec{b}) \equiv ab\cos\theta. \tag{A1-10}$$

With this definition, the scalar product of a vector with itself is just the square magnitude of the vector, since $\theta = 0$ in this case, and therefore $\cos\theta = 1$:

$$(\vec{a} \cdot \vec{a}) = a^2. \tag{A1-11}$$

Using the definition (A1-10) and (A1-11), the trigonometric equation (A1-9) arises then from the following steps:

$$\begin{aligned}
c^2 = (\vec{c} \cdot \vec{c}) &= (\vec{b} - \vec{a}) \cdot \vec{c} = (\vec{b} \cdot \vec{c}) - (\vec{a} \cdot \vec{c}) \\
&= \vec{b} \cdot (\vec{b} - \vec{a}) - \vec{a} \cdot (\vec{b} - \vec{a}) \\
&= (\vec{b} \cdot \vec{b}) - (\vec{b} \cdot \vec{a}) - (\vec{a} \cdot \vec{b}) + (\vec{a} \cdot \vec{a}) \\
&= b^2 - 2ab\cos\theta + a^2.
\end{aligned}$$

We see that the "derivation" of this equation between the sides of a triangle appears almost trivial; it is very apparent, however, that in this process we have made use of the yet unproven *distributive* property of the scalar product. Generally stated, this is the property that if $\vec{u} = \vec{a} + \vec{b}$, then

$$(\vec{u} \cdot \vec{v}) \equiv (\vec{a} + \vec{b}) \cdot \vec{v} \quad \text{is equal to} \quad (\vec{a} \cdot \vec{v}) + (\vec{b} \cdot \vec{v}). \tag{A1-12}$$

Using the definition of the scalar product, the statement (A1-12) is true provided

$$uv\cos\theta = av\cos\theta_a + bv\cos\theta_b,$$

θ being the angle between $\vec{u}$ and $\vec{v}$, and θ_a between $\vec{a}$ and $\vec{v}$, θ_b between $\vec{b}$ and $\vec{v}$. To validate (A1-12), we must therefore prove that

$$u\cos\theta = a\cos\theta_a + b\cos\theta_b.$$

The correctness of this equation is indeed easily read off from Figure A1.5. In this figure the shaded triangle is perpendicular to the direction of $\vec{v}$. The distributive

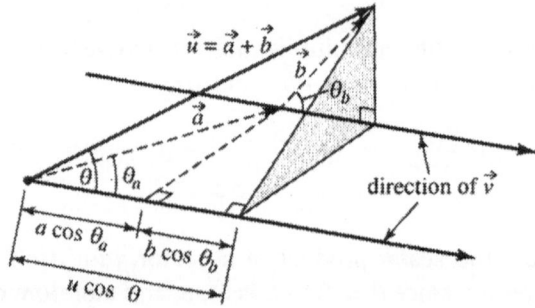

Figure A1.5. Proof of distributive law for scalar product.

property of the scalar product of two vectors $\vec{a}$ and $\vec{b}$ enables us to express its value in terms of the Cartesian components of these vectors. Using the representative (A1-8) of $\vec{a}$ in terms of the unit vectors, $\hat{\imath}, \hat{\jmath}, \hat{k}$, we may now write

$$(\vec{a} \cdot \vec{b}) = (a_x\hat{\imath} + a_y\hat{\jmath} + a_z\hat{k}) \cdot \vec{b}$$
$$= a_x(\hat{\imath} \cdot \vec{b}) + a_y(\hat{\jmath} \cdot \vec{b}) + a_z(\hat{k} \cdot \vec{b}).$$

But $(\hat{\imath} \cdot \vec{b}) = \hat{\imath} \cdot (b_x\hat{\imath} + b_y\hat{\jmath} + b_z\hat{k}) = b_x$ and, similarly, $(\hat{\jmath} \cdot \vec{b}) = b_y$, $(\hat{k} \cdot \vec{b}) = b_z$, so that

$$(\vec{a} \cdot \vec{b}) = a_xb_x + a_yb_y + a_zb_z. \tag{A1-13}$$

This formula could in fact be used as the basic, algebraic definition of the scalar product of two vectors $\vec{a}$ and $\vec{b}$. The previous geometric definition has the great advantage that it displays a property not so easily recognized in (A1-13), namely, the *invariance* of $(\vec{a} \cdot \vec{b})$. In fact, according to the definition (A1-10), $(\vec{a} \cdot \vec{b})$ has a value which depends only on the *relative* orientation of the two vectors, as expressed by the angle θ between them. Its value can therefore be ascertained independently of any coordinate system. This property is not manifest in the algebraic expression (A1-12) which makes use of the values of the Cartesian vector components in a specified frame of reference.

(7) ***The cross product of two vectors.*** Let us introduce this by an example: The rotation of the Earth about its axis (Figure A1.6). This rotation can be described by means of a vector $\vec{\omega}$ (the angular velocity vector), whose direction is the same as the axis of rotation, and whose magnitude is the *rate* of rotation, that is, the rate of change of the angle ϕ:

$$\omega = \frac{d\phi}{dt}.$$

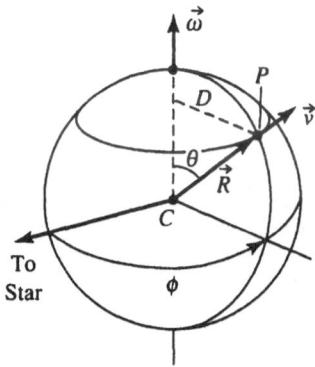

Figure A1.6. Rotation of the Earth.

We may then inquire about the actual velocity $\vec{v}$ of a point P on the Earth's surface, located by its position vector $\vec{R}$ relative to the Earth's center. The magnitude v of $\vec{v}$ is

$$v = \omega D = \omega R \sin \theta$$

and the direction of $\vec{v}$ is perpendicular to both the vector $\vec{R}$ and the vector $\vec{\omega}$. $\vec{v}$ is called the *cross product* of $\vec{\omega}$ and $\vec{R}$, and written as

$$\vec{v} = (\vec{\omega} \times \vec{R}).$$

Let us extract from this example the defining properties of the cross product of two vectors $\vec{a}$ and $\vec{b}$:

(i) $\vec{c} = (\vec{a} \times \vec{b})$ is a *vector*, perpendicular to the plane defined by $\vec{a}$ and $\vec{b}$.

(ii) The magnitude c of $\vec{c}$ is given by

$$c = |\vec{a} \times \vec{b}| = ab\sin\theta, \tag{A1-14}$$

θ being the angle subtended between $\vec{a}$ and $\vec{b}$.

(iii) In the plane containing $\vec{a}$ and $\vec{b}$, a positive sense of rotation is defined by turning $\vec{a}$ into the direction of $\vec{b}$. The tip of $\vec{c}$ then points in the direction of advance of a right-handed screw turned in a positive sense.

It follows from this definition that

$$(\vec{a} \times \vec{b}) = -(\vec{b} \times \vec{a}). \tag{A1-15}$$

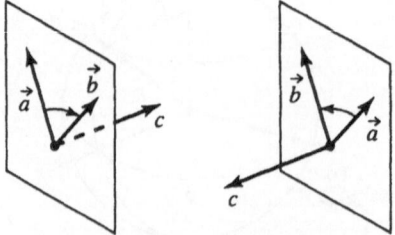

Figure A1.7. Cross product $\vec{c}$ of two vectors $\vec{a}$ and $\vec{b}$.

The components of $\vec{c} = (\vec{a} \times \vec{b})$ can be expressed in terms of the components of $\vec{a}$ and $\vec{b}$. Let us consider this for the special case that $\vec{a}$ and $\vec{b}$ lie in the x–y plane. In this case $\vec{c}$ is parallel to the z-axis, and we have (see Figure A1.8)

$$
\begin{aligned}
c_z = ab\sin\theta &= ab\sin(\theta_b - \theta_a) \\
&= ab(\sin\theta_b \cos\theta_a - \cos\theta_b \sin\theta_a) \\
&= (a\cos\theta_a)(b\sin\theta_b) - (a\sin\theta_a)(b\cos\theta_b) \\
&= a_x b_y - a_y b_x. \tag{A1-16}
\end{aligned}
$$

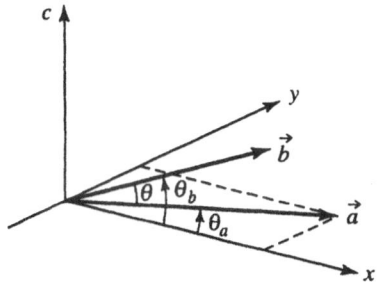

Figure A1.8. Cross product of two vectors $\vec{a}$ and $\vec{b}$ in the x–y plane.

One can show that this expression for the z-component of $\vec{c}$ is generally correct, even in the case that $\vec{a}$ and $\vec{b}$ are not in the x–y plane. Equation (A1-16) and its generalization for the components of any vector $\vec{c} = (\vec{a} \times \vec{b})$:

$$c_x = a_y b_z - a_z b_y,$$
$$c_y = a_z b_x - a_x b_z, \qquad \text{(A1-17)}$$
$$c_z = a_x b_y - a_y b_x,$$

allow an easy demonstration of the distributive property of the cross product. If

$$\vec{c} \equiv (\vec{u} + \vec{v}) \times \vec{w} = (\vec{u} \times \vec{w}) + (\vec{v} \times \vec{w}),$$

then

$$c_x = (u_x + v_x)w_y - (u_y + v_y)w_x$$
$$= (u_x w_y - u_y w_x) + (v_x w_y - v_y w_x), \text{ etc.}$$

1.5   References and Supplementary Reading

[1] *Foundations of Modern Physical Science* by G. Holton and D. Roller. Addison Wesley, Reading, MA (1958). This fine book contains a penetrating analysis of Galileo's contributions to kinematics.

[2] *Calculus* (Vol. I) by T. M. Apostol. Blaisdell, New York (1967). You will find here a lucid and rigorous mathematical analysis of the racetrack paradox of Zeno.

[3] *Episodes from the Early History of Mathematics* by A. Aaboe. Random House, New Mathematical Library, New York (19xx). A slim paperback which contains fascinating and penetrating descriptions and discussions of some important elements in Babylonian and Greek mathematics. It is rich in historical detail. The author discusses briefly, but very clearly, the paradoxes of Zeno (pp. 41 and 42).

[4] "A Transcutaneous, Ultrasonic Blood Velocity Meter," by H. F. Stegall, R. F. Rushme, D. W. Baker. *Journal Applied Physiol.* **21**(2), 707 (1966).

[5] "Ultrasonic Doppler Shift Blood Flowmeter," by D. L. Franklin, W. A. Schlegel, N. W. Watson. *Biomedical Sci. Instrum.* **1**, 309–315 (1963).

[6] "Technique for Radio Telemetry of Blood-Flow Velocity from Unrestrained Animals," by D. L. Franklin, N. W. Watson, K. E. Pierson, and R. L. van Citters. *Amer. J. Medical Electronics* **5**, 24 (1966).

[7] *Physical Principles of Ultrasonic Diagnosis* by P. N. Wells. Academic Press, New York (1969).

[8] "Laser Doppler Measurements of Blood Flow in Capillary Tubes and Retinal Arteries," C. Riva, B. Ross, and G. B. Benedek. *Investigative Ophthalmology* **2**, 936–944, November (1972).

[9] *An Introduction to Mechanics* by D. Kleppner and R. Kolenkow. McGraw-Hill, New York (1973).

[10] *Dialogue on the Great World Systems*, in the Salusbury translation, revised, annotated and with an Introduction by Giorgio de Santillana. University of Chicago Press, Chicago (1953).

[11] Two New Sciences by Galileo Galilei; translated by H. Crew and A. De Salvio under the title *Dialogues Concerning Two New Sciences*, Macmillan (1914), (1933); Northwestern University Press (1936), (1946), (1950); Dover, New York, (1952).

1.6   Problems

(These problems are arranged within each group in order of increasing difficulty.)

Uniform Motion

1. In an amusement park there is a slide for children in the form of a cylindrical track (see Figure).

Because of friction the children slide down at uniform speed $v = 3$ m/s. Given $H = 6$ m, and $R = 2$ m:

(a) How long does the ride take?

(b) What is the slope of the track?

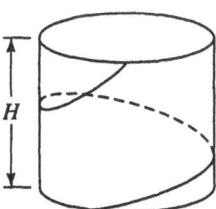

2. (a) Two cars A and B enter the tollbooth of the Massachusetts Turnpike one-half hour apart in time. The first car travels at 50 mph. What speed must the second car maintain to overtake the first in Albany, a distance of 200 miles away?

(b) Reformulate this problem to cover the more general case that the two cars have speeds V_A and V_B, and the second car, B, is delayed a time Δt relative to the first. What should be the speed of B to overtake A after a distance d?

(c) Illustrate your solution graphically by plotting x versus t for both vehicles.

3. (a) Two cars A and B enter the Massachusetts Turnpike at 9:00 A.M. The first (A) enters in Boston and drives at 60 mph. The second enters the Turnpike at Worcester a distance of 25 miles further west of Boston. If both cars meet in Albany (200 miles from Boston), what is the speed of the second car and when do they meet?

(b) Set this problem up in the general case that car A has speed V_A and car B has speed V_B. ($V_B < V_A$.) If they are separated at $t = 0$ by a distance d, at what point and at what time will the car with the head start be overtaken?

(c) Illustrate your solution graphically by plotting x versus t for both vehicles.

4. A weather balloon ascends into the sky from the ground with uniform speed V. The balloon is observed by means of a telescope a distance R from the launch site. The angular elevation θ of the balloon is recorded as a function of time.

(a) Obtain a formula for the elevation θ in terms of h and R.

(b) Obtain a formula for $(d\theta/dt)$ as a function of time.

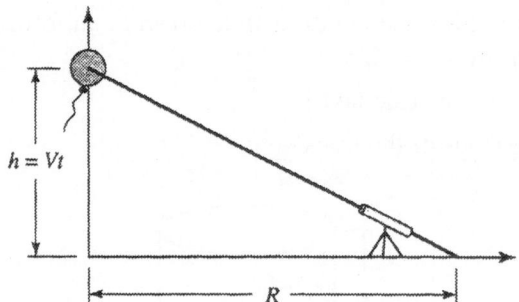

(c) Show that when the ascent time t is large compared to (R/V) that $(d\theta/dt)$ goes to zero as R/Vt^2.

Constant Acceleration in One and Two Dimensions

5. Show that a mass, if initially at rest, and subject to a constant acceleration a, reaches a speed that is proportional to the square root of the distance s traveled:

$$v = \sqrt{2as}.$$

(a) Assuming no loss due to air resistance, what is the speed, at street level, of a stone dropped from the Empire State Building? ($a = g = 10$ m/s)

(b) The gravitational acceleration g is really quite small. In the "electron gun" of a TV tube, electrons experience a's of the order of 10^{14} m/s^2:

  (i) What speed do they reach in 1 cm of travel?

  (ii) What time does it take an electron to travel from the nozzle of the electron gun to the TV screen (20 cm) in uniform motion?

6. A modern jet airplane can have an acceleration of about $(g/3)$. Assuming that this acceleration is a constant during takeoff, how long must the runway be in order that the plane achieve a takeoff velocity of 200 mile/h?

7. Suppose a chipmunk is 12 cm long. An eagle can see an object provided that it subtends an angle greater than one minute of arc at the eagle's eye:

(a) What is the maximum height above ground at which an eagle can still see a chipmunk?

(b) Suppose, on seeing the chipmunk, the eagle drops like a stone to attack it. How much time has the chipmunk got to retreat safely to cover if the eagle sees him at the height calculated in (a)?

(c) Suppose the chipmunk can run with a maximum speed of 2 m/s. What is the maximum safe distance he can venture to avoid capture by eagles?

8. *The old dollar bill game.*   As shown in the sketch, a student suspends a dollar bill by one end so that its midpoint is between the separated thumb and fore-finger of his roommate. The rules of the game are as follows: The roommate must keep his hand stationary, but as soon as the bill has been released, he may catch it between his thumb and finger; if he succeeds, he may keep the dollar:

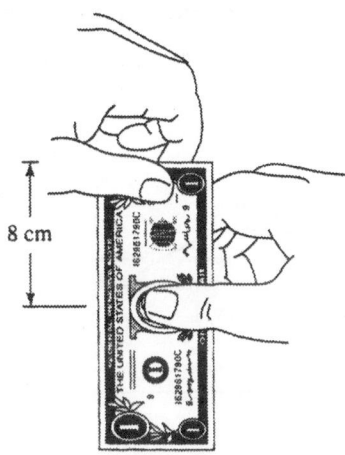

(a) How quickly must he react to get the dollar? (Ignore air resistance; it is only important for higher speeds of the falling bill. Also for simplicity take $g = 961 = (31)^2 \text{ cm s}^{-2}$.)

(b) The average person's reaction time is about $\frac{1}{5}$ s. If you owned the dollar, would you play the game allowing someone to place his thumb and finger near the bottom of the bill?

(c) How would the game be modified if played in an elevator ascending at a uniform speed of 3 m/s?

9. A stone falling with uniform acceleration g and initial velocity V_0 downward is photographed with a stroboscopic light which flashes at constant time intervals Δt. Let the first flash occur at the beginning of the fall:

(a) Show that the distance traveled between the nth and $(n + 1)$th flash (call this $d(n)$), is given by

$$d(n) = V_0 \Delta t + \frac{g}{2}(\Delta t)^2 (2n - 1).$$

(b) Show how this result can be used to measure g and to show that the acceleration is in fact constant.

10. *The monkey's problem.*   A monkey, pursued by a hunter with a rifle, reasons as follows: "As soon as I see the flash of his gun, I will let go of the limb to which I hang on, and thus, the bullet will pass over my head."

(a) Is he correct, assuming the hunter aims straight at him?

(b) Is the hunter correct in aiming straight at the monkey?

11. In the shot-put contest the shot is thrown by the athlete from a height h, at an angle α_0, with a speed v_0.

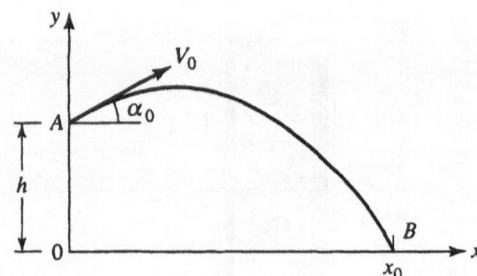

The shot lands at B, a horizontal distance x_0 from the point of departure A:

(a) Find expressions for the coordinates of the shot as functions of time $(x(t), y(t))$. (Uniform acceleration g in the $-y$-direction is assumed.)

(b) Find an expression for the distance x_0. *Hint*: Shot reaches x_0 at time T. Then

$$x(T) = x_0, \quad y(T) = 0.$$

(c) For those who want to tackle a harder problem than Galileo solved, try this: For a given h and v_0, how must the angle α_0 be chosen to yield the maximum distance x_0?

(d) The following case has been recorded:

$$h = 1.8 \text{ m},$$

$$v_0 = 13 \text{ m/s},$$

$$\alpha = 43°.$$

What is the value of x_0? In relation to problem (c), was this a good choice of α?

12. *Profile of water emerging from faucet.*   This problem is of an unusual type, as it deals with the kinematics of fluid flow. You have all observed that a thin stream of water, emerging at small speed v_0 from the nozzle of a faucet, has a profile as shown in the following figure:

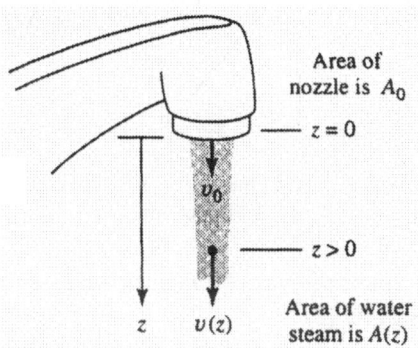

(a) Assuming that below the nozzle each element of fluid is in free fall, show that the velocity of the fluid, a distance z below the nozzle, is

$$v(z) = \sqrt{v_0^2 + 2gz}.$$

(b) What is the total outflow of water through the nozzle area A_0? (Flow measured in liters/second, for instance.)

(c) The total flow through the nozzle area A at $z = 0$ must be equal to the total flow through a horizontal plane a distance z below the nozzle. (It is important that you convince yourself of the correctness of this statement.) Express this flow in terms of $A(z)$ and $v(z)$. From equating the two expressions for the flow at $z = 0$ and $z > 0$, find a formula for the profile of the water stream as a function of z.

Nonuniform Motion

13. *The racetrack paradox revisited.* Zeno's analysis of the runner on the racetrack leads us to the erroneous conclusion that even though the runner's speed is constant he will never reach the end of the track.

   In this example you will discover that it may in fact take an infinite time for a runner to reach the end of the track even though he always runs in the same direction and stops only at the end of the track.

   In each of the three figures below we show possible variations of the runner speed as a function of his position along the track:

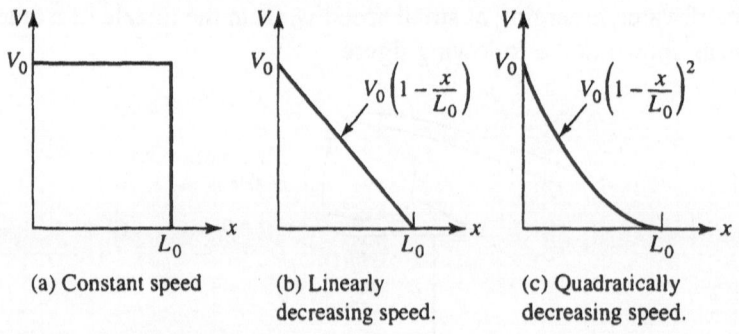

(a) Constant speed	(b) Linearly decreasing speed.	(c) Quadratically decreasing speed.

(i) Show that the runner never reaches the end of the track if he moves as described in (b) and (c).

(ii) Obtain a formula which gives the position of the runner as a function of time for case (b) and case (c) shown above.

(iii) Make a rough graph of each of the results you found in (ii).

14. Compute the acceleration of a runner as a function of his position x when his velocity is given either by

$$V(x) = V_0 \left(1 - \frac{x}{2}\right)$$

or

$$V(x) = V_0 \left(1 - \frac{x}{2}\right)^2.$$

Write down a general expression for $a(x)$ when $V(x)$ is given.

15. Imagine a race between two runners on a track of length L. The first runs with a constant speed V_0, and arrives at the end of the track in a time $T_0 = L/V_0$. The second starts with zero velocity, but increases his speed sinusoidally with time, so as to reach maximum velocity at the end of the track. Both runners reach the end of the track simultaneously. How fast is the second runner moving at this point?

Circular Motion

16. A wheel with 12 spokes turns with uniform angular velocity ω. The motion of the wheel is observed in a darkened room using a flashing (stroboscopic) light, which produces a brief burst of light 60 times each second:

    (a) Find the lowest angular velocity (ω_s) at which the spokes of the wheel appear to be stationary.

    (b) At what value of the angular velocity does the wheel appear to have 24 spokes?

    (c) At what angular velocities higher than ω_s will the wheel again appear stationary?

    (d) Suppose that the angular velocity of the wheel is close to ω_s, but differs from it by an amount $\delta\omega_s$, such that $|\delta\omega_s|/\omega_s \ll 1$. Show that the apparent angular velocity of the wheel is $\delta\omega_s$.

17. A movie camera takes pictures at a rate of 20 frames per second. Suppose one films a stagecoach chase in which the coach wheels have 18 spokes and a radius of 0.75 m. On showing the movie, one discovers that the wheels appear to be stationary:

    (a) How fast was the stagecoach moving during the filming?

    (b) Obtain a general formula for the speed at which the wheel appears not to rotate in terms of the number of spokes N, the radius of the wheel R, and the number of frames per second $(1/\Delta t)$.

18. In the drawing below we show the driving rod and piston of a steam engine acting on a drive wheel.

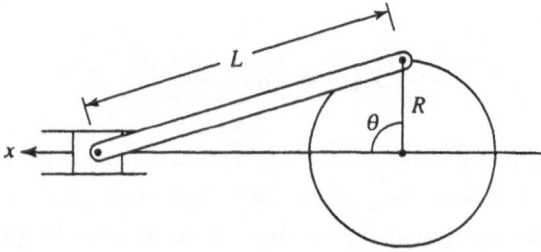

   Suppose that the engine shaft moves with constant angular velocity $\omega = d\theta/dt$:

   (a) Compute the position (x) of the piston center as a function of the angle θ.

   (b) Compute the velocity of the piston (dx/dt) as a function of time when ω is constant.

   (c) Show that when $R \ll L$ the piston will move back and forth sinusoidally.

19. Those of you who have experienced toy car racing know that unless the speed is carefully controlled the car can topple and fly off the track. In fact this is likely to occur when the centripetal acceleration is of the order of the acceleration of gravity g:

   (a) A toy car track generally has a minimum radius of curvature of 50 cm. At what speed will the centripetal acceleration equal g?

   (b) A VW microbus enters a circular cloverleaf exit at 100 m radius. What is a safe exit speed? (Use as your criterion for safety that the centripetal acceleration should not exceed $g/3$.) How does your numerical result compare with posted exit speed limits?

20. A wheel of radius R *rolls* on a horizontal surface at constant speed v_c (= speed of hub):

   (a) Find the coordinates $x(t)$ and $z(t)$ of the point P' on the rim of the wheel, which at time $t = 0$ touched the ground. (See the following figure.)

   (b) In a coordinate system moving with the wheel center, P' is in uniform circular motion. The net motion is the vector sum of the velocity as observed in this moving coordinate system, plus the vector velocity of the wheel center relative to the ground. Draw a vector diagram showing this sum and obtain the horizontal and vertical components of the velocity of P'.

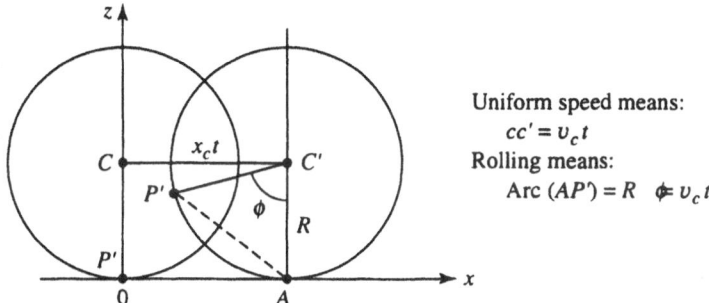

Uniform speed means:
$$cc' = v_c t$$
Rolling means:
$$\text{Arc } (AP') = R\ \phi = v_c t$$

(c) Determine the speed v of point P'. Show that it varies between 0 and $2v_c$.

(d) The velocity $\vec{v}$ of the point P' is perpendicular to the line AP'. Show this by using the scalar product.

(e) Show that the results of (b), (c), and (d) can be summarized by the statement that the velocity at any point of the periphery of the wheel is the same as if the wheel rotated about a fixed point A, with angular velocity $d\phi/dt = v_c/R$.

CHAPTER 2

Dynamics

2.1 Introduction

Whereas kinematics provides a mathematical description of motion, dynamics provides the source or "motive force" for that motion. For example, Galileo's kinematics describes that all bodies fall to Earth (in vacuo) with the same acceleration. Newton's dynamics tells us that this occurs because of the action of the "force of gravity." Furthermore, this self-same force of gravity can be used to *predict* the orbits and the motions of the planets through the sky.

The great advance from kinematics to dynamics was made by Isaac Newton (1642–1727). His three laws of motion are simple in statement yet profound in their implication. The power and utility of these three laws of motion can, in fact, only be appreciated after they have been applied to a wide variety of mechanical examples. Their proper application permits the solution of mechanical problems which at first sight seem impossible to analyze. Newton's laws of motion, when expressed in terms of conservation laws (such as conservation of momentum or energy), or when expressed in terms of the sophisticated mathematical language of "Hamilton's equations," provide a vital part of the foundations of thermodynamics and quantum mechanics. One can state Newton's three laws of motion and explain each, but their real significance and power comes to each of us only by applying them again and again to different problems in different branches of physics.

2.2 Newton's First Law: The Law of Inertia

This law is often expressed as follows. If a body has no *net* force acting upon it, it will not undergo a change in its state of motion. If stationary, the body will remain stationary; if moving the body will continue to move with constant velocity.

There are two difficulties with this form of the statement of the first law. The first difficulty is the use of the intuitive, but as yet undefined, notion of force. The second is that the coordinate system used to measure the motion is not specified. The term stationary, or uniformly moving, has meaning only in terms of measurements made in some particular coordinate system.

Because of these difficulties one can state Newton's first law in a second equivalent, but apparently more abstract, way (see [2]). Let us imagine an "isolated body" that is one which cannot interact with other bodies. Such an isolated body will clearly experience no "net force" upon it. In terms of such an "isolated body" we can express Newton's first law as follows. Coordinate systems exist in which an isolated body moves with constant velocity. Such a coordinate system is called an inertial coordinate system.

2.3   Newton's Second Law of Motion

This second law of motion relates the motion of a body to the "forces" acting upon it. Experimentally we may apply a force to a body by attaching to it a spring, and then pulling the spring out so that a definite extension of the spring is produced. If the extension of the spring is held constant during the motion, the force on the body will be constant. If we examine the motion of a body that is subject to such a single net force we would find the very simple result that the motion is one of constant acceleration; the velocity increases linearly with time. If we should apply the same force to a body of a different size we could find a different constant acceleration. In general, the "bigger" the body, the smaller the acceleration produced by a fixed force. From this we may conclude that for a fixed force the acceleration is proportional to the force. The constant of proportionality between the force and acceleration is called the "mass" of the body. We can assign a "mass" to each body by measuring its acceleration relative to a standard body of unit mass. If m_1 is the mass of the standard body, and m_2 the mass of a test body, then the mass m_2 can be related to m_1 by applying the same force to each body. The acceleration in each case is inversely proportional to the mass. Thus we have

$$a_1 = \frac{F}{m_1}, \qquad a_2 = \frac{F}{m_2}.$$

The mass of any test body can then be specified in terms of that of the standard body from the equation

$$m_2 = m_1 \left(\frac{a_1}{a_2} \right). \tag{2-1}$$

Mass then is a property of the body which measures its "inertia," i.e., the magnitude of its acceleration under the influence of a constant net external force.

One property of mass that emerges from experiments as described previously is that it is an "extensive" quantity: Imagine a piece of homogeneous material, out of which you cut two blocks, the second having double the volume of the first. Its mass, and hence its inertia, will be double that of the first block. So in a certain sense mass is a quantitative measure of the "amount of matter" in an object. But one must recognize that this is a loose way of speaking; two objects of different chemical composition have the same mass, if they experience the same acceleration under the same force. The primary definition of mass is that it is a measure of an object's inertia.

Let us now examine the effect of increasing the "force" when the mass of the test body is constant at say m_2. If we should, for example, stretch the spring to twice its original extension, or attach two identical springs and pull in the same direction, and with the same extension as with one alone, we would find that the acceleration would double. In fact, then the acceleration is directly proportional to the force when the magnitude of the force changes. Thus, in this case, one can write

$$\vec{F} = m\vec{a}.$$

One can use this as a means of determining force. By applying the force to a test body whose mass is known relative to some standard, one can determine the magnitude of the applied force by measuring the magnitude of the resulting acceleration. Also the direction of the acceleration produced in a body is the same as the direction of the force applied. This suggests that the relation $F = ma$ is a "vector equation," i.e., that F is a vector, a is a vector in the same direction, and m is a scalar. To determine whether this is the case, we must examine whether forces obey the laws of vector addition. By this we mean the following. Suppose a body is subject to several forces, each acting in a particular direction and with a particular magnitude. Is the resultant acceleration of the body the same or different from that which would obtain if that body were subject to a single force equal to the vector sum of the individual forces? It is by no means obvious that this should be the case. However, experimentally it is found to be true. Just as in the case of the superposition of velocities which applied in kinematics, so in the case of dynamics the direction and magnitude of the acceleration of a body is the vector sum or the vector superposition of the individual acceleration produced by each of the applied forces acting separately. The magnitude and direction of that acceleration is that which would be produced by a single force equal in magnitude and direction to the vector sum of the individual forces acting. Thus force can be regarded as a vector quantity in that several forces add like vectors in producing their net effect. Newton's second law can be stated as a vector formula

$$\sum_i \vec{F}_i = m\vec{a}. \tag{2-2}$$

The acceleration ($\vec{a}$) of a point body upon which several forces $\vec{F}_i$ act is directly proportional to the vector sum ($\sum_i \vec{F}_i$) of all the forces, and inversely proportional to the mass m of the body.

It is implicit in the statement of the second law of motion that the coordinate system in which the acceleration is measured must be an inertial coordinate system. Only in such a coordinate system will the acceleration be zero when the total force is zero.

2.4   Newton's Third Law of Motion: The Law of Action and Reaction

The forces that produce motion may be produced by gravity, or by magnetism, or by electrical charges, etc. In every case, however, we find that the force experienced by one body (call it i) is produced by another body or bodies (j). This force, however, is not produced solely by the action of body j on body i, but by the interaction between them.

We may inquire: If F_{ij} is the force acting on body i produced by some other body j, what is the reaction force F_{ji} produced on j by body i? When put in the form of this mathematical language we are in essence asking for the "symmetry properties" of the forces of interaction between bodies—regardless of the detailed origin of that force. Newton's third law tells us that the action and reaction forces are equal and opposite to one another, i.e., $\vec{F}_{ij} = -\vec{F}_{ji}$. In Newton's own words:

> To every action there is always opposed an equal reaction: or the mutual actions of two bodies upon each other are always equal, and directed to contrary parts [3].
>
> Whatever draws or presses another is as much drawn or pressed by that other. If you press a stone with your finger, the finger is also pressed by the stone. If a horse draws a stone tied to a rope, the horse (if I may so say) will be equally drawn back toward the stone; for the distended rope, by the same endeavor to relax or unbend itself, will draw the horse as much toward the stone as it does the stone toward the horse, and will obstruct the progress of the one as much as it advances that of the other.

A consequence of the third law of motion is that an isolated body cannot exert a force upon itself. If a body is observed to accelerate there are just two possibilities. The first is that the coordinate system in which the motion is viewed is not inertial. The second is that the body is acted on by some external force. If this is the case then the existence of that force can be checked by observing the reaction force acting between the accelerated body and that body with which it is interacting.

Newton's third law applies to point bodies as observed in inertial coordinate systems. If we consider an extended body, the internal forces which act between the parts of the body will, in accordance with the third law, add in pairs to give a net internal force of zero. The net internal force acting between all parts of the body is zero. This leads to the important principle of the conservation of momentum for extended bodies, as we shall see in Chapter 4.

These then are Newton's three laws of motion:

(1) that there are coordinate systems (inertial) in which a body subject to no external force moves with constant velocity;

(2) that the total force acting on a body produces acceleration in inverse proportion to the mass: $\sum_i \vec{F} = m\vec{a}$;

(3) that a force represents equal and opposite interaction between bodies ($\vec{F}_{ij} = -\vec{F}_{ji}$).

These three apparently very simple statements are the basis for the dynamics of moving bodies. If we add "constitutive relations," that i,s a description of the nature of the forces that act in particular cases, e.g., the form of electric force, the gravitational force, etc., we have a complete basis for classical mechanics.

2.5   The Fundamental Forces of Physics

To complete our formulation of the basic laws of mechanics, it is necessary to specify the nature of the forces that act between bodies. These forces can be divided into two classes: the fundamental forces and the derived forces. The fundamental forces represent the irreducible minimum of independent elementary interactions which permit the deduction of all the remaining "derived forces."

The fundamental forces of physics are four in number. These are

(1) the force of gravity;

(2) the electromagnetic forces;

(3) the "strong" nuclear force; and

(4) the "weak" nuclear forces.

The derived forces include such forces as friction, viscosity, surface tension, and elasticity. All these can be derived from the fundamental forces. In fact they are derived from just one of these, i.e., the electromagnetic force. We shall not at this time give a detailed description of each of these derived forces. As they arise in our consideration of physical situations we will explain the properties of each in turn.

Of the four fundamental forces only the first two will really be of concern to us. The two nuclear forces are of very short range and are zero when the interacting particles are separated by distances greater than about 10^{-13} cm. They determine the structure and stability of atomic nuclei; for example, that the Oxygen nucleus ^{16}O is stable, the Potassium isotope ^{40}K radioactive, and that Uranium ^{235}U can disintegrate by fission. But having acknowledged their role in determining that atomic species make up our observable world, we can dismiss them from our consideration.

The gravitational and electromagnetic forces are at the origin of the phenomena we shall be discussing in mechanics, statistical physics, electricity and magnetism, oscillations, waves, and fluid dynamics.

The most spectacular effects of the force of gravity are to be found in the motion of the astronomical bodies, the stars, planets, galaxies, etc. It is, of course, responsible for the weight of terrestrial bodies and the uniform gravitational acceleration found by Galileo.

The electromagnetic forces, when combined with the laws of atomic dynamics (quantum mechanics) are responsible for the structure of atoms, molecules, and solids, and for the interactions between them. It is interesting to realize that the dynamics of atomic particles are in many ways similar to the classical dynamics of Newton's laws. The essential features of indeterminacy and quantization and exchange symmetry, however, play a vital role in determining motion at the atomic level.

2.5.A.   The Gravitational Force

The form of the gravitational force was discovered by Isaac Newton, probably during the period 1666–1667 in which he discovered the three laws of motion. His discoveries were published only in 1687 as the result of the persuasion of his close friend Edmond Halley. Halley had come to Newton in 1684 for advice on the question as to what force could produce the elliptical orbits of the planets around the Sun. The detailed kinematics of planetary motion had been found by Johannes Kepler (1571–1630), after very careful analysis of the precise observational measurements made by Tycho Brahe (1546–1601) over the period of an entire lifetime. Newton told Halley that he knew the form of the law of force "and much other matter." Newton's extraordinary mathematical skill enabled him to prove quite rigorously that the force responsible for the planetary motion was one of attraction along the line joining the Sun and the planet, and also that the force had to vary inversely with the square of the distance between the Sun and the planet. He also reasoned that the force of attraction was proportional to the product of the masses of the two bodies. To Newton, whose mind saw unity where others saw diversity, it was clear that the force of attraction between the Sun and planets had to be the same as the force that attracts a falling body toward the center of the Earth. And so he was led to the formulation of the law of universal gravitation, namely, that between *any* two point bodies (1) and (2) there exists a force of gravitational attraction that acts along the line of centers, and whose magnitude is given by

$$F_{12} = G \frac{m_1 m_2}{r_{12}^2}.$$ (2-3)

The constant of proportionality is the so-called gravitational constant. Its magnitude was first measured by Henry Cavendish in 1771 by a direct measurement, using a torsion balance, of the force of attraction between massive balls. The present value for G is

$$G = (6.67 \times 10^{-8}) \frac{\text{cm}^3}{\text{gm s}^2}$$

in the CGS system of units. Force, which has the unit of gm cm/s^2 is given the unit of the dyne, 1 dyne being equal to 1 gm cm/s^2. In the MKS system of units in which length is in meters, mass is measured in kilograms, time in seconds, the combination 1 kg m/s^2 is called 1 Newton. And in these two systems we can write G as

$$G = 6.67 \times 10^{-11} \text{ N m}^2/\text{kg}^2$$

or   (2-4)

$$G = 6.67 \times 10^{-8} \text{ dyne cm}^2/\text{gm}^2.$$

It is interesting to observe that the "gravitational mass" which appears in the law of gravitational force could, in fact, be different from the "inertial mass" which appears in Newton's second law of motion. It is possible to measure experimentally the ratio of the gravitational and inertial masses of a body. Such experiments have shown that these two conceptually distinct masses are actually identical within one part in 10^{11} [4]. The identity of gravitational and inertial mass is an accident in classical mechanics. However, this identity is the experimental basis of the "principle of equivalence" which is one of the foundations of Einstein's theory of relativity.

Newton's law of universal gravitation provides an immediate explanation for Galileo's kinematics of falling bodies, i.e., the motion of a falling body is one of constant acceleration whose magnitude (g) is independent of the mass of the body. This is seen as follows: The gravitational force on a body of arbitrary mass m_1 is*

$$F_{\text{grav}} = G\frac{m_1 m_e}{r_e^2}. \tag{2-5}$$

Here m_e and r_e are the mass and radius of the Earth. According to Newton's second law, this force will accelerate the mass m_1 in accordance with the equation

$$F_{\text{grav}} = m_1 g, \tag{2-6}$$

where g is the acceleration. Clearly the force is directly proportional to the mass m_1, and the acceleration is inversely proportional to the mass m_1. Thus the acceleration is quite independent of the mass of the body. A massive body is pulled more strongly to Earth—it is "heavier"—than a body with small mass. However, the body actually accelerates under this force in inverse proportion to its mass. This leads to constant acceleration independent of mass. The gravitational acceleration g is then

$$g = \frac{Gm_e}{r_e^2} \tag{2-7}$$

at sea level and latitude 45° this is equal to 980.6 cm/s^2 or 9.806 m/s^2.

The constancy of the gravitational acceleration permits the simple determination of the mass of a body. All that is needed is to measure its "weight." The weight of a body is the *force* exerted on it by the gravitational pull of the Earth.

---

*The Earth is not a point body. This expression needs to be justified. See the discussion on page 54.

Having measured this with either a balance bar or a spring, one can immediately obtain the mass of the body from the equation, Weight $= mg$. The weight of a body is measured in CGS units of dynes, in MKS units in Newtons. The unit of force in the English units is one pound. The mass unit in English units is the slug. To obtain the mass in slugs from the weight in pounds one simply divides by the gravitational acceleration expressed in ft/s^2. In these units $g = 32.17$ ft/s^2, m (slugs) $= W$ (pounds)$/g$ (ft/s^2).

2.5.B.   Motion of a Satellite or the Moon Around the Earth

A combination of Newton's law of universal gravitation, the laws of motion, and the kinematics of circular motion, permits us to understand the motion of man-made satellites or the Moon about the Earth. In the general case, the orbit of such bodies is an ellipse; however, it is possible in the interest of simplicity to regard the motion as being perfectly circular. This will permit a sufficiently accurate analysis of the motion.

Let us begin by considering the speed which a body must be projected horizontal to the surface of the Earth so that it neither returns to the Earth, nor enters an orbit which will substantially increase its distance from the Earth. This is the speed with which a satellite must be injected into circular orbit around the Earth.

If there were no force acting on such a satellite, it would of course move in a straight line with constant velocity. In order to travel in a circle it must experience a force which constantly turns the velocity vector inward toward the Earth. That is, the satellite must experience a gravitational force just large enough to provide the centripetal acceleration associated with the circular motion. Newton's thinking on this point is given below:

> A centripetal force is that by which bodies are drawn or impelled, or in any way tend, toward a point as center.
>
> Of this sort is gravity, by which bodies tend to the center of the Earth; magnetism, by which iron tends to the lodestone; and that force, whatever it is, by which planets are continually drawn aside from the rectilinear motions, which otherwise they would pursue, and made to revolve in curvilinear orbits. A stone, whirled about in a sling, endeavors to recede from the hand that turns it, and by that endeavor, distends the sling, and that with so much the greater force, as it is revolved with the greater velocity, and as soon as it is let go, flies away. That force which opposes itself to this endeavor, and by which the sling continually draws back the stone toward the hand, and retains it in its orbit, because it is directed to the hand as the center of the orbit, I call the centripetal force.

And the same thing is to be understood of all bodies, revolved in any orbits. They all endeavor to recede from the centers of their orbits; and were it not for the opposition of a ... force which restrains them to, and detains them in their orbits, which I therefore call centripetal, would fly off in right lines with a uniform motion ... nor could the Moon without some such force be retained in its orbit. If this force were too small, it would not sufficiently turn the Moon out of a rectilinear course; if it were too great, it would turn it too much, and draw the Moon down from its orbit toward the Earth. It is necessary that the force be of a just quantity, and it belongs to the mathematicians to find the force that may serve exactly to retain a body in a given orbit with a given velocity."

Let us then be Newton's mathematicians and find the velocity needed to keep a man-made satellite in orbit not too far above the surface of the Earth. That is, the radius of the orbit is approximately the same as the Earth's radius. From our study of acceleration in circular motion we know that the centripetal acceleration of circular motion is related to the speed (V) of the satellite and the radius of the orbit (r_e) by the equation

$$a_{\text{centripetal}} = V^2/r_e. \tag{2-8}$$

The force of gravity must, in accordance with Newton's second law, be equal to

$$F_{\text{grav}} = ma_{\text{centripetal}} = \frac{mV^2}{r_e}, \tag{2-9}$$

where m is the mass of the satellite. The gravitational force acting on the satellite depends on the mass of the satellite and its distance r from the center of the Earth in accordance with the formula

$$F_{\text{grav}} = \frac{Gm_em}{r^2}. \tag{2-10}$$

Since $r \simeq r_e$ the force of gravitational attraction is about equal to the weight of the satellite, i.e.,

$$F_{\text{grav}} = mg \tag{2-11}$$

as also follows from (2-7). Using then (2-11) and (2-9) together we find

$$mg = \frac{mV^2}{r_e}. \tag{2-12}$$

The mass cancels on both sides of this equation, and we see that the velocity required for a stable satellite orbit is

$$V = \sqrt{g r_e}. \tag{2-13}$$

Putting in the known values of $g = 9.8$ m/s^2, $r_e = 6.37 \times 10^6$ m. We find $V \simeq$ 7900 m/s or $V \simeq 4.95$ mile/s $= 17,800$ mile/h. An object moving with this speed, horizontal to the surface of the Earth, will not return to the Earth, but will travel in an orbit around it.

We can easily compute the time required for such a satellite to travel around the Earth. The period of the orbit is

$$T = \frac{2\pi r_e}{V}, \tag{2-14}$$

i.e., the ratio of the circumference of the orbit to the speed. Using the numerical values given above, we find $T = 84$ min.

In the preceding calculation we have assumed that the *total* gravitational force exerted on the satellite by each element of mass in the Earth acts between the center of the Earth and the satellite. Newton's law of gravitation applies to the force of attraction between two point bodies. Newton surmised (correctly as it turned out) that a spherical body whose mass is distributed uniformly throughout the sphere exerts a force inversely proportional to the square of the distance from the *center* of the sphere to the small body outside it. This is actually difficult to prove, and this difficulty was in fact part of the reason for the delay in Newton's writing of *Principia*. In *Principia* Newton presents a proof, using geometrical methods, which is a remarkable demonstration of his extraordinary mathematical ingenuity.

The analysis just given for a man-made satellite traveling in an orbit whose radius is about equal to the Earth's radius can be easily extended to calculate the time needed for the Moon to orbit the Earth. Astronomers have found this to be equal to 27.3 days. (Since the Earth rotates about its own axis, the apparent period of the Moon is 2 days longer than the "sidereal period" which is the figure of 27.3 days.)

If we assume for simplicity that the Moon's orbit is a circle of radius r_m, the balance between centripetal and gravitational force gives the equation

$$\frac{G m_e m_m}{r_m^2} = \frac{m_m V^2}{r_m}. \tag{2-15}$$

Here m_m is the mass of the Moon. Remembering that the gravitational acceleration at the surface of the Earth is Gm_e/r_e^2, we can write the left-hand side of (2-15) as follows by multiplying the numerator and denominator by r_e^2:

$$m_m g \left(\frac{r_e}{r_m}\right)^2 = \frac{m_m V^2}{r_m}. \tag{2-16}$$

The left-hand side of this equation is seen essentially to be the weight of the Moon, since $g(r_e/r_m)^2$ is the gravitational acceleration that a body would have if dropped with zero velocity toward the Earth from the height equal to the Earth–Moon distance.

In (2-16) the mass of the Moon cancels on both sides, and we find that the velocity of the Moon required for a circular orbit is

$$V = \left(\sqrt{g r_m}\right) \left(\frac{r_e}{r_m}\right). \tag{2-17}$$

The period of the motion is

$$T = \frac{2\pi r_m}{V} = 2\pi \sqrt{\frac{r_m}{g}} \left(\frac{r_m}{r_e}\right).$$

Using the known values of the distance from the Earth to the Moon $r_m = 3.8 \times 10^5$ km, and the radius of the Earth $r_e = 0.635 \times 10^4$ km, we obtain $T = 2.35 \times 10^6$ s. Since 1 day contains 8.64×10^4 s, this time is equal to 27.3 days in very good agreement with the actual value for the Moon's period, and is a dramatic demonstration of the correctness of Newton's law of gravitation. Here is how Newton himself commented on this calculation. Speaking of the year of discovery, 1666, he said

> And the same year I began to think of gravity extending to the orb of the Moon, and . . . from Kepler's Rule . . . I deduced that the forces which keep the Planets in their orbs must [be] reciprocally as the squares of their distances from the centers about which they revolve: and therefore compared the force requisite to keep the Moon in her orb with the force of gravity at the surface of the Earth, and found them to answer pretty nearly.

2.6   Nonfundamental or Derived Forces

2.6.A.   Contact Forces

Whenever two objects are in contact—say a book lying on a table—the two objects exert a force on each other. Without the support of the table the book would fall under the pull of gravity. The table exerts an *upward* force on the book equal in magnitude to its weight. At the same time, the table experiences a force, vertically *downward*, equal again to the weight of the book.

These observations tell us nothing about the "nature" of these forces of contact. In the light of the atomic–molecular structure of matter, and the electronic structure of atoms, we must conclude that these forces are electrostatic in nature.

In the example of the book on the table, the force exerted by the table on the book is perpendicular (normal) to the table, provided the tabletop is horizontal. If we tilt the table by an angle α, the book will first stay put until the tilt angle is raised to a critical value α_c at which the book will begin to slip (see Figure 2.1). In the tilted case, the gravity force Mg can be regarded as having two components: One, $N' \equiv Mg \cos \alpha$, being perpendicular to the tabletop, and the other, $T' = Mg \sin \alpha$, being parallel to the tabletop (see Figure 2.1).

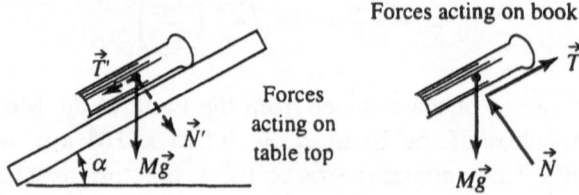

Figure 2.1. Object on an inclined plane surface.

If the book is at rest, the forces exerted by the tabletop must consist of two components: One (N) compensating N', and the other (T) compensating T'. Thus,

$$N = Mg \cos \alpha \qquad \text{and} \qquad T = Mg \sin \alpha. \tag{2-18}$$

The tangential component (T) of the force exerted by the table is called the force of friction.

Since we have observed that the book remains stationary until the angle of inclination α reaches a critical value α_c, we must conclude that the force of friction rises from $T = 0$ when $\alpha = 0$ to a maximum value $T = Mg \sin \alpha_c$ beyond which it cannot grow. It is possible to relate the maximum friction force to the magnitude of the normal force. This is done as follows. At the point of slipping, the normal force is $N = Mg \cos \alpha_c$ and $T = T_{max} = Mg \sin \alpha_c$, so that

$$T_{max} = N \tan \alpha_c. \tag{2-19}$$

The ratio of the maximum frictional force T_{max} to the normal force is defined as the coefficient of static friction μ_0:

$$\mu_0 \equiv \frac{T_{max}}{N}. \tag{2-20}$$

In the present example we see that one can determine μ_0 by measuring the angle at which slip occurs.

In general then we may characterize the force of friction as a quantity whose magnitude and direction are always such as to prevent slippage until the necessary friction force must exceed T_{max}. T_{max} is proportional to the normal force N, and its coefficient of proportionality is μ_0. μ_0 depends on the nature of the surfaces in contact. We give in Table 2.1 some typical values of μ_0 for different surfaces.

Table 2.1

Surfaces	μ_0
Wood on wood	0.25–0.50
Wood on brick	0.3–0.4
Ski wax on wet snow	0.1
Ski wax on dry snow	0.04
Rubber on solids	1–4

One might think, by a bit of clever generalization from the above experiment, that the coefficient μ_0 could also be measured by finding the force necessary to pull the book over the horizontal table at constant speed (see Figure 2.2). In order to have the book moving at constant speed v, all forces on the book must again cancel (otherwise, an acceleration would result by Newton's second law). This means that T would have to be equal to F in magnitude; in addition, since the book slips under the action of F, T is as large as possible, that is, equal to T_{max}. We might now think that in such a situation T is again related to N by the coefficient μ_0. However, experiment shows that while T_{max} is proportional to N, i.e.,

$$T_{max} = \mu N. \tag{2-21}$$

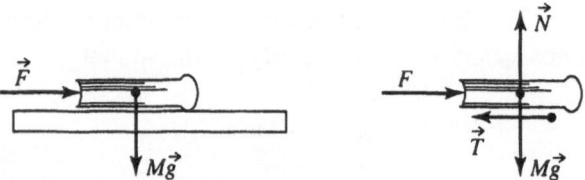

Figure 2.2. Sliding friction.

The coefficient μ now depends on the speed of the moving surface. Thus μ is called the coefficient of sliding friction. Its magnitude is generally smaller than μ_0, and it decreases with speed v. A typical variation of μ with v is shown in Figure 2.3.

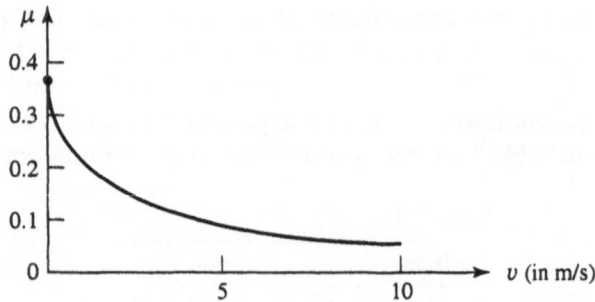

Figure 2.3. Coefficient of sliding friction μ, as a function of speed.

Returning now to the normal component of the force of contact (in our case, the force exerted by the table on the book), we may ask ourselves: What is the process by which this force always manages to have just the required magnitude, such as to compensate exactly the normal component of the weight of the supported object? If we imagine a support less rigid than a tabletop, the answer to this question emerges more easily (see Figure 2.4). A plank supported near the ends and loaded in the middle will bend. It will bend more the heavier the load. We see in this example that the manner in which the support "generates" the force N on the load is by means of an elastic deformation. N increases with deformation, and for any given load the plank will yield until a state of deformation is reached, for which N is equal to Mg in magnitude. Thus, $\vec{N}$ is an *elastic force*, generated by the deformation of the support. (Of course, this statement says nothing about what elasticity "really" is; again, the force generated by the elastic deformation of a body is, in the last analysis, electrostatic in origin.)

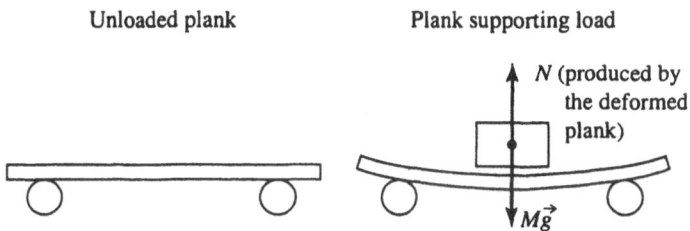

Figure 2.4. Unloaded and loaded plank.

Most forces of contact between solid bodies are such elastic forces; this fact is often hidden by the circumstance that for rigid objects, like a tabletop, the elastic deformation associated with N is so small as to escape casual detection. In the device known as a *spring*, by contrast, we have an elastic object that was designed to yield in response to a load in an observable and predictable way. Properly designed, it will undergo shortening or stretching in direct proportion to the applied load, and thus may be used to measure such loads, as in a spring balance.

The same elasticity, and elastic yield in response to an applied force, holds for a string or rope; it may not be easy to show this convincingly with a piece of parcel string, but anyone who has experience with a mountaineer's nylon rope has seen this point dramatically demonstrated.

2.6.B.   Drag Forces in Fluids (Liquids and Gases)

Any solid object moving through a fluid at speed v experiences a force in the direction opposite to its motion. This force is called the fluid drag, or simply drag. Except at very low speed, it is the net effect of phenomena of great complexity; it depends on the speed, size, and shape of the object, and on the density, viscosity, and compressibility of the fluid.

The resistance experienced by a fast-moving car, by an airplane, or by a falling raindrop, are examples of air drag. In these cases (excluding supersonic speeds), the resistance is roughly proportional to the square of the object's speed v, to the density ρ of air, and to the area A of the object in a plane perpendicular to the object's motion

$$F_{\text{Drag}} = \tfrac{1}{2} C_D A \rho v^2, \tag{2-22}$$

C_D is the (empirical) drag coefficient, and is a pure number of order 1.

The drag force in (2-22) is *independent* of the viscosity of the fluid. Even without any quantitative concept of viscosity, we know that honey is more viscous than water, and we know that a small pebble dropped into a glass of water falls faster than if dropped into a jar of honey. Indeed, for very small velocities (2-22) does not hold; in this case, the drag force *is* proportional to the viscosity of the fluid, and in addition depends linearly on the velocity. For the special case of a sphere of radius R, an exact expression for this force can in fact be derived from fluid dynamics:

$$F_D = 6\pi\eta Rv, \tag{2-23}$$

where η is the viscosity of the gas or liquid. (One may in fact then use (2-23) as a means of *defining* viscosity, if one has ascertained that F_D is in fact proportional to R and v.) Equation (2-23) is called Stokes' law. It applies to such cases as the settling of tiny water droplets in fog (drop diameter $< 10^{-2}$ cm), or to a grain of sand settling in water. But more importantly, it also applies to the determination of the sedimentation rate of macromolecules in an ultracentrifuge. (This will be discussed in Volume II.) From the measured rate it is possible—by using Stokes' law—to determine the radii of macromolecules. Furthermore, since the weight of a macromolecule is proportional to R^3, while the viscous resistance is proportional to R, molecules of different size sediment at differing rates. This fact is the basis of devices which provide the spatial separation of molecules of different molecular weight.

2.7   Newton's Laws Applied to Problems in Dynamics

2.7.A.   Introduction

Newton's laws are deceptively simple in statement. Their full meaning and power becomes clear only after much practice, applying them to a great variety of situations.

Only in the simplest case is it immediately clear what is to be done. If we have a single mass m, acted on by a known force, specified by both magnitude and direction, then the problem is reduced to one of mathematics: Solve the equation: $m\,d\vec{v}/dt = \vec{F}(\vec{r}, \vec{v}, t)$. This may or may not be an easy problem, but in any case the task is clear, and we will later give some examples of this type.

More generally, however, one deals with systems of several interacting bodies, some in contact with each other, others having their motion constrained by rails, planes, cables, etc.

Confronted with such a situation, the initial question is often: How do I apply Newton's laws; I don't know all the forces at the outset? In this introductory section we provide a procedure which, if followed, will permit a systematic analysis and the solution of apparently complex problems in mechanics: Here are the steps to be taken:

One: Decompose the total system into each of its individual parts.

Two: For each part draw a force diagram which represents *all* the force vectors acting on that part, and label each force, whether or not its value is known. In the labeling process it is important to identify those force pairs that are equal and opposite by virtue of Newton's third law.

Three: Choose a coordinate system relative to which you can conveniently express the components of the forces, velocities, and accelerations. Make sure that the coordinate system is inertial. (Do not center your coordinate system on an accelerating body.)

Four: Write down the conditions of kinematical constraint. Here are some examples of "kinematical constraints":

(a) Two coupled railroad cars both have the same velocity.

(b) Two weights attached at the ends of a rope around a pulley move with opposite velocities.

(c) The bob of a pendulum moves along the periphery of a circle.

Five: Write down Newton's second law of motion for each part of the system. Decompose each of the forces into components along the chosen coordinate axes. For each body the mass times acceleration in each coordinate direction is equal to the sum of the force components in that direction.

Also add to these the equations of kinematical constraint.

Examine the resulting set of equations and identify the number of unknowns. Make sure that the total number of equations is equal to the total number of unknowns.

If this procedure is properly followed, the set of equations obtained represent the statement of the original physical problem in mathematical language. From this point on, the solution is strictly a problem of mathematics. Probably the single most common error of beginning students is to try to guess at the very outset the magnitude of the forces acting between the parts of a mechanical system. Generally this is quite impossible.

The student should understand that the procedure presented above will be at the basis of the discussions of the examples which follow. However, he should not expect that each step will be made explicit in every example. Nevertheless, by studying the examples to come, and by solving homework problems, it should become second nature to the student to proceed along the lines presented in the prescription outlined above.

2.7.B.   Illustrative Examples

(i)   Freight Train

We begin with a very simple example of a Diesel engine pulling a freight car and a caboose (Figure 2.5). This latter is equipped with a brake; if applied, this brake will generate a frictional force opposing the pull of the engine. In this first example we will identify each of the five steps mentioned in the previous section.

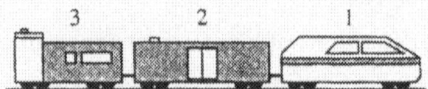

Figure 2.5. Train assembly.

Our *first* step is then to identify the system as consisting of three separate masses M_1, M_2, M_3, linked together by the couplings.

In the *second* step we represent each mass schematically, and draw all the forces acting on each, taking cognizance of Newton's third law (Figure 2.6). $\vec{F}_D$ is the driving pull of the engine; $\vec{F}_B$ is the braking force (friction) of the caboose. T_3 and T_2 are the tensions in the couplings between cars.

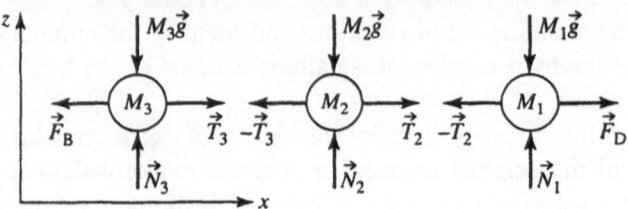

Figure 2.6. Forces on each of the cars in Fig. 2.5

Third, a convenient coordinate system is one with the x-axis in the horizontal, and the z-axis in the vertical direction.

As the *fourth* step, we recognize that due to the coupling of the cars, their velocity and accelerations must be the same. Also, none of the masses has a vertical acceleration.

In the *fifth* and last of the preliminary steps, we now state the equations, expressing Newton's second law and the equations of constraint: Beginning with the motion in the x-direction, we have the three equations

$$M_1 \frac{dv_{1x}}{dt} = F_D - T_2, \tag{2-24a}$$

$$M_2 \frac{dv_{2x}}{dt} = T_2 - T_3, \tag{2-24b}$$

$$M_3 \frac{dv_{3x}}{dt} = T_3 - F_B. \tag{2-24c}$$

It is useful to stick to the convention that symbols like $F, T, \ldots$ represent the magnitudes of the vectors $\vec{F}, \vec{T}, \ldots$, and hence are always positive. In (2-24a), for example, the x-component of $\vec{F}_D$ is F_D, that of $-\vec{T}_2$ is $-T_2$. The x-component of the motion of the three cars is constrained by the relation

$$v_{1x} = v_{2x}, \qquad v_{2x} = v_{3x}. \tag{2-25}$$

Proceeding to the z-components, we notice that the constraints $v_{1z} = v_{2z} = v_{3z} = 0$ also express vanishing acceleration in the z-direction; hence Newton's second law states that the sum of the force components in the z-direction vanish for each mass

$$-M_3 g + N_3 = 0,$$

$$-M_2 g + N_2 = 0,$$

$$-M_1 g + N_1 = 0.$$

These equations determine the forces $\vec{N}$ exerted by the rails on the cars as equal to each car's weight.

We now turn our attention to (2-24) and (2-25). We will assume that the engine pull F_D is known, as well as the masses. We have a total of five equations, and five unknowns $v_{1x}, v_{2x}, v_{3x}, T_2, T_3$. F_B, the force of braking, presents a more subtle problem; consider what was said about frictional forces in Section 2.6. To handle it properly, we must distinguish the two cases, one where $F_B^{max} > F_D$, and the train is prevented from moving, and the case where the train can move:

(a) The brake applied to the last car prevents the train from moving. In this case, all three velocities are zero. Instead of (2-25), we have actually now three constraint equations

$$v_{1x} = v_{2x} = v_{3x} = 0 \qquad (2\text{-}25\text{a})$$

together with (2-24). This is a set of six equations, which allows the determination of F_B as the sixth unknown. The solution of (2-24), using (2-25), is now

$$\left. \begin{array}{l} F_D = T_2 \\ T_2 = T_3 \\ T_3 = F_B \end{array} \right\} \quad \text{or} \quad F_B = T_2 = T_3 = F_D.$$

The tensions in the couplings between cars are *equal* to the engine pull F_D, and the braking force $\vec{F}_B$ is equal and opposite to $\vec{F}_D$.

Up to this point, any student could probably have produced these answers by "intuition," and the great detail mainly serves the purpose of illustrating the "prescriptions" in the simplest possible context.

(b) We now proceed to the case where the brake is released somewhat, so that F_B is smaller than F_D, and the train accelerates. F_B is now *not* an unknown any more: Since the train moves F_B assumes the maximum value consistent with the pressure applied on the brake shoes, and this pressure is controlled by the brakeman. Our unknowns are now indeed the five quantities v_{1x}, v_{2x}, v_{3x}, T_2, and T_3, for which we have just five equations.

At this point, the purely mathematical part begins. Equation (2-25) allows us to introduce a single unknown velocity $v_x = v_{1x} = v_{2x} = v_{3x}$, in terms of which we can write (2-24a,b,c) as

$$M_1 \frac{dv_x}{dt} = F_D - T_2, \qquad (2\text{-}26\text{a})$$

$$M_2 \frac{dv_x}{dt} = T_2 - T_3, \qquad (2\text{-}26\text{b})$$

$$M_3 \frac{dv_x}{dt} = T_3 - F_B. \qquad (2\text{-}26\text{c})$$

We have reduced the problem to three equations for the three unknowns v_x, T_2, T_3.

T_2 and T_3 can be eliminated by adding the three equations (2-26a,b,c):

$$(M_1 + M_2 + M_3)\frac{dv_x}{dt} = F_D - F_B. \tag{2-27}$$

This is the equation of motion for the train as a whole, and gives us its velocity in terms of F_D and F_B. There remains the task of finding T_2 and T_3. For this purpose, we solve (2-27) for dv_x/dt and insert into (2-26a) and (2-26c):

$$\frac{M_1}{M_1 + M_2 + M_3}(F_D - F_B) = F_D - T_2,$$

$$\frac{M_3}{M_1 + M_2 + M_3}(F_D - F_B) = T_3 - F_B,$$

from which simple algebra gives us

$$T_2 = \frac{M_2 + M_3}{M_1 + M_2 + M_3}F_D + \frac{M_1}{M_1 + M_2 + M_3}F_B, \tag{2-28a}$$

$$T_3 = \frac{M_3}{M_1 + M_2 + M_3}F_D + \frac{M_1' + M_2}{M_1 + M_2 + M_3}F_B. \tag{2-28b}$$

With this, the problem is solved, apart from the mathematical problem of finding v_x from the equation for dv_x/dt.

The results of (2-27) and (2-28) are worth some discussion: First, we observe that (2-27) states that the total mass of the train times its acceleration in the x-direction is equal to the x-component of the total net force on the train. We could have found this result more quickly by considering the train as a whole as a single object of mass $(M_1 + M_2 + M_3)$ by labeling only the external forces F_D and F_B on this single unit. This example then illustrates the important point of the *consistency* of results obtained by applying the above-mentioned set of prescriptions, *irrespective* of the manner in which a system is divided into elements. The student should also appreciate the role played by Newton's third law in guaranteeing this consistency.

What we *gain* by dividing the system into smaller units is information on the forces of interaction between these parts, in our case, the forces T_2 and T_3. The magnitudes of these forces are of some interest; let us inspect them in the case that the brake has been released completely, so that $F_B = 0$. In this case

$$T_2 = \frac{M_2 + M_3}{M_1 + M_2 + M_3} F_D,$$

$$T_3 = \frac{M_3}{M_1 + M_2 + M_3} F_D. \tag{2-29}$$

We see that T_2 is equal to the engine pull times the fraction of the total mass represented by the two cars behind the engine, whereas T_3 is F_D times the fraction of the total mass represented by the last car. Upon some reflection, this result appears "obvious." $\vec{T}_2$ must impart to the two cars $(M_2 + M_3)$ the same acceleration as $\vec{F}_D$ must do to the whole train $(M_1 + M_2 + M_3)$, and therefore represents that fraction of the total pull. The same applies to $\vec{T}_3$.

Looking at the general result for the tensions T_2 and T_3 between the cars of the accelerating train (2-28), the student will recognize that no amount of guesswork will have predicted the forces acting on each car. However, by following the systematic procedure we outlined, the entire solution emerges naturally.

(ii)   Block on an Inclined Plane

Consider a wedge, sliding without friction on a horizontal surface. On the inclined surface of the wedge a block is placed, which again may move along this inclined surface without friction. The problem we want to solve is: What horizontal force $\vec{F}$ must be applied to the wedge to keep the block stationary on the inclined plane? (See Figure 2.7.)

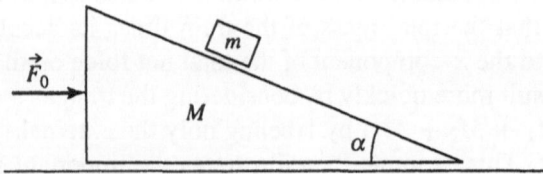

Figure 2.7. Block on a movable wedge.

This is a purely didactic example, which again, however, offers a great deal of insight into the meaning of Newton's laws, and shows the rewards of adhering to a systematic procedure in attacking the problem. The two components of this system are the wedge, of mass M, and the block, of mass m. Let us make a diagram showing all the forces acting on each part (Figure 2.8). In these diagrams, $\vec{P}$ is the force

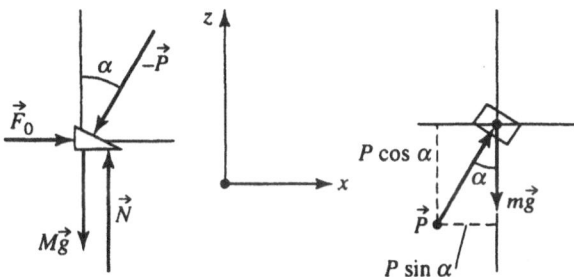

Figure 2.8. Force diagram for a wedge (mass M) and for a block (mass m).

exerted by the wedge on the block. It is normal to their common surface, since the frictional force is assumed to be zero. Since the wedge too slides without friction on the horizontal supporting surface, the force $\vec{N}$ exerted by this surface on the wedge is vertically upward. As a convenient coordinate system, we choose a horizontal x-axis, and a vertical z-axis. In this coordinate system the force components on the wedge and block are

On the wedge:
$$\begin{cases} F_x = F_0 - P \sin \alpha, \\ F_z = N - Mg - P \cos \alpha, \end{cases} \qquad (2\text{-}30\text{a})$$

On the block:
$$\begin{cases} F_x = P \sin \alpha, \\ F_z = -mg + P \cos \alpha. \end{cases} \qquad (2\text{-}30\text{b})$$

Next consider the constraints. The problem asks for the force $\vec{F}_0$ necessary to keep the mass m in a fixed position on the wedge. Their velocity components in the x-direction must therefore be the *same*; their velocity components in the z-direction must both be zero. Let $\vec{V}$ be the velocity of the wedge, $\vec{v}$ that of the block; the constraint equations then read as follows:

$$V_x = v_x, \qquad (2\text{-}31\text{a})$$

$$V_z = 0, \qquad (2\text{-}31\text{b})$$

$$v_z = 0. \qquad (2\text{-}31\text{c})$$

There are thus three equations of constraint; the four force components of (2-30) will give rise to four equations of motion, when Newton's second law is applied to them. So we will have a set of seven equations. The unknowns are: F_0, N, P, V_x, V_z, v_x, v_z, also seven in number. We thus have the correct number of equations.

The next and final preparatory step is to write Newton's second law for each of the four force components listed in (2-30). In doing so, for the z-components of $\vec{V}$ and $\vec{v}$, we recognize (2-31b and c), which require zero acceleration in the z-direction, and hence require that the net force on each body be zero:

$$N - Mg - P \cos \alpha = 0, \tag{2-32a}$$

$$-mg + P \cos \alpha = 0. \tag{2-32b}$$

The motion in the x-direction obeys the equations

$$M \frac{dV_x}{dt} = F_0 - P \sin \alpha, \tag{2-33a}$$

$$m \frac{dv_x}{dt} = +P \sin \alpha. \tag{2-33b}$$

Equations (2-32) give at once

$$N = (M + m)g$$

and

$$P = \frac{mg}{\cos \alpha}.$$

Inserting this into (2-33), and using (2-31a), we can write (2-33a) as

$$M \frac{dV_x}{dt} = M \frac{dv_x}{dt} = F_0 - mg \tan \alpha$$

and (2-33b) as

$$\frac{dv_x}{dt} = g \tan \alpha. \tag{2-34}$$

Thus, we know dv_x/dt from this last equation, and inserting its value into the equation containing F_0, we find

$$F_0 = (M + m)g \tan \alpha. \tag{2-35}$$

With this, all the unknowns are determined. We know F_0, the accelerations $dv_x/dt = dV_x/dt$, and the two forces of constraint, $\vec{N}$ and $\vec{P}$.

It is again worthwhile, in retrospect, to see whether these results make sense, from a more intuitive point of view. First, we observe the consistency of (2-34) with (2-35): Since the block does not slide on the wedge, the force F_0 must give a common acceleration dv_x/dt to a total mass $(M + m)$, and F_0 has indeed that value: $F_0 = (M + m)g \tan \alpha = (M + m) \, dv_x/dt$.

Another, generally useful, test is the inspection of limiting cases, here the cases $\alpha = 0$ and $\alpha = \pi/2$. For $\alpha = 0$, $F = 0$: One doesn't have to push to keep the block on its relative position on the wedge; for $\alpha = 90°$, $F_0 = \infty$: no finite push (no finite acceleration in the x-direction) can keep the block from falling. The limiting forms of the result make sense.

(iii) Harmonic Oscillation

A very important "simple" problem, with a single mass acted on by a specified force, is exemplified by the prototype of all elastic vibrations: A mass M, attached to a spring, which slides without friction on a horizontal plane (see Figure 2.9): (a) shows the mass at rest (the spring is relaxed); this is the equilibrium position, and we *choose* the coordinate origin so as to make this position coincide with $x = 0$. In (b) and (c), the spring is compressed (stretched), and the mass at position x. x thus measures the degree of deformation of the spring; for $x > 0$, the spring is stretched, for $x < 0$, it is compressed. The force exerted by the spring *on* the mass is given by an expression of the form

$$F_x = -kx. \tag{2-36}$$

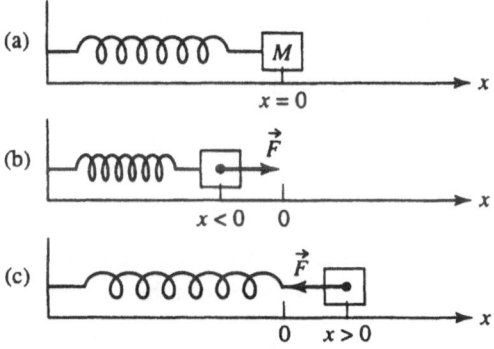

Figure 2.9. Mass attached to a spring.

The − sign in (2-36) expresses the fact that F_x is to the *left* when the spring is stretched (and hence x is positive), but to the right when the spring is compressed (and x is negative). No other force besides that of the spring is acting in the x-direction. (You will by now also understand why we can omit reference to the forces in the z-direction!)

The problem at hand is now to find out what the motion of the mass is, if at any one time it has been displaced from the equilibrium position $x = 0$. Newton's second law tells us that

$$M\frac{dv_x}{dt} = M\frac{d^2x}{dt^2} = F_x = -kx.$$

We thus have established the *equations of motion*

$$\frac{d^2x(t)}{dt^2} = -\left(\frac{k}{M}\right)x(t). \tag{2-37}$$

This equation is of a qualitatively different kind from those previously encountered: The force, although well *determined*, is *unknown* yet to us, since it depends on the position $x(t)$ at time t, which is just the position coordinate whose value as a function of time we want to establish. Equation (2-37) has the structure of a relation between the second derivative of an unknown function $x(t)$ and the function itself. Such equations are called differential equations. We can offer here no general methods for handling this type of equation. It will help, however, to state in words the question that is asked by (2-37): Find a function $x(t)$, whose second derivative is proportional to that function. It is perfectly legitimate, at this point, to use one's store of knowledge of functions and their derivatives. The student, who by now will have practiced some calculus, will in fact have run across such a function. He will have seen the relations

$$\frac{d\sin\phi}{d\phi} = \cos\phi, \qquad \frac{d\cos\phi}{d\phi} = -\sin\phi.$$

We can go from there. We may next consider the functions $\sin(c\phi)$ and $\cos(c\phi)$, with c a constant. Their derivatives follow from the chain rule

$$\frac{d\sin(c\phi)}{d\phi} = \frac{d\sin(c\phi)}{d(c\phi)} \cdot \frac{d(c\phi)}{d\phi} = \cos(c\phi) \cdot c,$$

so we have

$$\frac{d\sin(c\phi)}{d\phi} = c\cos(c\phi) \tag{2-38a}$$

and similarly

$$\frac{d\cos(c\phi)}{d\phi} = -c\sin(c\phi). \tag{2-38b}$$

Combining the two (2-38a and 2-38b), it follows that

$$\frac{d^2\sin(c\phi)}{d\phi^2} \equiv \frac{d}{d\phi}\left(\frac{d\sin(c\phi)}{d\phi}\right) = \frac{d}{d\phi}(c\cos(c\phi)) = c\frac{d\cos(c\phi)}{d\phi} = -c^2\sin(c\phi).$$

Repeating this with $\cos(c\phi)$, we have the two results

$$\begin{aligned} \frac{d^2\sin(c\phi)}{d\phi^2} &= -c^2\sin(c\phi), \\ \frac{d^2\cos(c\phi)}{d\phi^2} &= -c^2\cos(c\phi). \end{aligned} \tag{2-39}$$

We have now in our hands two functions which satisfy an equation of the type (2-37). Instead of ϕ we have the time t, and a comparison with (2-39) shows that the role of c^2 is taken by the quantity (k/M), for which we shall introduce the symbol ω^2:

$$\left(\frac{k}{M}\right) \equiv \omega^2, \tag{2-40}$$

so the functions $\sin(\omega t)$ and $\cos(\omega t)$ will both satisfy (2-37). In fact, we can multiply $\sin(\omega t)$ with a constant a, $\cos(\omega t)$ with a constant b, add them:

$$a\cos(\omega t) + b\sin(\omega t)$$

and still have a solution of (2-37):

$$\frac{d^2}{dt^2}(a\cos(\omega t) + b\sin(\omega t)) = a\frac{d^2\cos\omega t}{dt^2} + b\frac{d^2\sin\omega t}{dt^2}$$

$$= a(-\omega^2 \cos \omega t) + b(-\omega^2 \sin \omega t)$$
$$= -\omega^2 (a \cos \omega t + b \sin \omega t).$$

It is seen here that a and b can in fact be arbitrary constants. So, in the absence of any supplementary information, we conclude that the solution $x(t)$ of (2-37) is of the form

$$x(t) = a \cos(\omega t) + b \sin(\omega t) \qquad (2\text{-}41)$$

with

$$\omega = \sqrt{\frac{k}{M}}. \qquad (2\text{-}42)$$

At this point the obvious question arises: What is the meaning of the two constants a and b? Since they are arbitrary, how are they determined? Here we do well to remember that Newton's law relates the *rate of change* of the velocity to the force acting on a mass. Assume we connect the spring to the mass at time $t = 0$, and that at this time, the mass has position x_0, and a velocity v_0. For times $t > 0$, the change of v and x, starting from the initial values x_0, v_0 is then controlled by (2-37), but to know the actual motion, x_0 and v_0 must be known and specified. The two constants a and b in (2-41) can in fact be expressed in terms of these *initial values* x_0 and v_0. From (2-41) we find that the velocity $v_x = dx/dt$ can be written as

$$v_x = -a\omega \sin(\omega t) + b\omega \cos(\omega t). \qquad (2\text{-}43)$$

If we now assume that x_0 is the value of x at $t = 0$, v_0 is the value of v_x at $t = 0$, and observe that $\cos(0) = 1$, $\sin(0) = 0$, we find

$$x_0 = a, \qquad v_0 = b\omega.$$

In this way, a and b are related to the initial values, and the completely specified solution of (2-37) may now be written as

$$x(t) = x_0 \cos(\omega t) + \left(\frac{v_0}{\omega}\right) \sin(\omega t). \qquad (2\text{-}44)$$

We also write the expression for the velocity

$$v_x(t) = -\omega x_0 \sin(\omega t) + v_0 \cos(\omega t).\tag{2-45}$$

The motion described by (2-44) and (2-45) is called the *harmonic oscillation*. We shall now inspect this motion a bit more closely.

First we observe that both the sine and cosine functions are *periodic functions*; if their argument increases by 2π, or a multiple of (2π), the values of both the sine and the cosine are again the same

$$\sin(\alpha + 2\pi) = \sin\alpha; \qquad \cos(\alpha + 2\pi) = \cos\alpha.$$

Therefore, by (2-44) and (2-45), increasing ωt by (2π) leads to the same values of x and v_x:

$$x\left(t + \frac{2\pi}{\omega}\right) = x(t), \qquad v_x\left(t + \frac{2\pi}{\omega}\right) = v_x(t).$$

This identifies $(2\pi/\omega)$ as the *period T of oscillation*, T being defined as that time interval after which both position x and velocity v_x return to their initial values. We thus have the important relation

$$\text{period } T = \frac{2\pi}{\omega} = 2\pi\sqrt{\frac{M}{k}}.\tag{2-46}$$

The information worth remembering obtained in this result is that the period of oscillation is proportional to the square root of the mass, and inversely proportional to the square root of the spring *constant k* of the spring. See also Figure 2.8.

Two more quantities must be defined in relation to harmonic motion. The first is the concept of the frequency f of oscillation, that is, the number of oscillations per unit time. f is simply given by the reciprocal period

$$f \equiv \frac{1}{T} = \frac{\omega}{2\pi}.$$

This also settles the meaning of $\omega = \sqrt{k/M}$ as 2π times the frequency of oscillation. ω is also called the *angular* frequency of oscillation.

The second is that of the *amplitude* of the oscillation. Equation (2-44) can be rewritten by using the fact that *any* two constants a, b can be expressed in the form

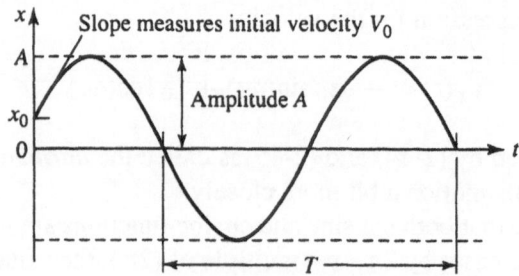

Figure 2.10. Graph of harmonic oscillation, displaying x_0, v_0, period T, and amplitude of oscillation.

$$a = \sqrt{a^2 + b^2} \cos \phi = A \cos \phi,$$

$$b = \sqrt{a^2 + b^2} \sin \phi = A \sin \phi,$$

$x(t)$, as given by (2-44), may therefore be rewritten as

$$x(t) = A \cos \phi \cos(\omega t) + A \sin \phi \sin(\omega t)$$

$$= A \cos(\omega t - \phi) \tag{2-47}$$

with A given by $A = \sqrt{x_0^2 + v_0^2/\omega^2}$. This way of writing the solution identifies the *amplitude A* of the oscillation. A measures the maximum value of the displacement of the mass from its equilibrium position. In the two special situations that either the initial velocity v_0 or the initial displacement are zero, we can write

$$x(t) = A \cos(\omega t), \qquad A = x_0, \; v_0 = 0,$$

or

$$x(t) = A \sin(\omega t), \qquad A = \frac{v_0}{\omega}, \; x_0 = 0.$$

(iv)   Spring Gun

In this device, a mass M_1 is attached to a spring; the compressed spring is then used to propel a projectile of mass M_2 placed alongside M_1 (Figure 2.11). The spring is first held in a compressed position, and initially both masses are at rest and in contact. As the trigger is pulled, the spring propels both masses; at a given time they separate and follow their independent courses. The problem is to determine the motion of both M_1 and M_2.

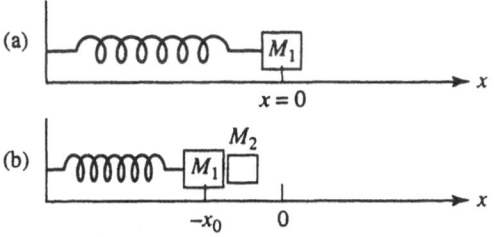

(a) Relaxed spring with mass M_1.
(b) Compressed spring ready to propel M_2.

Figure 2.11. Spring gun.

We have here a system of two components, M_1 and M_2. In labeling the forces on both masses, we shall omit here the forces of gravity and constraint acting in the z-direction. The two force diagrams are shown in Figure 2.12. $\vec{F}_s$ is the force exerted by the spring on M_1. The x-component of this force is

$$F_{s,x} = -kx.$$

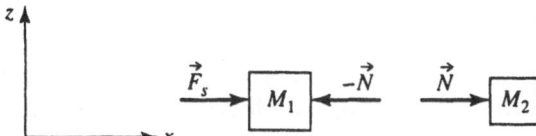

Figure 2.12. Forces on masses M_1 and M_2 in the spring gun.

The initial position of M_1 in the compressed spring is $x = -x_0$, and as long as the spring is compressed, the position x of M_1 is negative, and $F_{s,x}$ is positive, that is, the force pushes to the right.

It is important that we clearly distinguish two stages in the time sequence of events:

(a) As long as the spring is compressed, it will accelerate the mass M_1 to the right, and in this process will have to accelerate M_2 at the same rate. During this time interval we thus have a constraint condition

$$v_{1x} = v_{2x}. \tag{2-48}$$

(b) As soon as M_1 overshoots the position $x = 0$, the spring will pull M_1 to the left. However, the force of contact $\vec{N}_2$ on M_2 must push to the right, or else be zero, since the two masses are not glued together. The net force on M_1 will then point to the left, while on M_2 it will be either to the right, or be zero. This is enough to conclude that the masses will separate, and the force $\vec{N}$ will cease to exist. Therefore, in this second stage, which starts as soon as the position $x = 0$ has been reached by M_1, the force $\vec{N}_2$ will cease to exist, and M_2 will be in force-free motion along the x-axis.

Let us first take the time interval in which M_1 moves from $-x_0$ to 0 upon release of the trigger: Newton's law gives us

$$\text{applied to } M_1: \quad M_1 \frac{dv_{x_1}}{dt} = -kx - N, \tag{2-49a}$$

$$\text{applied to } M_2: \quad M_2 \frac{dv_{x_2}}{dt} = +N. \tag{2-49b}$$

In view of the constraint $v_{1x} = v_{2x}$, Equation (2-48), the same acceleration occurs in both equations, and upon addition, we have

$$(M_1 + M_2)\frac{dv_x}{dt} = -kx.$$

For x negative, dv_x/dt is > 0, hence the velocity increases in the $+x$-direction, and M_1 pushes on M_2. The motion is a harmonic oscillation, with angular frequency ω_{12}:

$$\omega_{12} = \sqrt{\frac{k}{M_1 + M_2}},$$

where

$$x(t) = -x_0 \cos(\omega_{12} t), \tag{2-50}$$

x starts from position $-x_0$ with zero initial velocity, and (2-50) conforms to the expression (2-44), which gives $x(t)$ in terms of the initial conditions.

The maximum velocity is reached at time

$$t_M = \frac{\pi}{2\omega_{12}}$$

when the cosine in (2-50) is zero (i.e., $x(t_M) = 0$), and the velocity

$$v_x(t) = \omega_{12} x_0 \sin(\omega_{12} t)$$

has the value

$$v_M = \omega_{12} x_0. \tag{2-51}$$

At this point, the two masses separate since M_1 is subsequently accelerated to the left, but cannot pull M_2 back. What are the subsequent motions of M_1 and M_2?

Consider first M_2. It is located at $x = 0$ at the time t_M, with velocity v_M, and free of any forces in the x-direction. So it will continue in uniform motion along the x-direction

$$x_2(t) = v_M(t - t_M) \qquad \text{for} \quad t > t_M. \tag{2-52}$$

Mass 1 is now subject to the spring force alone; its equation of motion is now

$$M_1 \frac{d^2 x_1}{dt^2} = -k x_1$$

and its motion is a harmonic oscillation with angular frequency $\omega_1 = \sqrt{k/M_1}$:

$$x_1(t) = a \sin(\omega_1 t) + b \cos(\omega_1 t). \tag{2-53}$$

Again one can determine a and b from that fact that at $t = t_M$, we *know* that $x_1 = 0$, and that $(dx_1/dt) = v_M$. Let us insert this into (2-53) and its derivative

$$v_1(t) = a\omega_1 \cos(\omega_1 t) - b\omega_1 \sin(\omega_1 t)$$

and obtain

$$0 = a \sin(\omega_1 t_M) + b \cos(\omega_1 t_M),$$
$$v_M = \omega_1 a \cos(\omega_1 t_M) - b\omega_1 \sin(\omega_1 t_M).$$

From these equations we may obtain a and b:

$$a = \frac{v_M}{\omega_1} \cos(\omega_1 t_M), \qquad b = -\frac{v_M}{\omega_1} \sin(\omega_1 t_M).$$

This can now be inserted into (2-53) with the result valid for $t > t_M$ only:

$$x_1(t) = \frac{v_M}{\omega_1} \big(\sin(\omega_1 t)(\cos \omega_1 t_M) - \cos(\omega_1 t) \sin(\omega_1 t_M) \big)$$

$$= \frac{v_M}{\omega_1} \sin \omega_1 (t - t_M). \tag{2-54}$$

Upon inserting the actual values of v_M and ω_1, we find a more transparent expression for the amplitude of this motion

$$v_M = \omega_{12} x_0, \qquad \omega_1 = \sqrt{\frac{k}{M_1}}, \qquad \text{so} \qquad \frac{v_M}{\omega_1} = \sqrt{\frac{M_1}{M_1 + M_2}} x_0.$$

So after firing off the mass M_2, the spring with mass M_1 attached oscillates with a reduced amplitude $\sqrt{M_1/(M_1 + M_2)}x_0$ (see Figure 2.13), but at a higher frequency $\omega_1 = \sqrt{k/M_1} > \sqrt{k/(M_1 + M_2)}$.

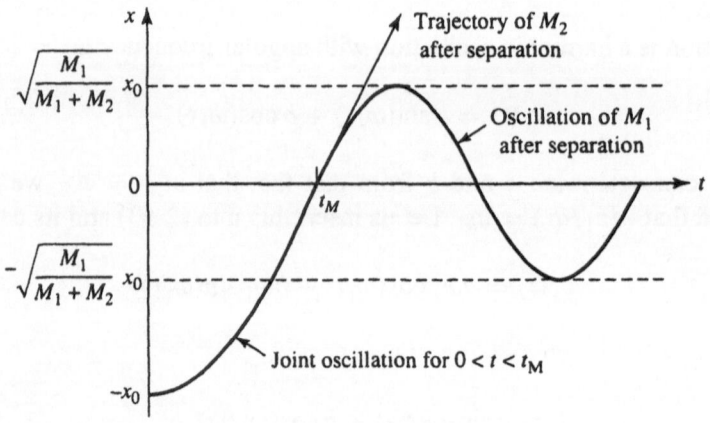

Figure 2.13. Trajectory of masses M_1 and M_2 in a spring gun.

(v) Viscous Damping Force. The Exponential Function

A body moving at low speed through a fluid experiences a viscous drag force, which is proportional to the velocity of the object relative to the fluid. In many cases of interest this drag force is accompanied by other forces acting on the mass; for instance, a small droplet of liquid ejected from a spray can move under the joint effect of gravity and viscous drag, etc.

In this example we discuss the motion of a mass m under the *sole* effect of such a velocity proportional drag force. There are then, in this problem, no preliminary steps of decomposing the system, no constraints, and we can proceed directly to the statement of Newton's second law. The force acting on the mass m has the form

$$\vec{F} = -\beta\vec{v}, \tag{2-55}$$

where β is a constant (which, for a spherical mass of radius R, moving in a fluid of viscosity η, has the value $\beta = 6\pi\eta R$, according to Stokes' formula (2-22)); Equation (2-55) states that the vector $\vec{F}$ is antiparallel to $\vec{v}$. Newton's law, as a *vector* equation, now states that the vector $d\vec{v}/dt$ is equal to the force vector divided by the mass m. In the present case then, $d\vec{v}/dt$ is proportional to $\vec{v}$ itself. Kinematically, this means that any change in $\vec{v}$ due to the force will be *parallel* to $\vec{v}$, and that the mass therefore moves in a straight line. A convenient coordinate system is then one with, say, the x-axis parallel to $\vec{v}$. Thus, we can write the equation of motion as

$$m\frac{dv_x}{dt} = -\beta v_x \qquad \text{or} \qquad \frac{dv_x}{dt} = -\left(\frac{\beta}{m}\right)v_x.$$

In this equation then, a single constant (β/m) appears. Comparing both sides of the equation, it is clear that (β/m) must have the dimension of a reciprocal time. We shall therefore express (β/m) in terms of a time constant τ_0 as $(\beta/m) \equiv 1/\tau_0$, and write the equation of motion in its final form

$$\frac{dv_x}{dt} = -\frac{1}{\tau_0}v_x(t). \tag{2-56}$$

Let us discuss this equation qualitatively at first. It states that v_x decreases with time at a rate given by the value of v_x itself. Assume that $v_x = v_0$ at time zero. After a small time interval Δt, v_x will have decreased to

$$v_x(\Delta t) \simeq v_0 - \Delta t \left(\frac{dv_x}{dt}\right)_{t=0} = v_0 - \Delta t \left(\frac{v_0}{\tau_0}\right). \qquad (2\text{-}57)$$

In Figure 2.14, we draw this initial rate of decrease of v, and also plot the subsequent behavior of $v_x(t)$. The initial slope of this curve at $t = 0$ is $-(v_0/\tau_0)$. If v_x would continue to decrease at this rate, then it would reach the value zero in time τ_0, as seen from (2-57); if we insert $\Delta t = \tau_0$:

$$v_0 - \tau_0 \left(\frac{v_0}{\tau_0}\right) = 0.$$

However, long before this happened, the slope of this curve has decreased. Consider the point P in Figure 2.14. $v(t)$ is smaller than v_0, hence the downslope of the curve at this point

$$\left(\frac{dv}{dt}\right)_{\text{point } P} = -\frac{v(t)}{\tau_0}$$

is less than the downslope at $t = 0$: The curve flattens out the more, the smaller v becomes, and v reaches zero only as t approaches infinity.

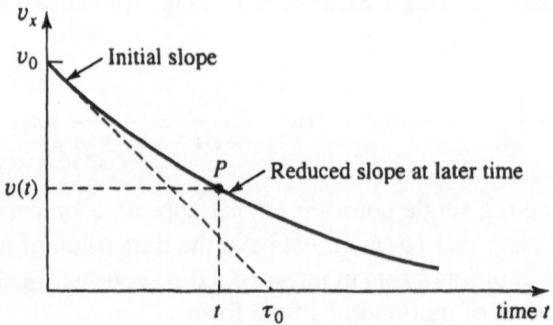

Figure 2.14. Motion of mass subject to drag force.

We now proceed to establish the shape of this curve of v versus time more precisely: The equation for $v_x(t)$ is what in mathematics is called a first-order differential equation because the value of the unknown function $v(t)$ is related to its *first* derivative. The question asked by this equation is then: What function is there which has a derivative proportional to the function itself? In case the student has heard already of the exponential function, he will be able to find the answer:

$$v(t) = v_0 e^{-t/\tau_0}.$$

It will, however, be worthwhile to proceed to a *construction* of this result. We have already seen that, starting at $t = 0$, the value of v after a small interval Δt (which, as we can see from Fig. 2.14, must be small compared to τ_0), the velocity has the value

$$v(\Delta t) = v_0 \left(1 - \frac{\Delta t}{\tau_0} \right). \tag{2-58}$$

Consider now the decrease of v in a second time interval of length Δt. v decreases to approximately

$$v(2\Delta t) \simeq v(\Delta t) + \Delta t \left(\frac{dv}{dt} \right)_{\Delta t} = v(\Delta t) - \Delta t \frac{v(\Delta t)}{\tau_0} = v(\Delta t) \left(1 - \frac{\Delta t}{\tau_0} \right)$$

and by using (2-58), we can write this as

$$v(2\Delta t) = v_0 \left(1 - \frac{\Delta t}{\tau_0} \right)^2. \tag{2-59}$$

Clearly, we can continue to proceed in this manner. After N steps we will have

$$v(N\Delta t) = v_0 \left(1 - \frac{\Delta t}{\tau_0} \right)^N. \tag{2-60}$$

At this point, we may well call the total elapsed time $N\Delta t$ by the letter t, which stands for elapsed time, and rewrite (2-60) as

$$v_x(t) \simeq v_0 \left(1 - \frac{t}{N\tau_0} \right)^N. \tag{2-61}$$

This is only an approximate answer because our steps Δt are finite. In order to reach a finite elapsed time t in steps $\Delta t = t/N$, the number of steps N must be the larger, the smaller we choose Δt, but the more accurate the result obtained will be. We will therefore have to choose N in (2-61) very large; we must "go to the limit

$N \to \infty$" as the mathematician would say. To see what happens in this limit, we first observe that since t and τ_0 are finite numbers, the number

$$n \equiv N \frac{\tau_0}{t}$$

will also get arbitrarily large as N goes toward infinity. In terms of n, (2-61) reads

$$v_x(t) \simeq v_0 \left(1 - \frac{1}{n}\right)^{n(t/\tau_0)}.$$

At this point we can make use of the algebraic rule that

$$A^{ab} = (A^a)^b$$

and rewrite $v_x(t)$ as

$$v_x(t) = v_0 \left[\left(1 - \frac{1}{n}\right)^n\right]^{t/\tau_0}. \tag{2-62}$$

This expression is now in a form in which we can let the number n go to infinity. The limit $\lim_{n\to\infty}(1 - 1/n)^n$ exists, and is equal to the inverse of the number e (the basis of natural logarithms)

$$\lim_{n\to\infty} \left(1 - \frac{1}{n}\right)^n = \frac{1}{e} = \frac{1}{2.7182\ldots}.$$

In terms of this number we may write the solution in its final form

$$v_x(t) = v_0 \left(\frac{1}{e}\right)^{t/\tau_0} \equiv v_0 e^{-t/\tau_0}. \tag{2-63}$$

This procedure then has shown us how the exponential function arises as the solution to (2-56). Since e is larger than 1, the negative exponent in (2-63) describes a function whose value decreases with time; in words, we say "$v_x(t)$ decreases or decays exponentially."

It should be clear that exponential decrease occurs for any quantity whose rate of *decrease* is proportional to that quantity itself. If the rate of *increase* is proportional to a quantity, we obtain exponential growth. We may now look again at the plot of v_x versus time. The solution (2-63) puts us now in a position to sharpen up the role of the characteristic time τ_0 (see Figure 2.15). At time $t = \tau_0$, v_x is reduced to $(1/e^2)v_0$, at $t = 2\tau_0$, to $(1/e^2)v_0$, at time $t = 3\tau_0$, to $(1/e^3)v_0$, etc. So once $t > \tau_0$, the velocity is pretty well on its way to the final value zero.

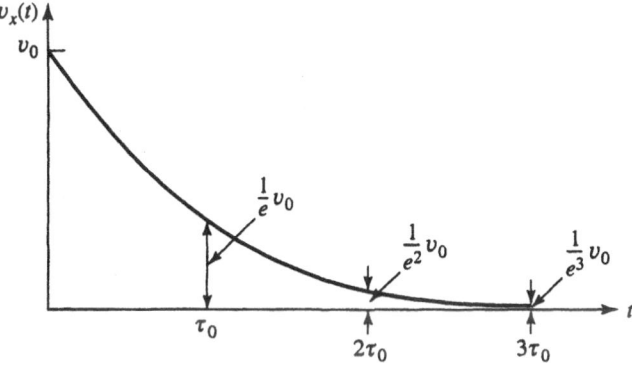

Figure 2.15. Exponential decay of velocity.

(vi)   Air Drag and Terminal Velocity

In this example we discuss the case of a mass in free fall, but subject to the air drag force, (2-22). The forces on the mass M are in Figure 2.16. $\vec{F}_D$ is the drag force of magnitude

$$F_D = \alpha v^2$$

and direction opposite to that of the velocity $\vec{v}$. We assume that the mass is released with zero initial velocity. Let $z = 0$ mark the point of release, the positive z-

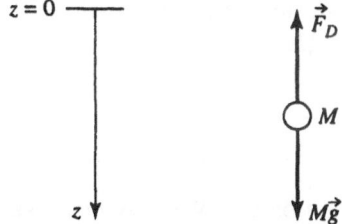

Figure 2.16. Forces on mass M falling in a fluid.

direction pointing vertically downward. The equation of motion for the velocity component v_z (v_x and v_y will remain zero) is

$$M\frac{dv_z}{dt} = Mg - \alpha v_z^2. \qquad (2\text{-}64)$$

Let us analyze this equation of motion qualitatively first: Just after release of the object, v_z is still very small and F_z is $\approx Mg$. The mass picks up speed in proportion to the elapsed time. As this happens, gradually the drag force $-\alpha v^2$ becomes more prominent; this will slow down the rate of acceleration. Eventually, v_z will have increased to a point where the drag force has caught up with Mg; F_z is then zero, no further acceleration takes place: The mass has reached *terminal speed*. This terminal speed, v_∞, is given by the condition $F_z = 0$:

$$v_\infty^2 = \left(\frac{Mg}{\alpha}\right). \qquad (2\text{-}65)$$

In order to get an idea of the numerical value of the terminal speed, we recall from (2-21) the value of the constant α in the drag force

$$\alpha = \tfrac{1}{2}c_D\rho A,$$

ρ is the density of air, 1.25 kg/m^3, A is the maximum area of the falling object in a plane perpendicular to its motion, and c_D is an empirical coefficient of order 0.5 for a spherical object, but larger, up to 1, for a more irregular shape. Then

$$v_\infty = \sqrt{\frac{2Mg}{c_D\rho A}}. \qquad (2\text{-}66)$$

We can at once see how terminal speed depends on the size of the object. Let L be a typical linear dimension, say the diameter, of the falling body. Then A will be proportional to L^2, M to L^3. Hence the terminal velocity increases with the square root of L:

$$v_\infty \sim \sqrt{L}.$$

Table 2.2 will have some data on v_∞. We may now ask how quickly this terminal speed is approached, or equivalently, what *distance* the mass must fall to begin to come close to terminal speed. This question can be qualitatively answered by

Table 2.2: Critical distance and terminal speeds for various objects ($c_D = 0.5$).

Object	Mass (kg)	Area A (m^2)	z_c (m)	v_∞ (m/s)	v_∞ (mph)
Rain drop (radius 0.1 cm)	4×10^{-6}	3×10^{-6}	2.1	6.5	14.5
Hailstone (radius 1 cm)	4×10^{-3}	3×10^{-4}	21	20	45
Spent bullet	1.4×10^{-2}	2×10^{-4}	110	47	105
Eagle in nosedive	20	0.25	130	51	115
Man	75	0.6	200	63	140
Bomb	500	0.2	4000	280	630

looking at the plot v_z^2 versus z in Figure 2.17. Clearly, the value of z, at which the free-fall line

$$v_z^2 = 2gz$$

intersects the terminal velocity line, defines a characteristic distance z_c, which is a measure of the distance, at which the object begins to come close to v_∞. This value z_c is defined by

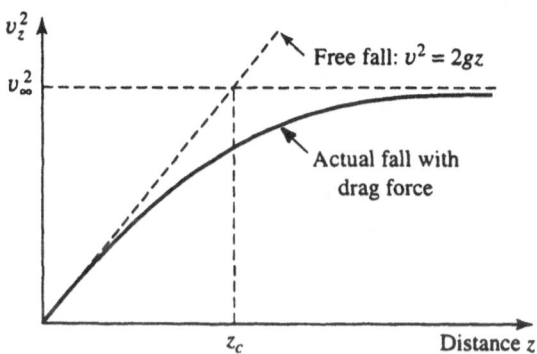

Figure 2.17. Plot of free fall and fall under drag force.

$$v_\infty^2 = 2gz_c \qquad (2\text{-}67)$$

or

$$z_c = \frac{v_\infty^2}{2g} = \frac{M}{c_D \rho A}.$$ (2-68)

It is not beyond our means to determine the exact shape of the curve drawn in Figure 2.17: Newton's law in this case takes the form

$$M\frac{dv_z}{dt} = Mg - \alpha v_z^2.$$

Upon dividing by M, and using (2-65), this may be written as

$$\frac{dv_z}{dt} = g\left(1 - \frac{v_z^2}{v_\infty^2}\right).$$ (2-69)

In Figure 2.17 we plotted v_z^2 against distance z, rather than v_z as a function of time. By using the identity

$$\frac{dv_z}{dt} = \frac{dv_z}{dz} \cdot \frac{dz}{dt} = \frac{dv_z}{dz}v_z = \frac{1}{2}\frac{dv_z^2}{dz},$$

we can transform (2-69) into an equation for v_z^2 as a function of z:

$$\frac{dv_z^2}{dz} = 2g\left(1 - \frac{v_z^2}{v_\infty^2}\right).$$ (2-70)

This equation can be brought into a familiar form: Call

$$u(z) = 1 - \frac{v_z^2(z)}{v_\infty^2},$$

then

$$\frac{du}{dz} = -\frac{1}{v_\infty^2}\frac{dv_z^2}{dz},$$

and we can write our equation as

$$\frac{du}{dz} = -\left(\frac{2g}{v_\infty^2}\right) u(z).$$

This you will now recognize as the equation for the decaying exponential discussed previously. Using the fact that $u(z = 0) = 1$, we have

$$u(z) = e^{-(2g/v_\infty^2)z} = e^{-z/z_c},$$

z_c being the length defined in (2-68). Converting back to $v_z^2(z)$, $v_z^2(z) = v_\infty^2(1-u)$, we get

$$v_z^2(z) = v_\infty^2\left(1 - e^{-z/z_c}\right). \tag{2-71}$$

This is the curve drawn in Figure 2.17.

(vii)   Damped Oscillation of a Pendulum

Every clock, from grandfather's pendulum clock, to an Accutron wristwatch, to a modern atomic clock, is based on the observation of periodic motion. It is the only way that we can give operational meaning to the concept of time: The period τ_0 of some reference oscillator defines a *unit* of time, and any statement about elapsed time is just a statement about a number of elapsed intervals τ_0. Modern civilization makes ever increasing demands on the accuracy, with which this standard interval can be defined.

   In this example, we analyze the motion of a simple pendulum, and then we will have a few comments on its performance as a timepiece. To simplify our analysis, we introduce the idealized, so-called "mathematical" pendulum: a "point"-mass suspended by a weightless string of fixed length ℓ, and moving in a vertical plane through the point of suspension. The mass is therefore constrained to move on an arc of circle with radius ℓ.

   In this problem we have a single body of mass M, constrained to move in a circle of radius ℓ about the center O. The force diagram displays the three forces acting on this mass: gravity $M\vec{g}$, the string tension $\vec{T}$, and a damping force $\vec{F}_D = -\beta\vec{v}$ (we make this assumption of proportionality of this force to $\vec{v}$ for reasons of simplicity). Choosing a coordinate system as shown, we have the force components

$$F_x = -T\sin\theta - \beta v_x, \tag{2-72a}$$

$$F_z = Mg - T\cos\theta - \beta v_z. \tag{2-72b}$$

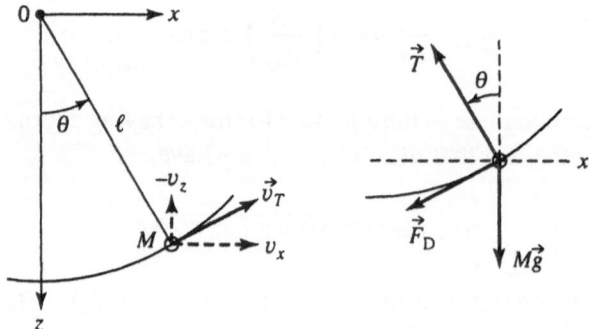

Figure 2.18. Pendulum and force diagram on mass M.

Let us now turn to the kinematical constraints. Most conveniently, we introduce here the concept of the tangential velocity component v_T; this quantity is equal to the magnitude v of $\vec{v}$ if the pendulum moves counterclockwise (θ increases), but equal to $-v$ if the mass moves clockwise (and θ decreases). The constraint that $\vec{v}$ is tangential to a circle of radius ℓ can then be expressed in the constraint equations

$$v_T = \ell \left(\frac{d\theta}{dt} \right), \tag{2-73}$$

$$v_x = v_T \cos \theta, \qquad v_z = -v_T \sin \theta. \tag{2-74}$$

(Notice that in Fig. 2.18, v_T is positive, v_z negative.) We can now proceed to the last step and write Newton's equation of motion for the x- and z-directions

$$M \frac{dv_x}{dt} = -T \sin \theta - \beta v_x, \tag{2-75a}$$

$$M \frac{dv_z}{dt} = Mg - T \cos \theta - \beta v_z. \tag{2-75b}$$

Together with the equations of constraint, we have five equations for the five unknowns v_x, v_z, T, v_T, θ.

A considerable simplification of this system of equations is achieved if we restrict ourselves to considering small amplitude oscillations of the pendulum, for which the maximum angle θ remains small: $\theta_{max} \ll 1$. In this limiting case

$$\sin \theta \simeq \theta \ll 1, \qquad \cos \theta = \sqrt{1 - \sin^2 \theta} \simeq 1, \tag{2-76}$$

v_x can therefore be approximated by v_T, whereas $|v_z|$ is small compared to $|v_T|$:

$$v_x \simeq v_T, \qquad v_z \simeq -\theta(t)v_T. \tag{2-74a}$$

With this approximation in mind we can rewrite (2-75) as follows:

$$M\frac{dv_T}{dt} + \beta v_T + T\theta(t) \simeq 0, \tag{2-77a}$$

$$M\frac{d}{dt}(\theta v_T) + \beta(\theta v_T) + Mg - T \simeq 0. \tag{2-77b}$$

In the second equation the terms can be regrouped after working out the derivative of θv_T:

$$\frac{d}{dt}(\theta v_T) = \theta\frac{dv_T}{dt} + v_T\frac{d\theta}{dt} = \theta\frac{dv_T}{dt} + \frac{v_T^2}{\ell},$$

it may now be written in the form

$$\theta(t)\left\{M\frac{dv_T}{dt} + \beta v_T\right\} + \left\{Mg + M\frac{v_T^2}{\ell} - T\right\} \simeq 0.$$

The first term in this equation is just $-T\theta^2(t)$, as is seen by consulting (2-77a); since we considered negligibly small all terms in θ^2, we drop this term in the equation, and are left with

$$T = Mg + M\frac{v_T^2}{\ell}. \tag{2-78}$$

This equation tells us that the string tension, for small angles θ ($\theta \ll 1$) is equal to the weight Mg of the mass plus a term Mv_T^2/ℓ. Where does this last term come from? We have seen in Chapter 1 that a mass M, moving with speed v on a circular track of radius R, undergoes a centripetal acceleration (directed *toward* the center) of magnitude v^2/R. There is therefore a force required to keep the mass on the circle; this force is Mv^2/R. In the case of a pendulum, M moves in a circle of radius ℓ, and the term Mv_T^2/ℓ represents this centripetal force: T must compensate Mg, and in *addition* supply this centripetal force. Numerically, the force Mv_T^2/ℓ is much smaller than Mg in the small angle approximation, and we shall henceforth approximate T by Mg.

We now turn to (2-77a) for v_T. Since θ enters into this equation, and θ is related to v_T by (2-73), we best express v_T in terms of θ; this gives the equation

$$M\ell\frac{d^2\theta}{dt^2} + \beta\ell\frac{d\theta}{dt} + Mg\theta(t) = 0,$$

or, after dividing by M:

$$\frac{d^2\theta}{dt^2} + \left(\frac{\beta}{M}\right)\frac{d\theta}{dt} + \left(\frac{g}{\ell}\right)\theta(t) = 0. \tag{2-79}$$

We see that this equation contains two independent constants (β/M) and (g/ℓ).

We also observe that, for $\beta = 0$, (2-79) would have the *same form* as (2-37) for the mass attached to a spring, with the constant (g/ℓ) taking over the role of (k/M); in the case that $\beta = 0$, the solution to (2-79) is therefore a harmonic oscillation

$$\theta(t) = a\cos(\omega_0 t) + b\sin(\omega_0 t), \tag{2-80}$$

ω_0 is the angular frequency of this oscillation, and is given by

$$\omega_0 = \sqrt{\frac{g}{\ell}}. \tag{2-81}$$

The period T_0 of this oscillation is then

$$T_0 = \frac{2\pi}{\omega_0} = 2\pi\sqrt{\frac{\ell}{g}}. \tag{2-82}$$

We notice the important facts that this period is:

(a) independent of the mass of the pendulum;

(b) independent of the amplitude; and

(c) proportional to the square root of the length of the string.

The second constant entering (2-79) is (β/M); it has the dimension of a reciprocal time, and we will introduce for it the symbols

$$\left(\frac{\beta}{M}\right) \equiv \gamma \equiv \frac{1}{\tau_0}. \tag{2-83}$$

Equation (2-79) may then be rewritten as

$$\frac{d^2\theta}{dt^2} + \gamma \frac{d\theta}{dt} + \omega_0^2 \theta(t) = 0 \tag{2-84}$$

and for $\gamma = 0$, the solution is given by (2-80).

We are now interested to investigate how the damping force, which provides the term $\gamma \, d\theta/dt$ in (2-84), affects the motion of the pendulum. We know from observation that for weak damping the pendulum oscillates with gradually decreasing amplitude, and eventually comes to a stop. Casual observation does not reveal whether the period T of oscillation is modified by damping.

In order to achieve a quantitative description of this damped motion, we will start with an educated guess for the form of $\theta(t)$. Choosing for the zero of time, an instant where the pendulum crosses the z-axis, so that $\theta(t = 0) = 0$, the motion of the *undamped* pendulum is described by

$$\theta(t) = A \sin(\omega_0 t),$$

A being the amplitude of the oscillation, and equal to the maximum angle $\theta_{\max}$ of deflection from equilibrium.

Damping will cause this amplitude A to decrease with time, and perhaps to modify the frequency too. We therefore guess that the motion of the *damped* pendulum will be given by an expression of the form

$$\theta(t) = A(t) \sin(\omega t). \tag{2-85}$$

To test the validity of this representation of $\theta(t)$, we simply insert it into (2-84) and check whether this equation can be satisfied, and for what values of ω and $A(t)$. This is a perfectly legitimate procedure, which is constantly used in physics; it simply consists of bringing to bear on the solution of a problem any insight and information one has already at hand.

To apply the proposed form of $A(t)$ to (2-84), we must calculate the two first derivatives of $\theta(t)$:

$$\frac{d\theta}{dt} = \frac{dA}{dt} \sin(\omega t) + \omega A(t) \cos(\omega t),$$

$$\frac{d^2\theta}{dt^2} = \frac{d^2 A}{dt^2}\sin(\omega t) + 2\omega\frac{dA}{dt}\cos\omega t - \omega^2 A(t)\sin(\omega t).$$

Inserting this, we then separately collect all the terms proportional to $\cos(\omega t)$ and $\sin(\omega t)$, and get an expression of type

$$\{\cdots\}\cos(\omega t) + \{\cdots\}\sin(\omega t) = 0,$$

the first bracket being

$$2\omega\frac{dA}{dt} + \gamma\omega A(t), \tag{2-86}$$

the second

$$(\omega_0^2 - \omega^2)A(t) + \gamma\frac{dA}{dt} + \frac{d^2 A}{dt^2}. \tag{2-87}$$

We will have solved the equation if these two expressions are both *zero*, and this in turn gives us just the two equations needed to determine $A(t)$ and ω.

From (2-86) we obtain

$$\frac{dA}{dt} = -\frac{\gamma}{2}A(t),$$

which we recognize as the equation for an exponential function. (See Examples (iv) and (v).)

$$A(t) = A_0 e^{-(\gamma/2)t} = A_0 e^{-t/2\tau_0}. \tag{2-88}$$

This solution also identifies $2\tau_0$ as the characteristic decay time of the amplitude (see Figure 2.19). The zero value of (2-87) will give us an expression for ω; after having replaced dA/dt by $-(\gamma/2)A$, $d^2 A/dt^2 = (\gamma^2/4)A$, we obtain the equation

$$\left(\omega_0^2 - \omega^2 - \frac{\gamma^2}{4}\right)A(t) = 0$$

and hence

$$\omega^2 = \omega_0^2 - \frac{\gamma^2}{4}. \tag{2-89}$$

This result implies that the period of the oscillation is increased by damping. This period T is

$$T = \frac{2\pi}{\omega} = \frac{2\pi}{\omega_0} \left/ \sqrt{1 - \frac{\gamma^2}{4\omega_0^2}} \right.$$

or

$$T = T_0 \left/ \sqrt{1 - \left(\frac{T_0}{4\pi\tau_0}\right)^2} \right. , \qquad (2\text{-}90)$$

T_0 being the period of undamped oscillation, and $\tau_0 = 1/\gamma$, the characteristic damping time. In Figure 2.19 we give a sketch of this damped oscillation, and label the two time constants T and τ_0. Notice that in constructing the curve for $\theta(t)$, we have multiplied the "envelope" $Ae^{-t/2\tau_0}$ with $\sin(\omega t) = \sin(2\pi t/T)$. The quantitative description of the damped oscillation which we now have in hand enables us to sharpen up the concept of weak damping: Damping is weak, if the

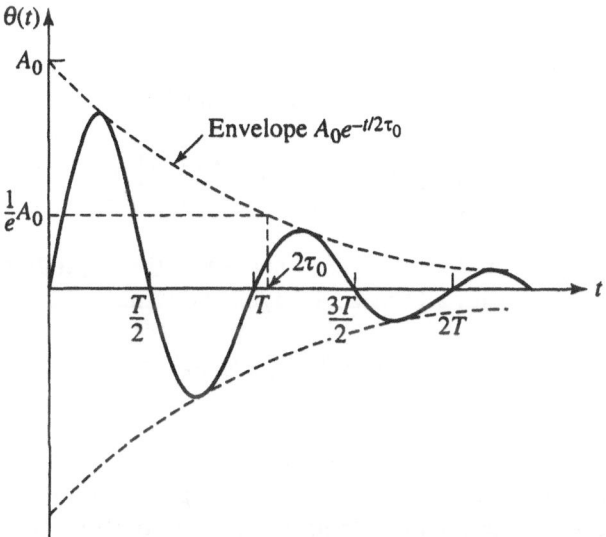

Figure 2.19. Damped oscillation

decay time constant τ_0 is large compared to the period T:

$$\frac{\tau_0}{T} \gg 1 \qquad \text{or, equivalently,} \qquad \frac{\gamma}{\omega_0} \ll 1.$$

In this weak damping case we see that the period T differs but little from the undamped period T_0, as can be inferred from (2-90). It is customary to call the ratio

$$\frac{\omega_0}{\gamma} = 2\pi \frac{\tau_0}{T_0} \equiv Q \tag{2-91}$$

the *quality factor* or simply "Q-value" of the damped oscillator. For weak damping Q is large compared to 1, and is a measure of the number of oscillations the pendulum (or any other oscillator) performs before its amplitude has died out.

Equation (2-89) shows that if the damping constant γ is large enough, ω^2 may approach the value zero, or even turn negative. The case where $\gamma = 2\omega_0$ is called the case of *critical damping*; ω is zero in this case. It is not immediately clear what becomes of expression (2-85) in this case. Some helpful information emerges, if we express $\theta(t)$ in terms of the initial angular velocity $(d\theta/dt)_{t=0}$. From the expression

$$\theta(t) = A_0 e^{-t/2\tau_0} \sin(\omega t)$$

we find $(d\theta/dt)_{t=0} = \omega A_0$, so that in turn we can write

$$\theta(t) = \left(\frac{d\theta}{dt}\right)_0 e^{-t/2\tau_0} \frac{\sin(\omega t)}{\omega}.$$

Assume now that ω is very small, due to near-critical damping; we can then approximate $\sin(\omega t)$ by ωt for all times $t \ll 1/\omega$, and write

$$\theta(t) = \left(\frac{d\theta}{dt}\right)_0 e^{-t/2\tau_0} t \tag{2-92}$$

and if ω actually is zero, this result is valid for all times.

We see then that in this case of critical damping the pendulum does not oscillate any more, but returns smoothly to its equilibrium position (Figure 2.20). This concludes our discussion of the damped oscillation of the pendulum. The detailed attention which we have given to this example is justified by the fact that (2-84) for damped harmonic motion describes a vast number of oscillatory phenomena in

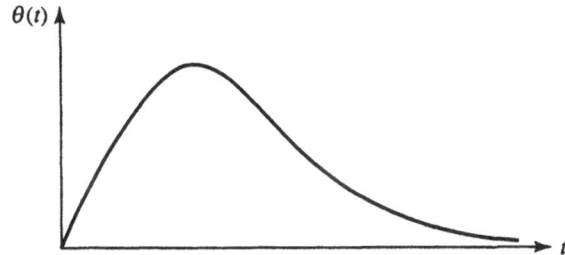

Figure 2.20. Amplitude of critically damped pendulum

all branches of physics, from particle mechanics to acoustics, electromagnetism, and atomic and molecular physics.

(viii)   The Centrifugal Pendulum

This example has a historic significance in that it describes the key part in a device called the "governor," used by Matthew Boulton and James Watt in 1789 to regulate the speed of their newly invented steam engine. The device, sketched in Figure 2.21, consists of a vertical shaft, rotating at an angular velocity Ω proportional to the engine speed. Mounted at the end of the shaft, and *forced to rotate with the shaft*, are two arms of length L, each carrying a mass M, the "fly ball," at its end. The arms are free to pivot about the joint J, in a vertical plane. An

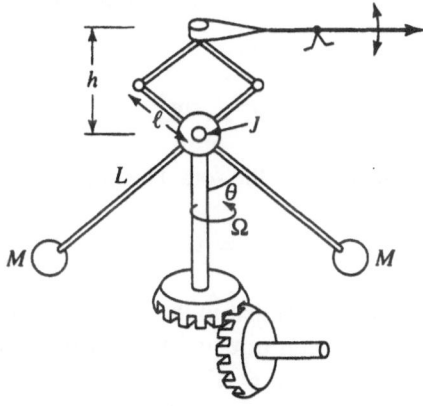

Figure 2.21. Fly ball governor (centrifugal pendulum).

extension of the arms provides the means to move a lever, attached at P, whose position regulates the pressure and flow of steam into the engine.

Our objective is to determine the equilibrium position θ_0 of the masses for a given angular velocity Ω of the shaft. Subsequently, we will ascertain that the masses may oscillate about that equilibrium position, and find the period of this oscillation. We begin by drawing a force diagram for one of the masses M, and the path of this mass. The forces on the mass are the weight $M\vec{g}$, and the tension $\vec{T}$ of the rod tying M to the pivot J (Figure 2.22).

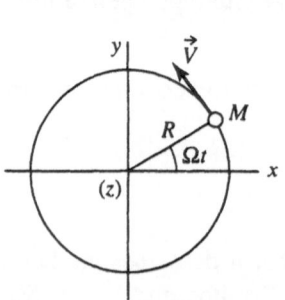
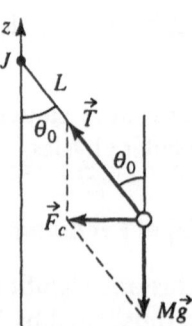

Top view: Circular path of mass M Side view in plane containing
Velocity: $v = R\Omega$ mass M and z-axis
Radius of circle $R = L \sin \theta_0$ $\vec{F}_c$ is the resultant of $\vec{T}$ and $M\vec{g}$

Figure 2.22. Path and force diagram for mass M.

In the steady state the mass describes a circular path in a horizontal plane; the radius of this circle is

$$R = L \sin \theta_0$$

θ_0 being the equilibrium position of the rod. The speed of M along this circle is

$$v = \Omega R = \Omega L \sin \theta_0.$$

Now, as we have seen in Chapter 1, that in the case of the uniform circular motion of a mass, there is an acceleration directed toward the center (centripetal acceleration), of magnitude $a_c = v^2/R = \Omega^2 R = \Omega^2 L \sin \theta_0$. The net force on M must therefore be such as to supply this acceleration. This net force must therefore be in a horizontal plane, directed toward the shaft of the pendulum, and have a magnitude

$$F_c = Ma_c = M\Omega^2 L \sin\theta_0. \tag{2-93}$$

Consulting the force diagram in Figure 2.22, we see that the z-component of the net force must vanish

$$T \cos\theta_0 - Mg = 0, \tag{2-94}$$

and the horizontal component must have the values given by (2-93):

$$T \sin\theta_0 = M\Omega^2 L \sin\theta_0. \tag{2-95}$$

These last two equations will supply us with both the values of the tension T and the equilibrium angle θ_0, which are our two unknowns in this case: Equation (2-94) tells us that

$$T = \frac{Mg}{\cos\theta_0} \geq Mg \tag{2-96}$$

(since $0 < \cos\theta_0 < 1$), whereas from (2-95) we have

$$T = M\Omega^2 L, \quad \text{provided} \quad \sin\theta_0 \neq 0. \tag{2-97}$$

(If $\sin\theta_0 = 0$, (2-95) says nothing about the value of T!)

The conclusions to be drawn from the above two expressions for T is that:

(a) either $\Omega^2 L > g$, in which case θ_0 must be $\neq 0$, and the two formulas (2-96) and (2-97) for T give

$$\frac{g}{\cos\theta_0} = \Omega^2 L,$$

that is,

$$\cos\theta_0 = \frac{g}{\Omega^2 L};$$

(b) or $\Omega^2 L \leq g$, in which case we must conclude that $\theta_0 = 0$, since otherwise (2-96) and (2-97) would be inconsistent. Therefore, $T = Mg$ in this case.

We have thus found that there exists a threshold value for the angular velocity Ω of the shaft, below which the equilibrium angle θ_0 is zero; for Ω's above this value this equilibrium angle is finite and increases with increasing Ω:

$$
\begin{aligned}
\theta_0 &= 0 && \text{for} \quad \Omega^2 \leq \left(\frac{g}{L}\right), \\
\cos\theta_0 &= \frac{g}{\Omega^2 L} && \text{for} \quad \Omega^2 > \left(\frac{g}{L}\right).
\end{aligned}
\tag{2-98}
$$

It is interesting to observe that the threshold angular velocity of the shaft is given by

$$
\Omega_{\text{threshold}} = \sqrt{\frac{g}{L}},
$$

which is just the angular frequency of oscillation of a pendulum of length L!

Let us now assume a case where $\Omega > \sqrt{g/L}$, so that the angle has the finite value given by (2-98). Should the mass, at a given time, be pushed to a position $\theta \neq \theta_0$, and then released, then the forces acting on M must of necessity drive the mass back toward the equilibrium position θ_0, whether θ be larger or smaller than θ_0. This indicates that the mass will actually oscillate about the position θ_0, and eventually return to θ_0 by virtue of the damping of that oscillation.

Conversely, if the shaft angular velocity Ω is suddenly increased to $\Omega' = \Omega + \Delta\Omega$, a new equilibrium position $\theta_0' = \theta_0 + \Delta\theta_0$ is defined by (2-98). The mass therefore finds itself in a position offset from this new equilibrium by an angle $\Delta\theta_0$, and will eventually settle into its new equilibrium position θ_0' by means of a damped oscillation.

It will be important, in the context of the role of the centrifugal pendulum in the "governor," to have an idea of the period T of this oscillation. (See Chapter 6 where the role of the period of oscillation in the feedback control system is illustrated.)

We can find the period in the following manner. Let Ω be constant, and let us apply to the mass M a small extra force $\Delta\vec{F}$, perpendicular to $\vec{T}$, which will hold the mass at an equilibrium position $\theta = \theta_0 - \Delta\theta$ (see Figure 2.23). We shall establish the relation between ΔF and the shift $\Delta\theta$ in the equilibrium position of the mass. This relation will be linear for small ΔF; we will write it in the form

$$
\Delta F = -(kL)\Delta\theta = -k(L\Delta\theta).
\tag{2-99}
$$

With (2-99) we have introduced a proportionality constant k which measures the ratio of force ΔF to the linear displacement $(L\Delta\theta)$. The role of k is then exactly

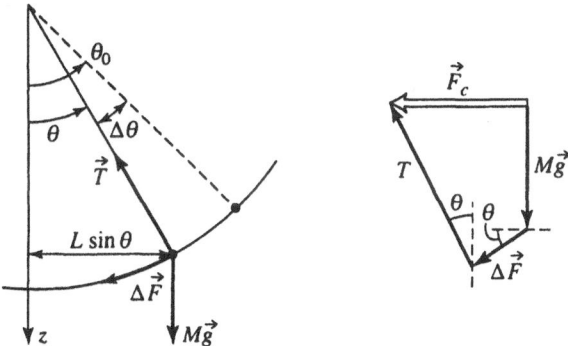

Figure 2.23. External force $\Delta \vec{F}$ and shift in equilibrium position.

that of the *spring constant* k in Example (iii), (2-36), and we can conclude at once that the period T and the angular frequency $\omega = 2\pi/T$ will be given by

$$T = 2\pi \sqrt{\frac{M}{k}}, \qquad \omega = \sqrt{\frac{k}{M}}, \qquad (2\text{-}100)$$

just as in the case of the spring, (2-46). To relate $\Delta \vec{F}$ to $\Delta \theta$, we consult the force diagram in Figure 2.23. The vector sum of the three forces must add up the horizontal, radially inward directed force $\vec{F}_c$ of magnitude $F_c = M\Omega^2 L \sin \theta$, needed to provide the centripetal acceleration. Writing out the vertical and horizontal components of these forces, we will have

$$Mg + \Delta F \sin \theta - T \cos \theta = 0, \qquad (2\text{-}101a)$$

$$\Delta F \cos \theta + T \sin \theta = M\Omega^2 L \sin \theta. \qquad (2\text{-}101b)$$

These are two equations for ΔF and T. T can be eliminated by multiplying (2-101a) by $\sin \theta$, (2-101b) by $\cos \theta$, and adding the two equations

$$Mg \sin \theta + \Delta F = M\Omega^2 L \sin \theta \cos \theta.$$

We now recall that, by (2-98),

$$Mg = M\Omega^2 L \cos \theta_0$$

so that ΔF may be written as

$$\Delta F = M\Omega^2 L \sin\theta (\cos\theta - \cos\theta_0).$$

θ and θ_0 differ only by the small angle $\Delta\theta$: $\theta_0 = \theta + \Delta\theta$; then $\cos\theta_0 = \cos(\theta + \Delta\theta) = \cos\theta \cos(\Delta\theta) - \sin\theta \sin(\Delta\theta) \simeq \cos\theta - \Delta\theta \sin\theta$ since $\sin(\Delta\theta) \sim \Delta\theta$, $\cos(\Delta\theta) \sim 1$. This gives us

$$\Delta F = M\Omega^2 L \sin^2\theta \, \Delta\theta \qquad (2\text{-}102)$$

and with this we have found the desired "spring constant" k, defined in (2-99):

$$k = M\Omega^2 \sin^2\theta. \qquad (2\text{-}103)$$

As long as this oscillation of the fly ball has small amplitude, k is nearly constant, and to calculate the angular frequency ω and period T of this oscillation, we are allowed to replace θ in (2-103) by its mean value θ_0. With the help of (2-100) we then conclude that the angular frequency ω of the fly ball oscillation about its mean position θ_0 is given by

$$\omega = \Omega \sin\theta_0 = \sqrt{\frac{g}{L}} \frac{\sin\theta_0}{\sqrt{\cos\theta_0}} \qquad (2\text{-}104)$$

and its period T by

$$T = \frac{2\pi}{\Omega} \frac{1}{\sin\theta_0}. \qquad (2\text{-}105)$$

Finally, an interesting question which arises is the choice of the optimum "operating" point θ_0, if this centrifugal pendulum is to be used in a governor. In this case one wants to optimize the change Δh in the position of point P in Figure 2.21 associated with a given change $\Delta\Omega$ of the shaft angular velocity, and at the same time keep the period T, (2-105), as small as possible. See the problem set at end of this chapter.

2.8   References and Supplementary Reading

[1] *P.S.S.C. Film on Newton's Laws of Motion* by E. M. Purcell. A perfectly lucid presentation of the essential experimental basis for Newton's laws of motion.

[2] *An Introduction to Mechanics* (Chapter 2) by D. Kleppner and R. Kolenkow. A carefully thought out and broad-ranging discussion of the meaning, origin, and implications of Newton's laws of motion with many clearly worked out and illuminating mechanical examples.

[3] *Mathematical Principles of Natural Philosophy and His System of the World (Principia)* by Sir Isaac Newton, revised by Florian Cajori, translated by Andrew Motte. University of California Press, Berkely, CA, (1934), (1962).

[4] "The Eotvos Experiment" by R. A. Dicke, *Scientific American* **205**, G4 (1961).

2.9 Problems

(The problems are arranged within each group, roughly in order of increasing difficulty.)

Constant Forces and Constraints

1. A block of mass m slides without friction on an inclined plane forming an angle α with the horizontal. Draw a diagram with all the forces acting on m. Choose a coordinate system as shown, and state the kinematical constraints. Determine the acceleration of m in the x-direction, and also the value of the force exerted by the plane on mass m.

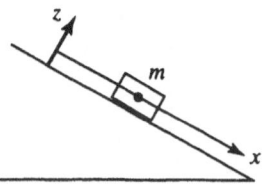

2. Atwoods Machine. Two masses, m_1 and m_2, are connected over a pulley with a rope of fixed length. Neglect friction, as well as the masses of pulley and

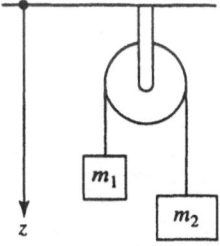

rope. Draw a force diagram for each of the masses; express the constraint conditions, and determine the accelerations dv_z/dt for both m_1 and m_2.

3. A freight train consists of N cars, of mass M each, pulled by an engine of mass M_e. As the train starts up, the engine exerts a pull F_e.

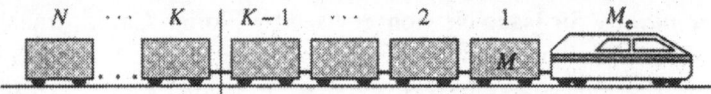

   (a) Find an expression for the acceleration of the train as a whole, neglecting friction.

   (b) Determine the tension T_K in the coupling between the Kth and $(K-1)$st car of the train during acceleration. Compare T_N with T_0.

4. An ocean-going research vessel uses a long steel cable to lower an instrument package to a depth L in the ocean. Neglecting the weight of the instrument package, the buoyancy corrections, find the tension T in the cable at point P, at a depth z below the pulley; for this, assume that w is the weight per unit length of cable, so that a piece of cable of length L has total weight $W = Lw$.

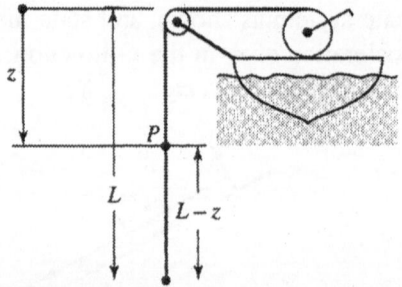

5. Consider an object of mass M, which by means of a handlebar can be pushed or pulled over a horizontal surface. The coefficient of sliding friction is $\mu < 1$. You are to explain the easily observed fact that it is easier to *pull* this object

than to *push* it, if in both cases the handlebar forms the same angle α with the horizontal. The object may be a toy, or a carpetsweeper, or a lawnmower, etc.

(a) Draw a diagram with all the forces acting on M, including T, the pull or push of the handlebar, the normal contact force N exerted by the floor, and the force of friction, N. Do this separately for the case of pushing and pulling.

(b) Show that the force T transmitted by the handlebar in the case of *uniform motion* of the mass (M) is

$$T = \frac{\mu M g}{\cos \alpha + \mu \sin \alpha} \qquad \text{(pull)},$$

$$T = \frac{\mu M g}{\cos \alpha - \mu \sin \alpha} \qquad \text{(push)}.$$

(c) T may become infinite in the case of pushing, if α is too large. At what angle α does this happen?

6. A person is jumping down from a ledge 10 ft high. Upon contact with the floor below, he goes into a kneebend, during which time his trunk decelerates uniformly.

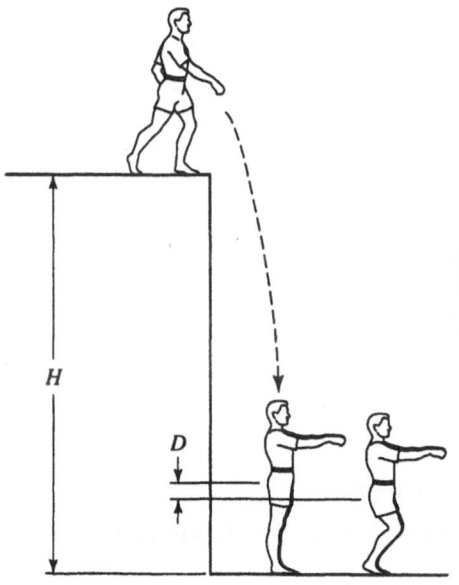

H = height of jump = 10 ft.
D = deceleration distance = 2 ft.
Mg = weight of trunk (including arms and head) = 120 lb.

(a) What is the person's velocity at the time that his feet make contact with the ground?

(b) What is the force exerted by the legs on the trunk during the deceleration phase?

(c) Can you estimate the force on the trunk if the person jumps with the knees locked straight?

7. In the figure below, the painter's assistant is painting a wall, while the painter holds the platform at a fixed height. Assuming that the weight of the assistant is W_A, that of platform plus lower pulley W_B, what is the minimum required weight of the painter, $(W_p)_{min}$, which is needed so that the assistant can do his job?

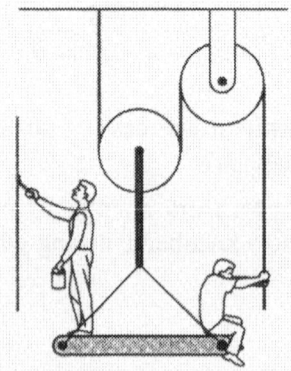

8. A mass m moves without friction on the inclined surface of a wedge of mass M. The wedge slides without friction on a horizontal surface. You are to determine the accelerations dV_x/dt, dV_z/dt of M, and dv_x/dt, dv_z/dt of m.

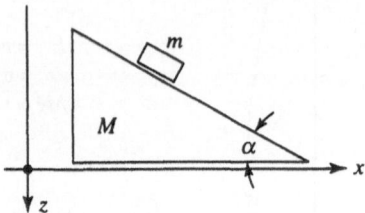

(a) Draw a force diagram for each mass.

(b) Determine the kinematical constraints. The figure below should help you to see that:

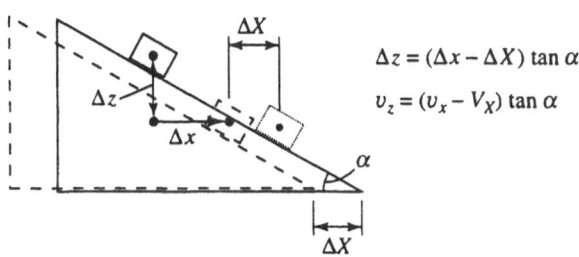

$$\Delta z = (\Delta x - \Delta X) \tan \alpha$$
$$v_z = (v_x - V_X) \tan \alpha$$

(c) Write the equations of motion relating the force components to the corresponding acceleration components.

(d) Check number of equations against number of unknowns.

Gravitational Forces and Circular Motion

9. *Geostationary satellite.* A satellite is launched into a circular, equatorial orbit. At the proper radius R_s of this orbit, the satellite has exactly the same angular velocity as the Earth, and seen from the Earth, will appear in a stationary position.

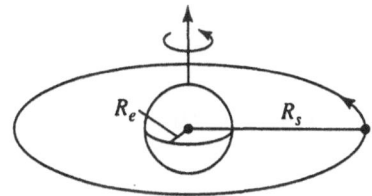

Using the fact that $GM_e/R_e^2 = g =$ gravitational acceleration at the Earth's surface $\simeq 10 \text{ m/s}^2$, $R_e = 6.3 \times 10^3$ km, and $\Omega_e = 2\pi/\text{day}$, find an expression for R_s in terms of g, R_e, Ω_e, and determine its numerical value.

10. A child is swinging a small pail filled with marbles in a vertical circle at constant speed. Assuming the radius of this circle to be $R = 50$ cm, what is the number of turns per second needed to prevent the marbles from dropping out of the bucket?

11. In an amusement park children enjoy a mock-airplane ride by riding in a gondola, lifted off the ground by the angular velocity Ω of the merry-go-round. Assume that $L = 5$ m. What is the angular velocity needed to lift the gondola to an angle $\theta = 60°$? What is the actual speed of the gondola in this case?

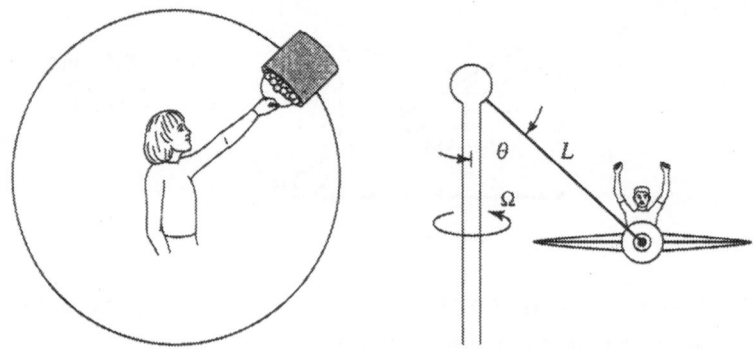

Velocity-Dependent Forces. Exponential Functions

12. The exponential functions e^x and $e^{-x} = (1/e)^x$:

  (a) The numbers e and $\frac{1}{e}$ are defined as

  $$e = \lim_{n \to \infty} \left(1 + \frac{1}{n} \right)^n = 2.71828\ldots,$$

  $$\frac{1}{e} = \lim_{n \to \infty} \left(1 - \frac{1}{n} \right)^n = 0.36788\ldots.$$

  Calculate the numerical values of this sequence of approximations to e and $1/e$, using $n = 1, 2, 4, 8, 16$.

  (b) Verify also that

  $$\left(1 + \frac{1}{n} \right)^n \cdot \left(1 - \frac{1}{n} \right)^n = \left(1 - \frac{1}{n^2} \right)^n$$

  approaches 1 as n gets large.

  (c) A loanshark will lend a customer the sum of \$1000 at a *daily* interest rate of $1/10\%$. What amount does the customer pay back at the end of 1 year? [Use

  $$\left(1 + \frac{1}{n} \right)^m = \left(1 + \frac{1}{n} \right)^{n(m/n)};$$

  and a table of exponential functions, or a pocket calculator.]

(d) A depositor puts $1000 into a savings account, which earns 5% in annual interest. Inflation reduces the buying power of the dollar by 6% annually. After 15 years the customer withdraws his money. What is the buying power of his money compared to that of the $1000 15 years earlier?

(e) e^x is the limit, as $n \to \infty$, of $[1 + (x/n)]^n$. For any integer n, you may use the binomial formula

$$(1 + a)^n = 1 + na + \frac{n(n - 1)}{2} a^2 + \cdots .$$

Apply this to $[1 + (x/n)]^n$, and show that as n gets very large, this leads to the series

$$e^x = 1 + x + \tfrac{1}{2}x^2 + \cdots .$$

If $|x| \ll 1$, we can therefore approximate e^x by $(1 + x)$.

13. A small water droplet in fog has a radius R of 10^{-3} cm. It settles in air at a uniform (terminal) speed determined by Stokes' law. Given the density ρ of water as 1 g/cm^3, the viscosity η of air as 10^{-4} g/cm s, find the terminal speed of the droplet. Determine also the characteristic time τ_0 with which terminal speed is approached by a droplet this size, starting from rest.

14. Consider an object of mass m, traveling with uniform velocity v_0. At time $t = 0$, it enters into a region where it experiences a viscous force $\vec{F} = -\beta\vec{v}$:

   (a) Give an expression for the velocity of the object for times $t > 0$.

   (b) Determine the distance D which the object moves until it finally comes to rest. For this, describe the position $x(t)$ of the mass by the educated guess $x(t) = a + be^{-t/t_0}$. Determine a, b, t_0 by requiring that $dx/dt = v_x$, and fit to your result in (a) and the initial condition. Then express the distance by means of a and b: $D = x(t = \infty) - x(t = 0)$.

15. A spherical object of diameter D, moving through air at speed v, experiences at low speed the viscous force

$$F = 3\pi D\eta v, \qquad \eta = \text{viscosity of air}$$

and at high speed it experiences the drag force

$$F = \tfrac{1}{2}C_D\rho\tfrac{\pi D^2}{4}v^2, \qquad \rho = \text{density of air}.$$

The force which actually describes the resistance is the one that is the larger, in any particular case. Show that the viscous ($\sim v$) force dominates if

$$\frac{Dv\rho}{\eta} < R_c$$

and the drag force ($\sim v^2$) dominates if

$$\frac{Dv\rho}{\eta} > R_c.$$

R_c is a pure number (critical Reynolds number). Determine its value. (Assume $C_D = 0.5$.)

16. A safe landing speed for a parachute jumper is $v = 15$ ft/s. Using the value $C_D = 1$ for the parachute's drag coefficient, and $\rho(\text{air}) = 1.25$ kg/m^3, what should be the minimum diameter of his open parachute? Assume the jumper's mass to be 80 kg.

Harmonic Oscillations

17. (a) Suppose that the position $x(t)$ of a mass in harmonic oscillation is expressed by the equation

$$x(t) = A \sin \omega(t - t_0).$$

Determine the position x_0 and velocity v_0 at $t = 0$, in terms of A and t_0.

(b) Suppose that the velocity $v_x(t)$ of a mass in harmonic oscillation is given by

$$v_x(t) = V \sin \omega(t - t_0).$$

   (i) Find an expression for $x(t)$. (Assume $x = 0$ at equilibrium.)

   (ii) Determine the position x_0 and velocity v_0 at $t = 0$.

   (iii) Find an expression for the amplitude of this oscillation.

18. *Accuracy of a pendulum clock.* The period of a pendulum of length L is given by

$$T = 2\pi \sqrt{\frac{L}{g}}.$$

The length L of the bar (a metal) is slightly dependent on temperature: A change of the temperature by a small amount Δt changes the length by a small amount $\Delta L = \alpha L \Delta t$. For most metals α is of order 10^{-5}/degree centigrade. What are the tolerable limits of temperature variation if the clock is to mark time with an error no larger than 1 s/day?

19. A mass m and weight mg are suspended on a spring:

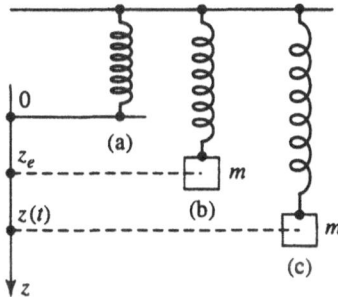

(a) Shows the relaxed spring,

(b) shows the spring with mass in equilibrium; and

(c) shows the oscillating spring.

0 is the position "zero" on the z-axis.

  (i) Make a diagram of all the forces acting on m. The force of the spring on m is $F_z = -kz$.

  (ii) Determine the equilibrium position z_e.

  (iii) Write Newton's equations of motion for v_z.

  (iv) Show that the motion is a harmonic oscillation about the equilibrium position z_e, and that the period of this oscillation is independent of the force of gravity.

20. Show that the position h of point P in Figure 2.21 (the centrifugal governor) responds to a small change $\Delta\Omega$ of the shaft angular velocity Ω by changing by the amount

$$\Delta h = -4\ell \cos\theta_0 \frac{d\Omega}{\Omega}.$$

Within what angular range should the operating point be chosen if, on the one hand, Δh should be made as large as possible, but on the other hand, the period T of the fly ball's oscillation be kept as small as possible?

21. *Oscillations and walking speed.* Has it occurred to you that the normal pace of walking is determined by the function of the leg as a pendulum?

    (a) Determine approximately the period of the leg–pendulum. To do so, you may make a cardboard model of the leg. In doing this, consider the questions:

        (i) Does it matter that this model does not have the same weight as the actual leg?

        (ii) The model obviously must be "similar" to the leg. What does this mean, exactly?

        (iii) Can you make a small size (scaled down) model, and then scale up the result? If yes, how is this done?

    (b) Compare the results so obtained with the period of relaxed walking.

22. A spring balance, as used in the kitchen, gives you a weight reading on an angle scale. When a weight W is placed on the platform, the pointer oscillates with decreasing amplitude, and finally comes to rest at the position θ determined by $W = k_\theta \theta$. Let ω_0 be the oscillation frequency of the unloaded scale. It is then found that for a certain weight W_1, placed on the scale, the oscillation frequency is reduced to $\frac{1}{2}\omega_0$. At what weight W_2 will the frequency be reduced to $\frac{1}{4}\omega_0$? (Express W_2 in terms of W_1.)

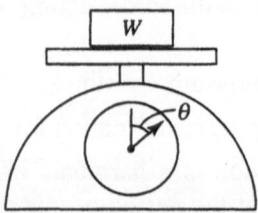

23. *Safety rope in mountaineering.* A rather recent invention is the elastic rope, which stresses by an amount proportional to the applied tension T:

$$T = \gamma \frac{\Delta L}{L_0} \equiv k\Delta L,$$

L_0 being the length at zero tension, ΔL the increment in length.

    Assume the climber has climbed to a height L_0 above the last piton, where the rope is anchored. In an accident he will then fall a height $2L_0$ before the rope begins to catch him:

(i) Find the velocity v_0 of the climber at this point. Subsequently, choose this point as the zero of time.

(ii) Choose a z-axis vertically downward, with $z = 0$ the point where the rope begins to stretch. Draw a force diagram for the falling body (for $z > 0$), and write the equation of motion.

(iii) Show that the motion for $z > 0$ is a harmonic oscillation

$$z(t) - z_0 = a \cos \omega t + b \sin \omega t;$$

give the values of z_0 and ω.

(iv) Determine a and b from the initial conditions $z(t = 0) = 0$, $v_z(t = 0) = v_0$ (see (i)).

(v) Determine the amplitude of this oscillation.

(vi) From (iv) and (v), determine the maximum deceleration the climber experiences, and the maximum value of the rope tension T.
Show that this maximum T is *independent* of the height of the fall, $2L_0$.

(vii) A typical 11 mm safety rope has a γ of 10^4 lb. What is the maximum T for a 150 lb climber?

24. Show that if a mass attached to a spring is damped by a constant force of friction (constant in magnitude, direction opposite to velocity), then the oscillation comes to a halt in a finite time. *Hint*: Show that one has two equations of motion

$$m\frac{dv_x}{dt} = -k(x + a) \qquad \text{for} \quad v_x > 0,$$

$$m\frac{dv_x}{dt} = -k(x - a) \qquad \text{for} \quad v_x < 0,$$

where $ka = F_f$ is the force of friction. Draw a diagram showing successive half-periods of motion, for which v_x has a given sign.

CHAPTER 3

Static Equilibrium and the Forces Acting on Muscles and Bone within the Human Body

3.1 Introduction

In the previous two chapters we have presented the mathematical and physical bases for mechanics. We saw how to describe the motion of a point body (kinematics) and how to predict such motion from a knowledge of the forces acting on the body (Newton's laws of motion). In the present chapter we consider what, at first sight, might seem to be a trivial topic—namely, the forces acting within and upon an extended body which is *not* moving. This topic, the static equilibrium of extended bodies, enables us to calculate the forces at work on the beams of a house, or on the girders in a bridge, or on the joints in a human body. The forces that act in a human body subjected to the force of gravity are surprisingly large and can be responsible for a number of pathologic effects, such as "slipped discs" and improper bone growth. By combining the laws of mechanics with accurate numerical anatomical data on the geometry of bones and the position and angle of the attachment of muscles, we can gain considerable insight into clinical symptoms, such as the "antalgic" gait of a person suffering from pain in the hip joint. We can, furthermore, use the principles of this chapter to understand the use and function of simple orthopedic devices. We can begin to understand the nature of the stress distribution in bones that lead to fracture, for example, in skiing accidents.

3.2 Conditions of Static Equilibrium

To proceed we must first obtain the conditions that must be satisfied by the forces acting on a rigid extended body in order that it be stationary. These conditions are not trivial because the body is not a point. If it were, the equilibrium condition would simply be that the sum of all the forces acting on it must be zero. An extended body, however, can have a net force equal to zero acting upon it and yet it will move. A very simple example of such a situation is seen in Figure 3.1. Here we see a rod upon which is exerted equal and opposite forces at the two ends. The vector sum of the forces is equal to zero, yet the body will, in fact, rotate. Its "center of mass" will remain stationary, but the rod will rotate about the center. Clearly, the requirement that the total force be zero is not enough to ensure the static equilibrium of an extended body. The additional condition is that the sum of the "torques" acting on the body must also be zero. This condition was discovered by the extraordinary Greek mathematician, astronomer, mechanician, and engineer—Archimedes of Syracuse who lived from 287 B.C. to 212 B.C. and was killed by a Roman soldier during the sacking of Syracuse at the end of a long Roman siege. The meaning and derivation of this second condition for equilibrium follows directly from Archimedes' law of the lever which appears in his book, *On the Equilibrium of Plane Figures* [1]. We may derive this law of the lever using intuitive symmetry arguments given by E. Mach [32] following the thoughts of Archimedes.

Let us imagine that we have a rod (of length L), and that we attach beneath its full length a weighty rectangular sheet as indicated in Figure 3.2. Clearly, the system of rod and weight will be in static equilibrium and will not rotate if it is supported at its center as indicated in the diagram. Now let us imagine that the rectangular sheet is cut into two pieces of lengths Q and P. Nothing will be changed, and the system will remain in equilibrium. Furthermore, we now may imagine that each segment of the rectangle is rotated about its center into the position indicated

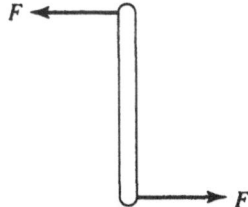

Figure 3.1. Solid object subject to a torque.

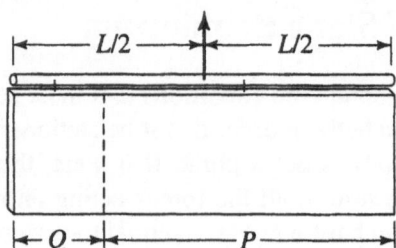

Figure 3.2. Rectangular sheet of length $L = P + Q$ attached to a rod of the same length.

in Figure 3.3. In this new situation we expect that the system remains in equilibrium because, intuitively, it is apparent that the weight of each section acts as if concentrated at its center, and this center does not move in the rotation process. Let us now examine the forces that act on the rod when it is in equilibrium. We do this in the force diagram in Figure 3.4.

We note first that the force on the support holding the center of the rod must be equal and opposite to the sum of the weight forces associated with sections P and Q. These weight forces are proportional to the lengths P and Q; i.e., $F_Q = CQ$ and $F_P = CP$. C is the constant of proportionality that relates the weight of each section F_Q or F_P to its length Q or P. Notice now that whereas we can vary the forces F_Q and F_P as we like, by deciding where to cut the section, it is necessary that these forces act at specific places along the rod to maintain static equilibrium. Calling the distance $(L - Q)/2$ the "lever arm" of F_Q, and calling $(L - P)/2$ the

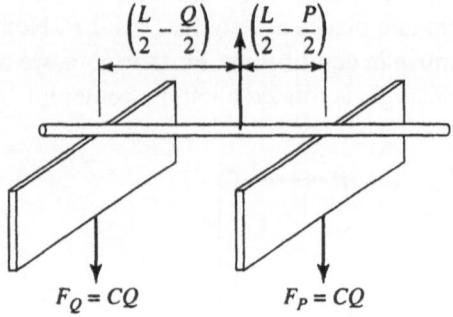

Figure 3.3. Rectangular sheet cut into parts of length Q and P, respectively, and rotated at their midpoints.

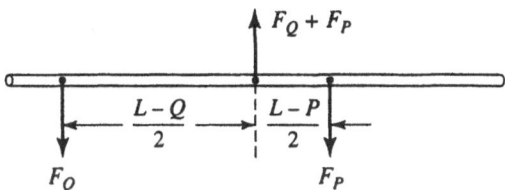

Figure 3.4. Force diagram for rod subject to the attached weights F_Q and F_P.

"lever arm" of F_P, we see that the locations at which F_P and F_Q must act are in the ratio

$$\left(\frac{L-Q}{2}\right) \Big/ \left(\frac{L-P}{2}\right).$$

But, since $(Q + P) = L$, we see that

$$\left(\frac{L-Q}{2}\right) \Big/ \left(\frac{L-P}{2}\right) = \frac{P}{Q}.$$

But the ratio P/Q is exactly equal to the ratio of F_P/F_Q. We have therefore that the ratio of the lever arms of the forces must be related to the ratio of the forces applied in accordance with the formula

$$\left(\frac{L-Q}{2}\right) \Big/ \left(\frac{L-P}{2}\right) = (F_P/F_Q)$$

or

$$F_Q \left(\frac{L-Q}{2}\right) = F_P \left(\frac{L-P}{2}\right).$$

This is the law of the lever. In words, it states that for the rod to be in equilibrium the force F_Q times its lever arm must be equal to the force F_P times its lever arm. We define the quantity "torque" as the product of the force and its lever arm. In the present case then, the torque associated with F_Q, i.e., $F_Q(L - Q)/2$, must be equal to the magnitude of the torque $(F_P(L - P)/2)$ associated with F_P. We see even in this simple example that one must associate a direction with the torque, i.e., that the torque should be a vector quantity because the torque associated with F_Q twists the rod in a direction opposite to the torque associated with F_P. We may, in fact,

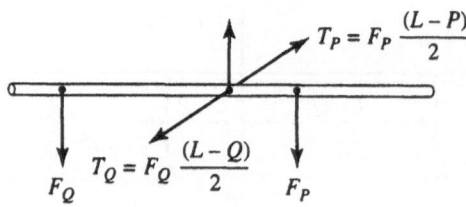

Figure 3.5. Torques T_Q and T_P resulting from the forces F_Q and F_P.

take the torque as a vector whose magnitude is equal to the product of the force and the lever arm, and whose direction is that in which a right-hand screw would advance if rotated in the same direction as the rotation of the rod. Thus the torque T_Q produced by F_Q can be represented by a vector pointing out of the paper, and that produced by F_P pointing into the paper. The condition for equilibrium then, is that the vector sum of the torques acting on the rod must be zero, and of course that the vector sum of the forces acting on the rod must be zero. This, in fact, is the complete statement of the conditions required for the static equilibrium of a rigid body. However, before it can be applied we must make quite clear and general the definition of the torque when the force is not perpendicular to the lever arm (see Figure 3.5).

Suppose the forces acting on a rod are as shown in Figure 3.6: upward force $\vec{F}_U$ and two forces $\vec{F}_Q$ and $\vec{F}_p$ acting on the rod but in arbitrary directions relative to the axis of the rod.

The equilibrium of the center of mass of the rod requires that the total force be zero, i.e.,

$$\vec{F}_U + \vec{F}_Q + \vec{F}_P = 0.$$

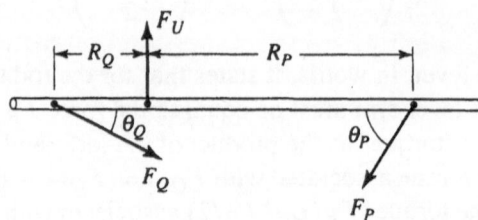

Figure 3.6. Equilibrium for the rod in the case of an arbitrary orientation of the forces $\vec{F}_Q$ and $\vec{F}_P$.

The law of the lever, when applied to this situation, clearly requires that the components of the force perpendicular to the rod must satisfy the condition that the torques associated with these components must add to give zero, i.e.,

$$(F_P \sin\theta_P) R_P = (F_Q \sin\theta_Q) R_Q.$$

We introduced previously in the Appendix to Chapter 1, Vectors (7) The cross product of two vectors. This concept enables us to express the torque quite simply so that it does reflect the fact that only the normal component of the force is used in the definition of the torque. The vector cross product $\vec{r} \times \vec{F}$ is equal in magnitude to

$$\left| \vec{r} \times \vec{F} \right| = rF \sin\theta.$$

The direction of $\vec{r} \times \vec{F}$ is that of the advance of a right-hand screw rotated in the direction of θ. The direction of θ is that obtained by rotating $\vec{r}$ toward $\vec{F}$. In Figure 3.7, $\vec{r} \times \vec{F}$ is a vector pointing out of the paper with magnitude $rF \sin\theta$.

With these considerations in mind we can define the vector torque quite generally about any chosen point by the formula

$$\vec{T}_0 = (\vec{r} \times \vec{F})$$

The torque about the origin of force F is just $\vec{r} \times \vec{F}$. If several forces act, the sum of the torques about the same origin must be equal to zero for static equilibrium. Mathematically then, for an extended body the conditions for static equilibrium are:

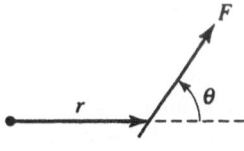

Figure 3.7.

The sum of all forces equals zero:

$$\sum_i \vec{F}_1 = 0. \tag{3-1}$$

The sum of all torques equals zero:

$$\sum_i \vec{T}_i = 0. \tag{3-2}$$

It is important to point out that if the net force on a rigid body is zero, and if the net torque about one particular point in the body is zero, then the net torque about *any* point will also be zero. One may choose *any* point as the origin for the computation of torques in a rigid body at equilibrium. The net torque will be zero regardless of the choice of origin. The proof of this remarkable result is left to the Homework Problems.

Having obtained the conditions for static equilibrium using Archimedes' law of the lever, it may be of some interest to recount a few stories connected with the life of this extraordinary man. Much about Archimedes can be found in the chapter on Marcellus (the Roman general in command of the forces that took Syracuse in 212 B.C.) in Plutarch's *Lives of the Noble Grecians and Romans*. Also a short biography and a detailed analysis of some of Archimedes mathematical discoveries are presented in the very fine paperback book, *Episodes from the Early History of Mathematics* by A. Aaboe [2].

Archimedes' discoveries of the magnification of force that could be achieved by using levers and compound pulleys led to the famous saying attributed to him "Give me a place to stand and I shall move the Earth." A more specific account of one of his demonstrations in this connection is to be found in Plutarch:

> Archimedes, however, in writing to King Hiero, whose friend and near relation he was, had stated that given the force, any given weight might be moved, and even boasted, we are told, relying on the strength of demonstration, that if there were another earth, by going into it he could remove this. Hiero being struck with amazement at this, and entreating him to make good this problem by actual experiment, and show some great weight moved by a small engine, he fixed accordingly upon a ship of burden from out of the king's arsenal, which could not be drawn out of the dock without great labor and many men; and, loading her with many passengers and a full freight, setting himself the while far off, with no great endeavor, by only holding the head of the pulley in his hand and drawing the cords by degrees, he drew the ship in a straight line, as smoothly and evenly as if she had been in the sea.

It is clear from his mathematical discoveries that Archimedes had tremendous powers of concentration. He could focus his full attention on a problem for long periods of time. The following story illustrates this graphically. Plutarch tells that, when engaged in his thinking, Archimedes would

> ...neglect his person to that degree that when he was occasionally carried by absolute violence to bathe or have his body anointed, he used to trace geometrical figures in the ashes of the fire, and diagrams in the oil on his body, being in a state of entire preoccupation, and, in the truest sense divine possession with his love and delight in science.

The ability to analyze and reduce to simple terms the most complex of questions is widely appreciated today as the hallmark of a great mind. Let us hear what Plutarch in the latter half of the first century A.D. said of Archimedes' mathematical works:

> It is not possible to find in all geometry more difficult and intricate questions, or more simple and lucid explanations. Some ascribe this to his genius; while others think that incredible effort and toil produced these, to all appearances, easy and unlabored results.

In closing the present section it is important to recognize that we have implicitly used in our analysis the concept of the center of gravity of a rigid body. We presumed that the weight of a rectangle acts effectively at the geometrical center of the rectangle. If the rigid body is an irregular figure the position of the center of gravity is not generally obvious. Nevertheless, as we shall discuss rigorously in Chapter 4, any distribution of matter in a rigid body has associated with it a point called the center of gravity (or the center of mass). This point has the property, that if a force F', equal and opposite to the weight of the body, is exerted at this point, and if the body is subject to no further external forces, then it will be in equilibrium; i.e., the body will be "in balance" and will experience no net force or no net torque. In essence then, the center of gravity is a mathematical point at which the weight of the body is effectively concentrated. In our subsequent examples, and in the problems at the end of this chapter, we will use this notion of the "center of gravity." In Chapter 4 we will provide a rigorous discussion of the concept of the center of gravity. We will show how one computes mathematically its location, and justify rigorously the fact that a body, under gravity, is in equilibrium when supported by a force equal to its weight at its center of gravity. For a symmetrical body, the center of gravity of a body is at its geometrical center.

3.3   Applications of Statics

3.3.A.   The Board Resting Against a Wall

As a simple first example of the applications of the laws of statics, let us think about the problem of resting a board with one end against a wall, and the other end on the floor. This is shown in Figure 3.8. Intuitively, we know that it is necessary to rest the board so that it makes a steep angle with the floor. If we would, by degrees, move the end of the board away from the wall, there will come a point that the board will slide along the floor and fall. We will now examine how steep the angle θ must be in order that the board not fall. This problem is quite similar to that of placing a ladder against a wall in such a way that the base does not slide when the ladder is occupied.

The problem can be analyzed with the aid of the free-body diagram (see Figure 3.9), showing the forces which act on the board. Five forces act upon the board: its weight W; the normal force at the floor F_{1N} and the friction force F_{1f} which acts tangential to the floor; and finally the normal force and tangential friction force at the wall F_{2N} and F_{2f}. The condition that the total force be zero in both the horizontal and vertical directions gives us the two equations

$$F_{2N} - F_{1f} = 0, \tag{3-3}$$

$$F_{2f} + F_{1N} - W = 0. \tag{3-4}$$

We may obtain one more equation by using the condition that the sum of the torques must be zero. We choose to compute the torques around the contact point

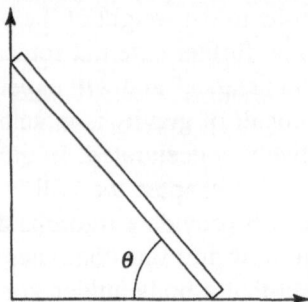

Figure 3.8. Inclined board resting against a wall.

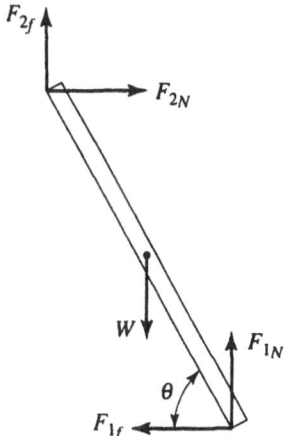

Figure 3.9. Forces acting on a board leaning against a wall.

at the top of the board. If the length of the board is ℓ, we find

$$F_{1N}\ell \cos \theta - F_{1f}\ell \sin \theta - W \left(\frac{\ell}{2} \cos \theta \right) = 0. \qquad (3\text{-}5)$$

We observe that these represent three equations in four unknowns F_{1N}, F_{2N}, F_{1f}, F_{2f}. The solution at this point is indeterminate in the sense that it is impossible to determine unique values for each of the four unknowns. The reason for this situation is that we have not yet specified the nature of the friction force with sufficient precision. Let us remedy this difficulty. The friction force that a surface exerts is tangential to the surface. Its magnitude can take any value from zero up to a maximum which is proportional to the magnitude of the normal force. The coefficient of proportionality between the friction force and the normal force is called the coefficient of friction (μ). Thus the friction forces F_{1f} and F_{2f} are always in the range

$$0 \leq F_{1f} \leq \mu_1 F_{1N},$$

$$0 \leq F_{2f} \leq \mu_2 F_{2N}.$$

Here μ_1 and μ_2 are the coefficients of friction appropriate, respectively, to the floor and wall. Before slipping the friction force takes on the value needed to make the net tangential force on the body equal to zero.

This specification of the frictional forces is still not sufficient to enable us to determine uniquely each force for any value of the angle θ. However, it is sufficient to permit us to calculate the point at which the board will start to fall. That is because at this point the frictional forces take on their maximum values; namely, $F_{1f} = \mu_1 F_{1N}$, $F_{2f} = \mu_2 F_{2N}$. Under these conditions we must look upon the three equations as consisting of three unknowns F_{1N}, F_{2N} and $\theta \equiv \theta_s$ the angle at which slipping begins to occur. With this understanding we can now solve the three equations for each of these three unknowns. We use first

$$F_{1f} = \mu_1 F_{1N}, \tag{3-6}$$

$$F_{2f} = \mu_2 F_{2N}. \tag{3-7}$$

in (3-3) to (3-5). Equations (3-3) and (3-4) then are two equations in the two unknowns F_{1N} and F_{1N}, viz.:

$$F_{2N} - \mu_1 F_{1N} = 0,$$

$$\mu_2 F_{2N} + F_{1N} = W.$$

Solving for F_{1N} and F_{2N} gives immediately

$$F_{2N} = \left(\frac{\mu_1}{1 + \mu_1\mu_2} \right) W, \tag{3-8}$$

$$F_{1N} = \left(\frac{1}{1 + \mu_1\mu_2} \right) W. \tag{3-9}$$

Furthermore, we can now obtain the third unknown θ_s. Using (3-6) and setting $\theta = \theta_s$ and dividing by $\ell \cos\theta$ in (3-5), we find that this equation gives

$$\tan\theta_s = \frac{1}{\mu_1}\left(1 - \frac{1}{2}\frac{W}{F_{1N}} \right). \tag{3-10}$$

Now (3-9) gives us F_{1N} and on substituting this into (3-10) we find

$$\tan\theta_s = \frac{1}{2\mu_1}\left(1 - \mu_1\mu_2 \right). \tag{3-11}$$

The magnitude of the coefficient of friction depends, of course, in detail on the surfaces in contact. For wood in contact with wood with both surfaces dry μ is in the range $0.25 < \mu < 0.50$. Metal to metal surfaces that are dry have $0.14 < \mu < 0.20$. A greased or oiled surface can have $\mu \sim 0.03$. In any case, the friction force never exceeds the normal force so that μ is always less than unity. Taking, for convenience, $\mu \sim \frac{1}{3}$ for our problem, we see at once a very interesting result. The slipping angle depends primarily on the coefficient of friction (μ_1) between the floor and the board. The frictional resistance to falling provided by the contact with the wall has only a weak effect. This is because the coefficient μ_2 enters the problem only in the product $\mu_1\mu_2$. Since μ_1 and μ_2 are, say, about $\frac{1}{3}$ their product is $\frac{1}{9}$, and this means that $\mu_1\mu_2$ has only about a 10% effect on the value one would obtain for $\tan\theta_s$ by ignoring μ_2 entirely. Numerically then, taking $\mu_1 \sim \frac{1}{3}$ and ignoring the $\mu_1\mu_2$ term we find

$$\tan\theta_s \cong \tfrac{3}{2}$$

or

$$\theta_s \cong 56°. \tag{3-12}$$

The board must be placed at a steeper angle than this, if it is not to fall. In fact, you can try yourself to measure the slipping angle. By this means you will, in fact, be able to deduce the coefficient of friction between the board and the floor.

3.3.B. Forces Acting at the Hip Joint

We can apply the conditions of static equilibrium to determine the magnitude and direction of the force acting on the hip abductor muscles, and the force which the acetabulum (hip socket) exerts on the head of the femur. We shall do this for certain important geometrical cases. For example, the case in which all the weight is on one foot, as occurs during walking. The second case we will consider is that in which a cane is used on the nonweight-bearing side of the body. The computed magnitude and direction of the forces acting on the femur and pelvis helps us to understand a number of clinically important observations. We will be able to understand the "antalgic gait" of a person with weakened abductor muscles or painful hip joint. In this gait the patient surprisingly throws his weight *over* the painful joint or weakened muscle. We will come to understand how this gait in turn leads to the pathological bone growth called coxa valga. The two cases we will analyze will also show us how the use of crutches or a cane after surgery can substantially reduce the risk of epiphysiolysis and avascular necrosis of the head of the femur. We shall also see why it is very important that surgical intervention at the hip joint must avoid weakening the hip abductor muscles.

To carry out such calculations it is necessary to have detailed and accurate information on the effective positions of attachment of muscles to the femur and the angles at which these forces act. In the case of the hip joint such information was first obtained, and the problem we shall consider analyzed by V. T. Inman [3]. A quantitative analysis of the forces acting at other joints and on other bones in the body, along with literature references, can be found in the very useful book *Biomechanics of Human Motion* by M. Williams and H. R. Lissner [4].

(i)   Elementary Anatomy of the Femur and Hip

Let us begin by discussing briefly some of the elementary anatomy of the hip and the femur (thigh bone). In Figures 3.10 and 3.11 we see the right hip bone. This bone consists of three parts which join together at the socket called the acetabulum.

The hip has three parts: the *ilium* which is the large upper part; the *ischium* which is the lower hind part—we rest on this in the sitting position. Finally, the third part is the front or anterior part called the *pubis*.

The right and left hip bones join together at the back with the *sacrum* and the *coccyx* at the sacro-iliac joints. They also join at the front at the pubic symphysis (symphysis means union). The entire unit is called the pelvic girdle. This is shown in Figure 3.12.

The thigh bone (femur) extends from the hip to the knee. At the upper end, the curved head of the femur articulates or fits into the acetabulum to form the hip

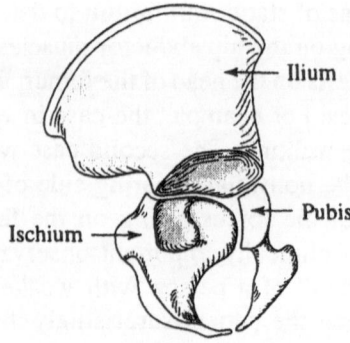

Figure 3.10. A young hip bone before fusion of sections. (Redrawn from Grant [6].)

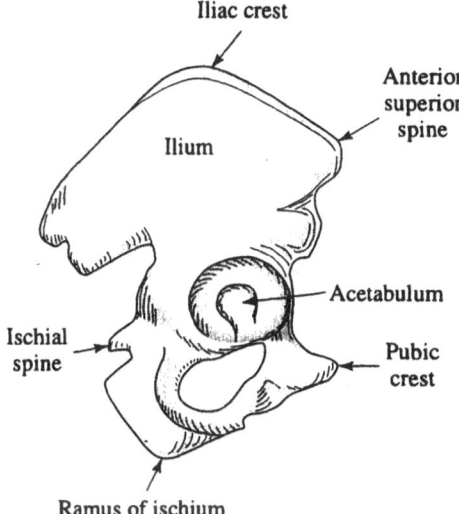

Figure 3.11. Right hip bone in an adult (from the side). (Redrawn from Cunningham [5].)

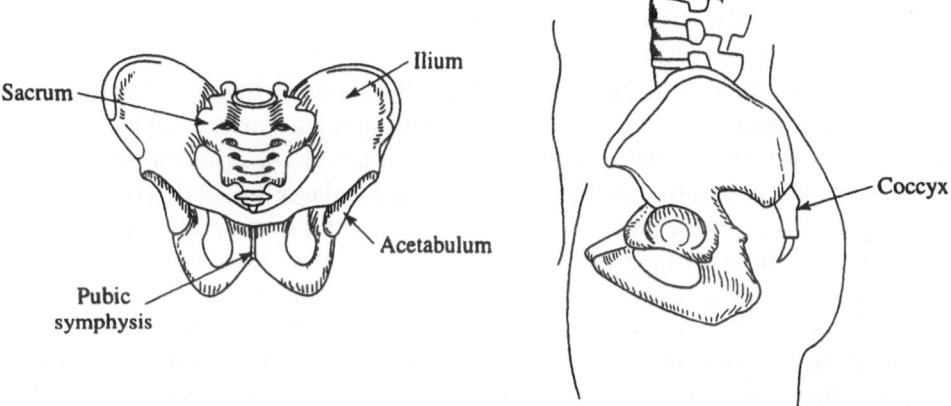

Figure 3.12. Front and side views of the hip. (Redrawn from Foster [14].)

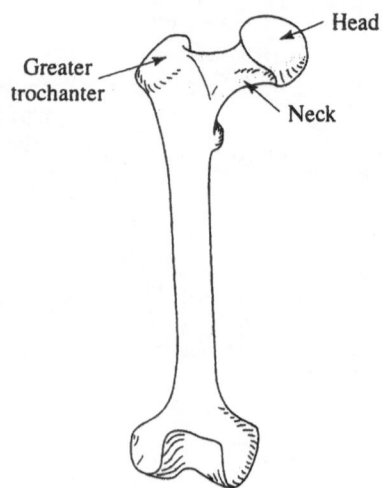

Figure 3.13. Right femur, anterior. (Redrawn from Cunningham [10].)

joint. We show in Figure 3.13 a diagram of the right femur as seen from the front (anterior aspect).

Capping the head of the femur is an *epiphysis* which is normally rigidly attached to the underlying bone. It fits the femur like a cap on a spike (see Figure 3.15).

In certain cases in which ossification between the epiphysis and the femur is not complete, the epiphysis can slip when subject to a force that is not perpendicular to the femur. The dissolution of the attachment of the epiphysis from its bone is known as epiphysiolysis.

The femur contains on its surface many protuberances that are important for muscle attachments. For our purposes the most important point of attachment on the femur is the greater trochanter (Figures 3.13 and 3.14). The tendons of five muscles are connected to this traction epiphysis. Insofar as abduction (movement from or toward the body axis) of the leg is concerned the two most important attachments are those of the *gluteus medius* and the *gluteus minimus* muscles. The other ends of these muscles are spread out and are attached to the ilium. The muscles called piriformis, obturator internis and its gemelli, the obturator externus, and the quadratus femoris are also attached in the vicinity of the greater trochanter; however, their function is to rotate the pelvis and they play a minor role in the abduction of the leg. The third important muscle, as far as our hip joint analysis is concerned, is the tensor fasciae femoris (latae). This muscle does not attach at the

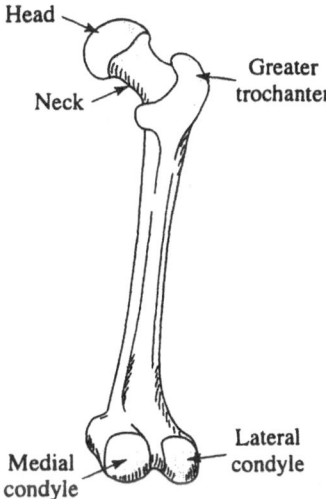

Figure 3.14. Right femur, posterior. (Redrawn from Cunningham [10].)

greater trochanter. It attaches below this point, and line of action is slightly medial (or toward the center) from the greater trochanter. Its other end interdigitates with the fascia lata and the iliotibial tract which runs from the lower part of the femur near the knee to the ilium.

The relative importance and effective line of action of these three muscles has been studied carefully by V. T. Inman [3], using x-ray photographs of cadavers. His

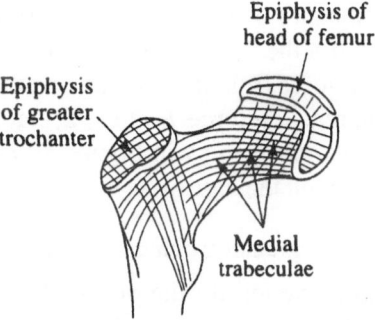

Figure 3.15. Cross section of head and neck of femur before ossification of epiphysis. (Redrawn from Grant [6].)

work indicates that the relative muscle masses of the tensor fascia femoris, gluteus minimus, and gluteus medius are 1:2:4, and he assumes that the force exerted by each is proportional to its mass. He was then able to deduce the line of action of an effective hip abductor force acting at the position of the greater trochanter. His data indicates that this lies about 70° above the horizontal even when the pelvis is elevated 15° above or below the horizontal plane. With this essential anatomical information available, Inman could calculate the forces acting on the head of the femur and on the acetabulum when a person of weight W rests upon his right foot. The right side is then the weight-bearing side.

(ii)   Calculation of the Force on the Head of the Femur and in the Hip Abductor Muscles

Having at our disposal the anatomical data described above, we may now apply the conditions for static equilibrium to calculate the forces acting at the hip joint. We begin by considering the situation in which the entire weight of the body is supported by one foot: the right foot. In this position, if we look upon the entire body as the system under consideration we see that the foot must lie directly under the center of gravity of the body as a whole. In this way the reaction force from the ground on the foot is along the same line as the weight force on the body, and the sum of the forces and torques on the entire body is zero when the reaction force from the ground is equal to the body weight.

    To calculate the force on the head of the femur and the force in the hip abductor muscles, we must consider the balance of forces and torques acting on the leg. In Figure 3.16 we show in an anatomical sketch the leg and hip bones and indicate the forces acting. In this figure we also indicate the relevant dimensions of the leg, showing in particular the distance between the greater trochanter and the center of the head of the femur (2.75 in), the distance between the greater trochanter and the line of action of the reaction force from the ground, and the line of action of the hip abduction muscles and the reaction force R at the acetabulum. To make perfectly clear the statics of the situation, we show in Figure 3.17 the equivalent free-body diagram of the forces at work on the leg:

F_1 is the force exerted on the greater trochanter by the hip abductor muscles.

R is the force of reaction exerted by the acetabulum on the epiphysis of the femur. The components of R are R_x and R_y.

N is the normal force exerted by the ground on the leg. It is equal to the weight W of the body $N = W$.

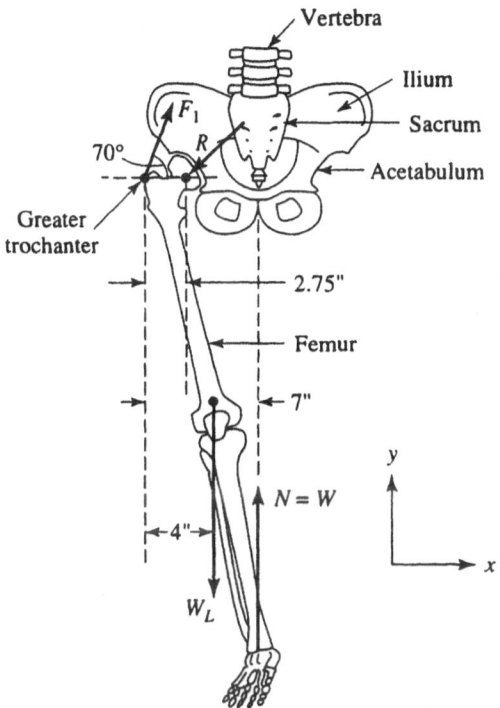

Figure 3.16. Anatomical sketch of leg and hip showing forces and dimensions. R is the force that the acetabulum exerts on the head of the femur. F_1 is the net force exerted by the hip abductor muscles. (Redrawn from Williams and Lissner, [4].)

W_L is equal to the weight of the leg. We shall take this as equal to one-seventh the body weight ($W_L = \frac{1}{7}W$) and it will act as shown in Figure 3.17 at the center of gravity of the leg, i.e., at the point shown slightly above the knee.

We can now compute the magnitude of $\vec{F}_1$, and the magnitude and direction of $\vec{R}$, the reaction force from the conditions of static equilibrium. Since all the forces are acting in the vertical plane we have a two-dimensional problem and we can take components along the x- and y-directions. The condition that the total force on the free body must be zero gives us the following equations for the x- and y-components of the forces:

y-direction:

$$F_1 \sin 70° - R_y - \tfrac{1}{7}W + W = 0, \qquad (3\text{-}13)$$

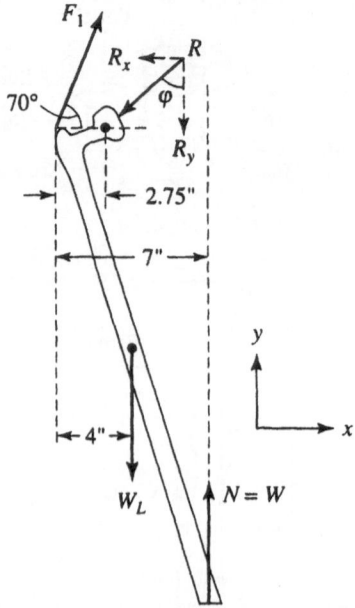

Figure 3.17. Forces on the resting leg bone.

x-direction:

$$F_1 \cos 70° - R_x = 0. \qquad (3\text{-}14)$$

The third equation can be obtained by computing the total torque about any point. We choose this point to be the center of the head of the femur. The reaction force from the acetabulum passes through this point and will therefore not appear in this equation. This choice of origin has the great merit that it will permit the direct determination of the force F_1. Then R_x and R_y can be obtained in terms of W directly from (3-13) and (3-14). The torque condition is seen to be

$$(F_1 \sin 70°)(2.75 \text{ in}) + \left(\tfrac{1}{7}W\right)(1.25 \text{ in}) - W(4.25 \text{ in}) = 0. \qquad (3\text{-}15)$$

We can draw a number of interesting conclusions from this equation before proceeding to the complete solution. From (3-15) we observe that the magnitude of the hip abductor force F_1 is determined largely by the magnitude and "lever arm" of the reaction force from the ground on the foot. In this equation this is the term W (4.25 in). 4.25 in is the distance between the line of action of the reaction force and the center of the head of the femur. As we shall see, the use of a cane on the left side results in a decrease in the size of this lever arm because the foot need not then be under the center of gravity of the body. On the other hand, if the left arm is carrying a weight, the condition for equilibrium of the body taken as a whole would increase the lever arm, and would also increase the reaction force N resulting in an increase in the force F_1. We also see that, to reduce the force in the hip abductor, it is advantageous to place the foot under the head of the femur, as can occur when standing on both legs.

Let us then complete the solution of the problem at hand and compute F_1 and R when the weight is supported by the right foot. From (3-15) we find

$$F_1 = W \frac{(4.25 - 0.179)}{2.58} = 1.58W. \tag{3-16}$$

Thus the force in the hip abductor muscles is about 1.6 times as large as the weight of the entire body W.

We now go on to compute the x- and y-components of the reaction force R on the head of the femur. From (3-14) and (3-16) we see that

$$R_x = 1.58W \cos 70° = 0.540W. \tag{3-17}$$

From (3-13) and (3-16) we find that

$$R_y = W \left(1 - \tfrac{1}{7}\right) + 1.58W \sin 70°,$$
$$R_y = (0.857 + 1.48)W,$$
$$R_y = 2.34W. \tag{3-18}$$

We see from (3-17) and (3-18) that the horizontal and vertical components of the reaction force at the acetabulum are, respectively, $0.54W$ and $2.34W$. The reaction force points very much downward. The magnitude and direction of the reaction force is computed below. (See Figure 3.17.)

$$\tan \phi = \frac{R_x}{R_y} = \frac{0.540}{2.34} = 0.231,$$
$$\phi \cong 13°. \tag{3-19}$$

The direction of the reaction force is about 13° away from the $-y$-direction. The magnitude of the reaction force is

$$R = \sqrt{R_x^2 + R_y^2} = W\sqrt{(0.54)^2 + (2.34)^2}, \tag{3-20}$$

$$R = 2.4W. \tag{3-21}$$

The head of the femur is subjected then to a force equal to 2.4 times the weight of the entire body. In the case of a 200 lb man this force can be nearly 500 lb!

(iii)   Clinical and Anatomical Implications

Having now determined the direction and magnitude of the hip abductor force and the force produced by the acetabulum, let us examine the significance of these results in terms of clinical and anatomical findings.

The first observation made by Inman was in connection with the direction of the reaction force R. He was able to show by careful x-ray measurements that R acts directly along the axis of the neck of the femur. Indeed, if one examines carefully with x-rays the bone structure of the head and neck of the femur, one finds it to be composed of a sponge-like yet ordered network called the medial trabeculae. The x-ray figures show that the axes of this network (see Figure 3.18) are arranged in such a way that they lie directly along the line of the force R. This observation alone suggests clearly that the growth of this bone is directed in part by the force applied to it. By studying the direction of R when the pelvis is rotated up and down about the horizontal plane, Inman found that R remained along the same direction, i.e., along the medial trabeculae.

By examining x-ray photographs of the femur at various ages in the development of the bone, it was found that the epiphysial line rotates as the bone grows so that it is always perpendicular to the medial trabeculae [3]. In this position the epiphysis is always normal to the force acting upon it and there is no shear stress on the cartilaginous base of the young growing epiphysis.

In case the hip abductor muscle is damaged or paralyzed, interesting changes occur. It is clear that when the abductor muscle does not function it is impossible to achieve equilibrium with the foot placed under the center of the body. The patient instinctively corrects this by shifting his weight so that his center of gravity lies above the center of the head of the femur. In so doing he exhibits what is known as an antalgic gait—the body tilts toward the side with the weakened muscle. With this gait he arranges that the foot, the center of gravity of the body, and the center of the head of the femur are in a straight line. The torque equation is satisfied

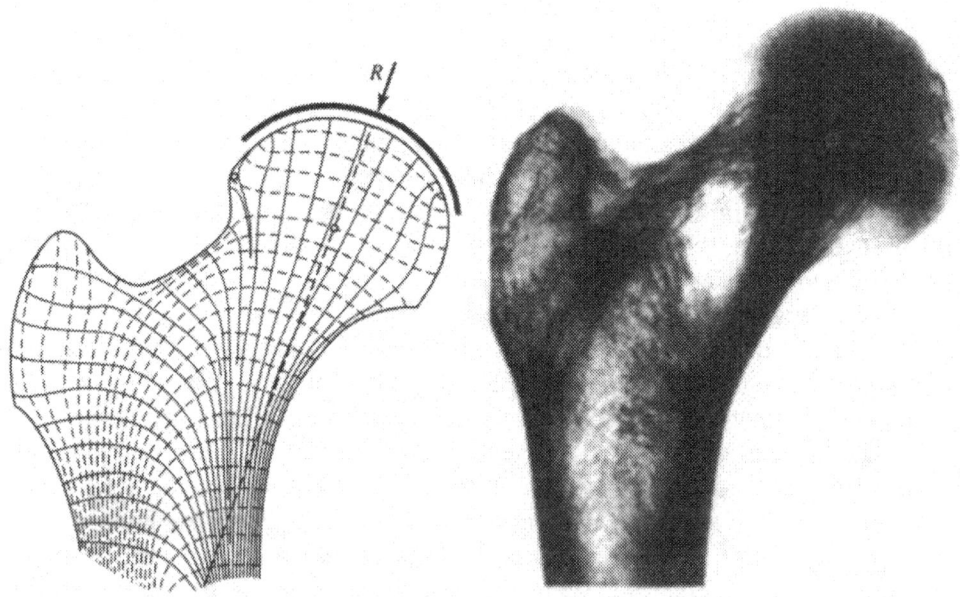

Figure 3.18. X-ray photograph of human femur.

with $F_1 = 0$. No tension is required in the hip abductor muscle. However, the reaction force at the acetabulum now points vertically downward. Its magnitude is reduced to about the value of the body weight. A heavy price, however, is paid for this adjustment in patients who suffer weakness of the abductor muscles early in life. The shift in direction of $\vec{R}$ results in a rotation of the epiphysial cartilaginous plate toward the horizontal plane. The bone then grows upward in response to the applied vertical force, and a condition known as coxa valga results. The neck of the femur gets larger and turns upward. As a result, one femur gets longer than the other and the pelvis is twisted even when standing on both feet. This in turn can lead to spinal curvature.

It is clear from these considerations that surgical approaches to the hip joint which weaken the hip abductor muscles will result in a change of magnitude and direction in the reaction force from the acetabulum, and can markedly affect the subsequent growth and development of the head and neck of the femur.

(iv)   Effect of a Cane on the Forces Acting at the Hip Joint

We saw in the previous discussion that the force in the hip abductor muscles are determined in a very important way by the torque exerted by the force on the foot about the center of the head of the femur. If the lever arm of this reaction force is reduced by bringing the foot closer to the vertical projection of the center of the femoral head, the hip abductor muscle force can be considerably reduced. The use of a cane permits this reduction of the force, without the twisting of the body toward the hip joint. We shall see in the discussion below that even a light pressure on a cane, carried in the hand opposite the affected side, can play an important role in lessening the pain in the weakened hip joint and can markedly aid in post-operative recovery after hip surgery.

When the weight of the body is borne on one foot, this foot must rest under the center of the sacrum so that the body as a whole will be in static equilibrium. Let us examine the position of the reaction force on the foot when part of the weight of the body is supported by a cane on the side contralateral to the weight-bearing side. The three forces acting on the entire body are the weight force (W), the normal force upward at the foot (N), and the upward force transmitted to the body through the cane (C). A man weighing 180 lb can rest 30 lb on a supporting cane without undue effort of the arm muscles. As this is equal to one-sixth the body weight, let us assume that the force C is equal, regardless of the actual weight of the person, to $1/6W$. We shall also assume that the cane rests on the ground at a position 12 in distant from the middle of the sacrum. Under these conditions

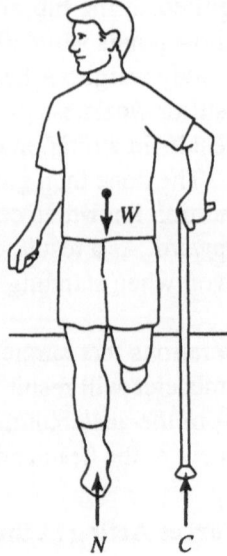

Figure 3.19. Forces on a person walking with a cane.

we can compute the location of the foot relative to the center of the head of the femur. Examining the free-body diagrams (Figures 3.19 and 3.20), we can apply the conditions of static equilibrium to the rigid body consisting of the cane and the right foot supporting the body.

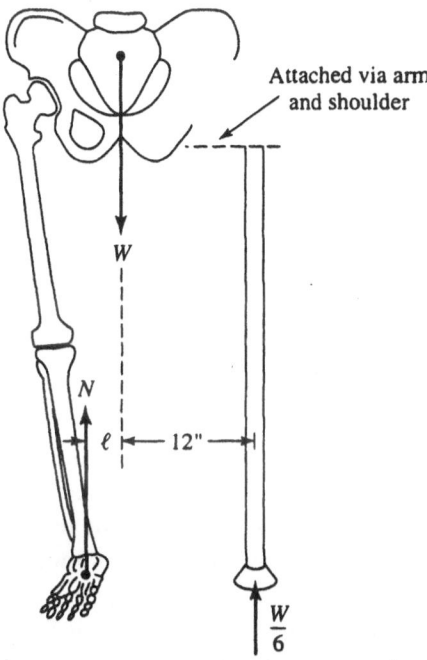

Figure 3.20. Force diagram for a person leaning on a cane for partial support.

The condition that the total force in the vertical direction is equal to zero gives

$$N + \frac{W}{6} - W = 0. \tag{3-22}$$

Thus the normal force on the foot is

$$N = \tfrac{5}{6}W. \tag{3-23}$$

The torques taken about the center of gravity gives the following equation for the position ℓ of the foot

$$N\ell - \frac{W}{6}(12 \text{ in}) = 0, \tag{3-24}$$

i.e.,

$$\ell = \left(\frac{W}{6N}\right) 12 \text{ in}. \tag{3-25}$$

Using (3-23) in (3-25) gives us the following result for

$$\ell = \frac{12 \text{ in}}{5} = 2.4 \text{ in}. \tag{3-26}$$

Thus the cane makes it possible to place the base of the foot 2.4 in away from the midline of the body. This, of course, considerably reduces the torque exerted by the normal force $\tilde{N}$ about the center of the head of the femur. We can now compute the actual magnitude of the forces in the hip abductor muscles and the reactor force at the acetabulum. We show in Figure 3.21 the free-body diagram of the leg in the position we have just computed. The weight of the leg $W_L = \frac{1}{7}W$ is located at the center of gravity of the leg which is located as shown in the figure. Note that now W_L is almost exactly under the center of the head of the femur. The normal force is now located on a line 1.85 in away from the center of the femur. Without the cane this distance was 4.25 in.

With this free-body diagram the equations for static equilibrium are:

Sum of forces in the y direction = 0:

$$F_1 \sin 70° - R_y - \frac{W}{7} + \frac{5}{6}W = 0. \tag{3-27}$$

Sum of forces in the x direction = 0:

$$F_1 \cos 70° - R_x = 0. \tag{3-28}$$

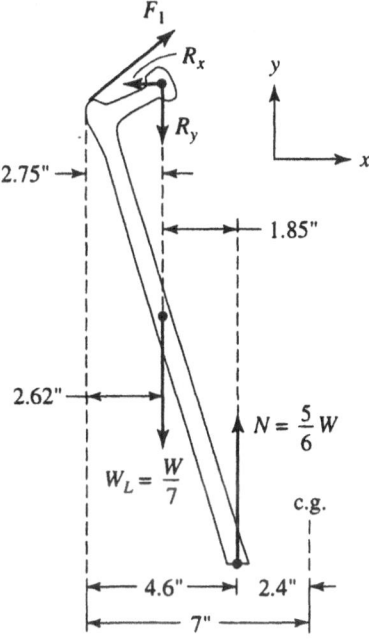

Figure 3.21. Free-body diagram of the leg.

Sum of torques about the center of the head of the femur $= 0$:

$$F_1 \sin 70°(2.75 \text{ in}) - \left(\frac{W}{7}\right)(0.13 \text{ in}) - \frac{5}{6}W(1.85 \text{ in}) = 0. \qquad (3\text{-}29)$$

From (3-29) we may immediately compute F_1 as the hip abductor force

$$F_1 = 0.60W. \qquad (3\text{-}30)$$

This placement of the foot therefore reduces the force in the hip abductor muscle to $0.60W$ compared with $1.6W$ without the cane. For a 200 lb man the hip abductor force is reduced from 320 lb to 120 lb simply by resting 33 lb ($\frac{1}{6}W$) of the weight on the cane. The reduction of 200 lb of force in the abductor muscle clearly will reduce pain if these muscles have been injured.

We can now also compute the components and magnitude of the reaction force at the acetabulum. From (3-28) we have

$$R_x = 0.20W. \tag{3-31}$$

Also the vertical component of R is obtained from (3-27) using the value of F_1 obtained in (3-30) viz.:

$$R_y = 1.25W. \tag{3-32}$$

The total reaction force at the head of the femur is

$$R = \sqrt{R_x^2 + R_y^2} = W\sqrt{(0.2)^2 + (1.25)^2} \tag{3-33}$$

or

$$R = 1.26W. \tag{3-34}$$

Thus the cane has permitted a reduction of the reaction force from $R = 2.34W$ which obtains without the cane to $R = 1.26W$ which we now find with the cane. The difference is a force equal approximately to the entire weight of the body, i.e., 200 lb for the man we have been using as an example. We see from these computations that when the cane is used to help support the body on the side opposite the affected hip joint, it is possible to reduce the force in the hip abductor muscle and the reaction force at the acetabulum by an amount roughly equal to the body weight even if the force on the cane is only one-sixth of the body weight. The use of the cane following hip surgery can play a very important role in the recovery of the patient. We quote below the observations on this point made by Dr. Blount, President of the American Association of Orthopedic Surgeons [7].

The patient of the wise orthopedic surgeon walks with crutches for six months after a fracture of the neck of the femur. He uses a stick for a longer time—the wiser the doctor, the longer the time. If his medical adviser, his physical therapist, his friends, and his pride finally drive him to abandon the cane, while he still needs one, he limps. He limps in a subconscious effort to reduce the strain on the weakened hip. If there is restricted motion he cannot shift his bodyweight, but he hurries to remove the weight from the painful hip joint when his pride makes him reduce the limp to a minimum.

The excessive force pressing on the aging hip takes its toll in producing degenerative changes. He should not have thrown away the stick.

The degenerative changes mentioned by Dr. Blount include a condition known as "avascular necrosis." In this condition the blood vessels which normally interweave within the bony structure of the head of the femur lose their blood supply: the blood vessels die and the bone tissue dies. The condition of avascular necrosis can be corrected by reducing the force on the femoral head. This can be done very effectively using crutches and later a cane. In his article [7] Blount describes a case in which a patient first suffered a fracture of the femur. Six months after the fracture she discontinued the use of crutches. Two years after that avascular necrosis appeared in the neck of the femur. To deal with this, the patient resumed the use of the crutches for one year, at which time the head of the femur had become revascularized. After that, the patient used a cane on the side opposite the affected hip for two years without any return of the avascularization.

3.3.C. Forces Acting on the Lumbar Vertebrae. Low Back Pain. Disease of Vertebral Discs

As a second example of the application of mechanics, we will examine the forces acting between the vertebra when the body is in some simple positions, such as occur in bending and lifting. The stresses which the muscles apply to the discs between the vertebrae can be quite large and these stresses can, along with a degeneration of the structure of the discs produce pain, muscle spasm, and immobilization of the lower back. Lower back pain is a very common ailment. Eighty percent of all of us suffer from it at one time or another in our lives. An understanding of the origins of this affliction depends in large measure on a grasp of the anatomy of the spinal column and the individual functional units that constitute it. Let us then begin with a brief review of the rudiments of the anatomy of the spine.

(i) Elementary Anatomy of the Spine, Vertebrae, and Back Muscles [6][8]–[10]

We show in Figure 3.22 a diagram of the vertebral column of man in the erect position. Directly above the sacrum and its curved "tail" the coccyx, rest five lumbar vertebrae. The one closest to the sacrum is the fifth and the furthest is the first lumbar vertebra. The curve of the line of these vertebrae is called the lumbar lordosis.

The curve of the vertebral column starts to reverse with the thoracic vertebrae of which there are twelve. The lowest thoracic vertebra is the twelfth and the top-

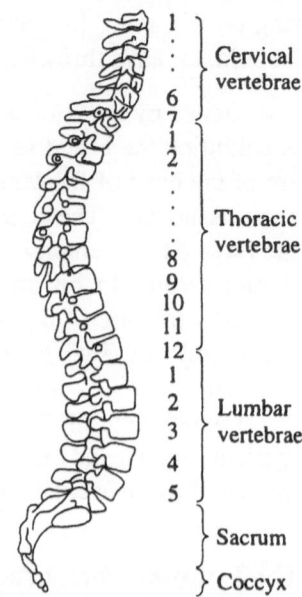

Cervical
vertebrae

Thoracic
vertebrae

Lumbar
vertebrae

Sacrum

Coccyx

Figure 3.22. The vertebral column.

most thoracic vertebra is the first. The curve of these thoracic vertebra which is convex posteriorly is called the dorsal kyphosis. The curvature here is less than the lumbar lordosis. Atop the first thoracic vertebra is the seventh cervical vertebra. The seven cervical vertebrae produce a curve which is convex anteriorly (anterior—toward the front of the body) and is called the cervical lordosis.

The actual curvature of the lumbar lordosis plays an important role in low back pain. This curvature in turn is determined by the lumbo-sacral angle, which is shown in Figure 3.23. The sacrum is rigidly attached to the pelvis. The fifth lumbar vertebra is separated from the sacrum by a disc. If the pelvis is allowed to slump forward because of weak abdominal muscles or weak hip flexor muscles, the lumbo-sacral angle is larger than normal ($\gtrsim 30°$) and a marked lumbar lordosis or sway back results. The relation between the lumbo-sacral angle and the lumbar curvature is shown schematically in Figure 3.24.

Clearly the angle at which the pelvis is held by the complex of muscles attached to it critically determines the curvature of the lower back.

Let us now examine the anatomy of the functional units of the spine: the vertebrae. In Figure 3.25 we see in side view two vertebrae in place above one another.

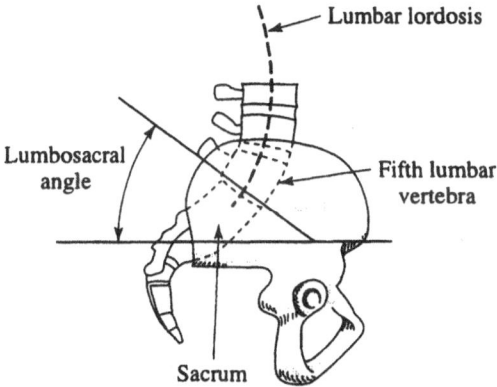

Figure 3.23. The lumbo-sacral angle is the angle between the horizontal and the top surface of the sacrum. (Redrawn from Cailliet [8].)

The anterior part of each vertebra is the part that bears the weight of the body. This is accomplished with the aid of a disc which extends between the upper part of one vertebral body and the base of the next vertebral body. This disc is a self-contained fluid system which absorbs shock, transmits the pressure between vertebra uniformly over the vertebral body, permits deformation of the intervertebral

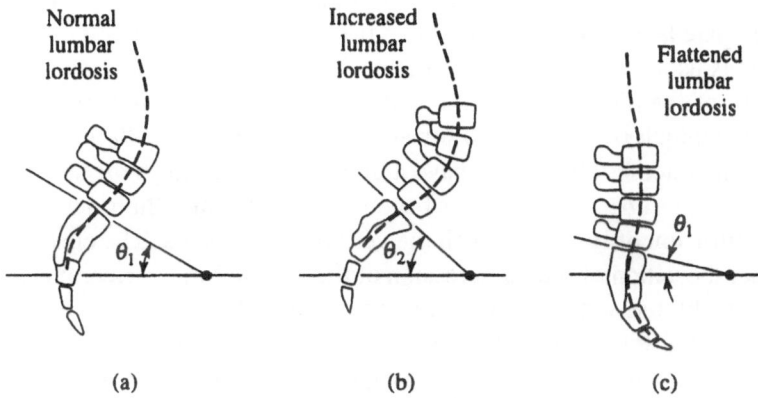

Figure 3.24. Lordosis. (a) Normal sacral angle and normal lumbar lordosis. $\theta_1 \approx$ 30°. (b) Forward tilting increases lumbo-sacral angle and lumbar lordosis. $\theta_2 >$ 30°. (c) Backward tilting decreases lumbo-sacral angle and lumbar lordosis. $\theta_3 <$ 30°. (Redrawn from Cailliet [8].)

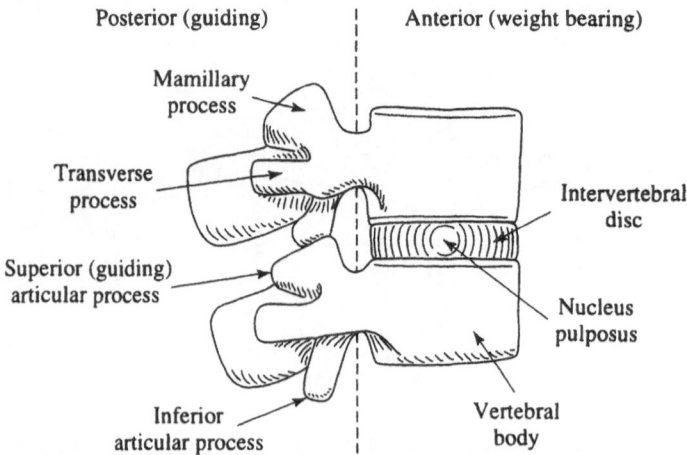

Figure 3.25. Two vertebrae in cross section. (Redrawn from Cailliet [8] and Cunningham [10].)

spacing, and therefore allows motion. The upper and lower plates of the disc are cartilages which are integral with the bone of the vertebral body. The wall of the disc is called the annulus fibrosis, which is composed of about a dozen layers of collagen fibers. These fibers are attached to the upper and lower cartilage plates, and in alternate layers the fibers cross like the limbs of the letter X. Clearly this arrangement is useful in giving the disc some strength against shearing movements. The matrix of the disc is called the nucleus pulposus and this consists of a viscoelastic gel which is normally 80% water and contains, as well, protein polysaccharides. The intervertebral discs have a blood supply through the second decade of life. In the third decade, the discs become avascular, and their nutrition is provided by diffusion of lymph through the vertebral end plates. The elasticity of the discs under the action of the compression of the surrounding vertebral is provided, in large part, by the annulus fibrosis. Degeneration of these fibers can occur as a result of aging or repeated injuries associated with overloading. Herniation of the annulus fibrosis can occur with the consequent extrusion of the nucleus pulposus through the posterior part of the disc or it can extrude into an adjacent vertebral body. This extrusion can cause pain by pressing on nerve endings in the ligamentous layers surrounding the vertebrae, or on the spinal nerve [8]. The narrowing of the intervertebral space can also cause the posterior facets of the vertebra to press against one another producing a crushing of their synovial capsule and consequent pain. We shall soon examine the anatomy of the posterior part of the vertebrae to see the location of these facets and the synovial capsule.

For our purposes the quantitative response of the intervertebral discs under the action of compressive loading is important. In Figure 3.26 we show data [11],[12] on the fractional compression of a lumbar disc as a function of the loading of the vertebra. We observe from this graph that the wet lumbar disc ruptures at a loading of about 1500 kg. The contraction at that point is about 35% of the original disc thickness. We note also that the disc is "elastic," i.e., the contraction is linear in the loading only up to about 100 kg. Above this loading, marked distortion of the disc is occurring as the stress–strain relation becomes nonlinear.

It is also of interest to note that the compressive loading (L_m) required for rupture of the disc decreases as one moves up the spiral column: for lower thoracic vertebrae $L_m = 1150$ kg, for the upper thoracic vertebrae $L_m = 450$ kg, and for the cervical vertebrae $L_m = 320$ kg. The area of each of these intervertebral discs decreases in such a way that the actual pressure or stress in kg/mm^2 required for rupture is the same for all the intervertebral discs, i.e., about 1.1 kg/mm^2.

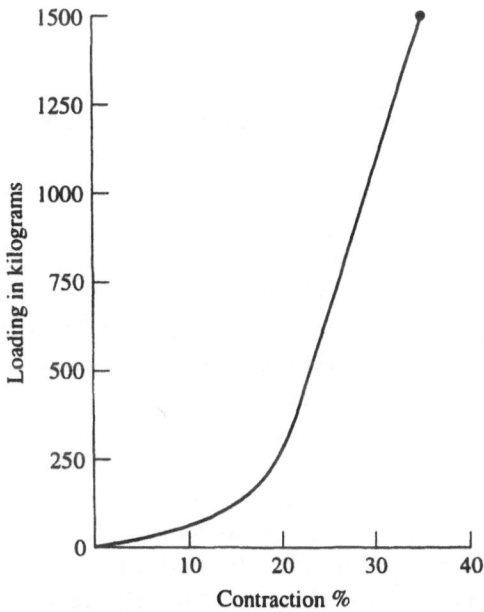

Figure 3.26. Loading versus strain in compression of wet lumbar vertebral discs of persons 40 to 59 years of age. (Redrawn from Yamada [11].)

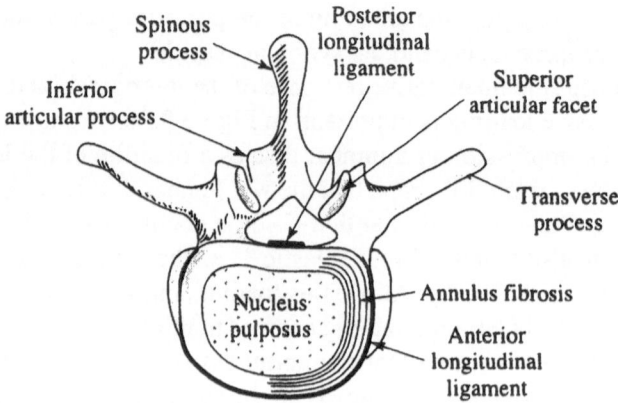

Figure 3.27. Vertebral segment viewed from above. (Redrawn from Cailliet [8] and Cunningham, [10].)

We may now complete our discussion of the elementary anatomy of the vertebral column by describing the posterior portion of the vertebrae and the location of the associated ligaments and some muscles. In Figure 3.25 we saw the posterior guiding portion of the lumbar vertebra in side view. In Figure 3.27 we show a schematic diagram of the top view of lumbar vertebra. The essential functional elements of the posterior segment are the superior and inferior articular processes. The faces of these processes are called facets. The inferior articular process of one vertebra is juxtaposed in the superior process of the vertebra below. These processes serve as guides or "stops," and they limit the transverse motion of the vertebra. The facets are lined with synovial tissue and separated by synovial fluid contained within an articular capsule. In Figure 3.28 we see these elements from the side in a normal vertebral column, and we also see their appearance when the disc has lost its elasticity or is squeezed by great stress. Clearly this will result in synovial irritation, and nerve root impingement.

Returning to Figure 3.27, we note that muscles are attached to the spinous and transverse processes, and these turn the vertebra in the vertical and horizontal plane, respectively. Further details can be seen in Figure 3.28. The space behind the vertebral body is called the vertebral foramen or vertebral canal. It is through this canal that the nerve roots pass from the brain downward. The vertebral foramen is lined with vertebral ligaments. These serve to help stabilize the vertebra against stress. The anterior longitudinal ligament is a tough glistening band that surrounds the front of the vertebral body and helps support the disc anteriorly. The posterior longitudinal ligament supports the back of the vertebral body. This support is not

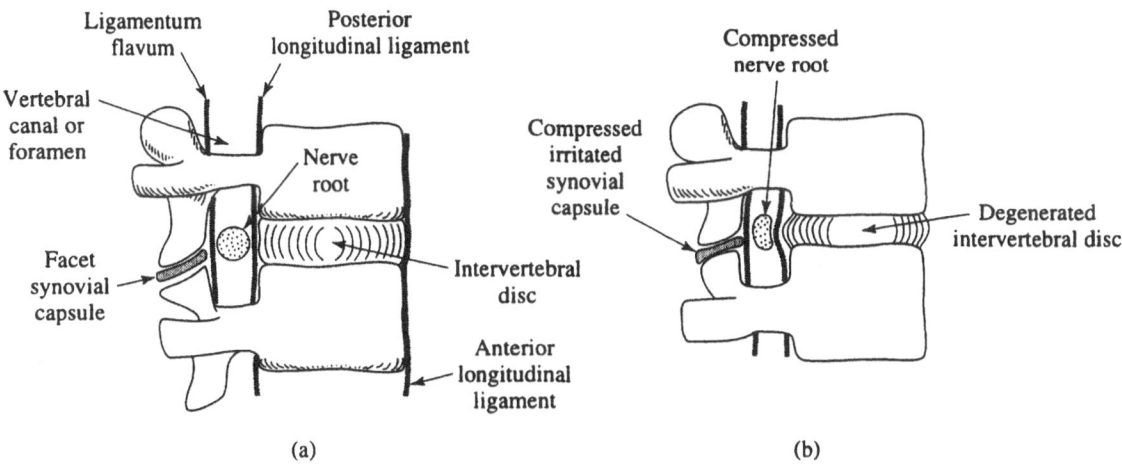

Figure 3.28. Cross-sectional views of two vertebrae. (Redrawn from Cailliet [8].) (a) Normal vertebral discs with normal intervertebral spacing. (b) Disc degeneration and narrowing, resulting reduction in diameter of the vertebral canal, compression of the nerve root; compression and irritation of the synovial capsules between the posterior articulations.

very complete and grows less substantial as the ligament runs down the vertebral column. When it reaches the first lumbar vertebra, the posterior longitudinal ligament starts to narrow. By the time it reaches the fifth lumbar vertebra, where it is very badly needed to help encapsulate and support the important lumbo-sacral disc, its width is only about one-half its original value. Clearly this is a weak point in the anatomy of the vertebral column particularly since the anulus fibrosus is also weaker posteriorly. In fact, herniation and extrusion of the disc does occur frequently in this less supported lumbar region. If the extruded disc presses on the nerve root, pain originating in the back and radiating down the buttock and leg will result [8]. Figure 3.28 shows the comparison between normally spaced vertebra and those with degenerated discs.

When bending the back or lifting objects from the floor, the principal muscles which lift the back are the erector spinae or sacrospinal muscles. These muscles attach between the ilium and the lower part of the sacrum with their upper parts attached to the spinous processes of all the lumbar and four thoracic vertebrae. On bending the back over as shown in Figure 3.29, the x-ray measurements by Inman [9] show that the gross mechanical effect of the erector spinae muscles can be approximated by describing them as a single cord which acts on the vertebral

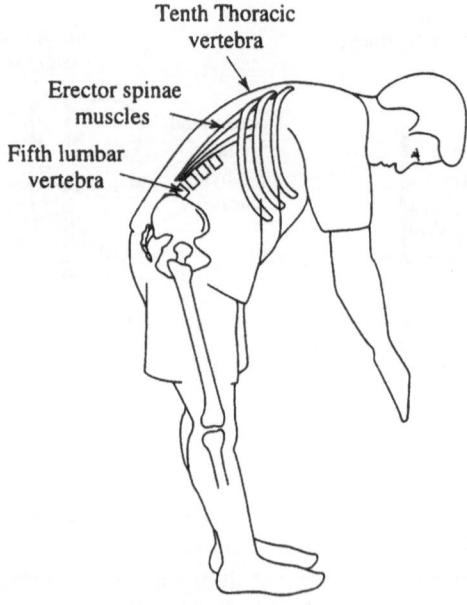

Figure 3.29. Diagrammatic representation of muscles used in raising the trunk. (From Strait et al. [9].)

column considered as a rigid body at a point which is two-thirds of the distance from the center of gravity of the head and arms. The angle of insertion of this effective cord is at an angle of about 12° relative to the axis of the vertebral column.

(ii)   Force on the Fifth Lumbar Vertebra on Bending and Lifting [9]

In accordance with the anatomical information so far discussed, we may draw the free-body diagram shown in Figure 3.30 to represent the forces acting on the vertebral column when the body is bent over. We imagine the spine as a rigid body hinged at the bottom at the lumbo-sacral disc.

The forces that act on the vertebral column are: W_2, the weight of the arms and head, plus the weight of any object being lifted by the arms in this bent position; W_1, the weight of the trunk; F_e, the force exerted by the erector spinae muscles; and, finally, the reaction force R exerted by the top of the sacrum pushing on the base of the lumbo-sacral disc. We seek to calculate from this free-body diagram the force F_e, and the reaction force R for specified values of the forces W_1, W_2, and specific angles θ between the axis of the back and the horizontal.

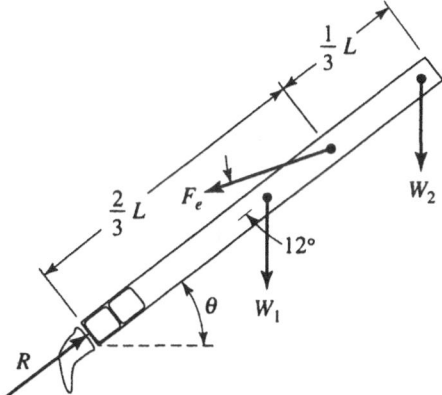

Figure 3.30. Free-body diagram of the vertebral column.

We consider first the least weighty case. The person simply bends over, with his arm hanging down straight, and he holds no weight in his hands. Let us assume, for example, that θ is 30°. Anatomical measurements show that

$$W_2 = 0.2W, \tag{3-35}$$

where W is the weight of the entire body, and that

$$W_1 = 0.4W. \tag{3-36}$$

The trunk, extending from the hips upward but excluding head and arms, is 40% of the body weight. This force acts effectively in the center of the vertebral column. The free-body diagram now looks as is shown in Figure 3.31: The problem has three unknowns. The x- and y-components of of the reaction force, R_x and R_y, and the magnitude of the force F_e in the erector spinae muscle. To obtain these we have three equations: the torque equation and equilibrium of the forces in the x- and y-directions. Taking the origin for the calculation of the torque at the lumbosacral disc, we obtain immediately an equation for the force in the erector spinae muscle, viz.:

$$F_e \sin 12° \left(\frac{2L}{3} \right) - 0.4W(\cos 30°)\frac{L}{2} - 0.2W(\cos 30°)L = 0. \tag{3-37}$$

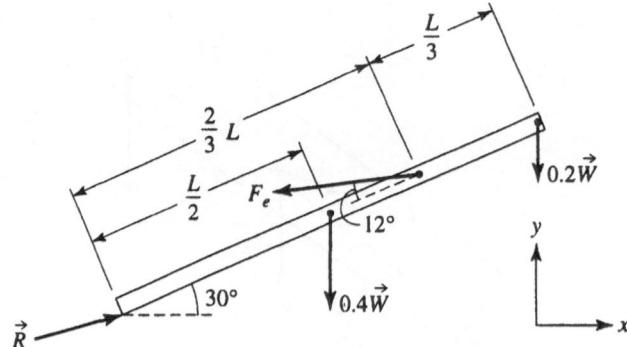

Figure 3.31. Free-body diagram of vertebral column.

The condition that the net force in the y-direction is zero gives

$$R_y - F_e \sin 18° - 0.4W - 0.2W = 0. \qquad (3\text{-}38)$$

The condition that the net force in the x-direction is zero gives

$$R_x - F_e \cos 18° = 0. \qquad (3\text{-}39)$$

These equations can be solved as follows. First, we obtain F_e from (3-37). This gives

$$F_e = 2.5W. \qquad (3\text{-}40)$$

The force in the erector spinae muscles must be two and a half times the total body weight! For a 200 lb man this corresponds to 500 lb in these muscles. The reason for this very large force is clear. The erector spinae muscle attaches to the vertebral column at a very shallow angle. Because of this, its lever arm about the lumbo-sacral disc is small. (This lever arm in fact is just ($\frac{2}{3}L \sin 12°$) which is equal to 0.14L.) Since its lever arm is small, the force F_e must be correspondingly large to counterbalance the weight forces W_1 and W_2.

We can obtain the x-component of the reaction force from (3-39) as

$$R_x = (2.5) \cos 18° W$$

or

$$R_x = 2.38W. \tag{3-41}$$

R_y then follows from (3-38) as

$$R_y = 1.37W. \tag{3-42}$$

We can now obtain the magnitude and direction of $\vec{R}$, from Figure 3.32:

$$\tan \phi = \frac{R_y}{R_x} = \frac{1.37}{2.38} = 0.575,$$

$$\phi = 35°. \tag{3-43}$$

Thus the reaction force points about 5° above the vertebral column. The magnitude of R is

$$R = \left(\sqrt{(1.37)^2 + (2.38)^2}\right)W = 2.74W. \tag{3-44}$$

Thus the reaction force at the lumbo-sacral disc is 2.7 times the total body weight. Again for a 200 lb man this is equal to 540 lb. The compressive force on the disc is equal to the component of R along the spinal column. This is equal to $(2.74)\cos 5° W$; i.e., it is almost identical to R itself. The compressive force in this orientation of the body is $2.75W$.

We can compare this result with the data mentioned previously on the compression of the discs. Remembering that 1 kg = 2.2 lb we see that for a 200 lb man just bending over the stress is 546 lb or about 250 kg. From the data in Figure 3.26, we see that this produces about a 20% contraction of the intervertebral spacing in a normal disc. Since the total volume of the disc must be regarded as constant, this corresponds to a 10% increase in the radius of the disc. Clearly, the disc will bulge

Figure 3.32. Orientation of the reaction force $\vec{R}$ relative to the vertebral column.

very substantially into the vertebral foramen. Also the narrowing of the disc brings the anterior guiding facets into closer apposition, squeezing the synovial capsule. Note that these estimates of disc compression are based on the assumption that the person is simply bending over with legs straight and arms hanging down. Suppose instead that one lifts something in this position, or worse yet, suppose he tries to lift a weight with the arms advanced in front of the head! The torque here will be very great indeed. This latter position is not too unusual; it occurs in lifting children out of cribs and carriages.

Let us compute the reaction force and the force in the erector spinae muscles when the arms hang down as in the free-body diagram in Figure 3.31, and the arms are lifting a weight. Let the weight lifted be equal to say one-fifth the body weight. In this case the diagram of Figure 3.30 has

$$W_2 = 0.2W + 0.2W,$$
$$W_1 = 0.4W,$$
$$\theta = 30°. \tag{3-45}$$

The three equilibrium conditions expressed in (3-37), (3-38), and (3-39) now become

$$F_e(\sin 12°)\frac{2L}{3} - 0.4W(\cos 30°)\frac{L}{2} - 0.4W(\cos 30°)L = 0, \tag{3-46}$$
$$R_y - F_e \sin 18° - 0.4W - 0.4W = 0, \tag{3-47}$$
$$R_x - F_e \cos 18° = 0. \tag{3-48}$$

Solving (3-46) gives

$$F_e = 3.74W. \tag{3-49}$$

Solving (3-47) gives

$$R_x = 3.56W. \tag{3-50}$$

Solving (3-48) gives

$$R_y = 1.96W. \tag{3-51}$$

The magnitude of the reaction force R is

$$R = \sqrt{R_x^2 + R_y^2} = W\sqrt{(3.56)^2 + (1.96)^2}$$
$$R = 4.07W. \tag{3-52}$$

When lifting a weight of one-fifth the body weight, the reaction force at the sacrum is increased from $R = 2.74W$ (with no weight) to $R = 4.07W$. Thus we have an increase of about $1.3W$ because basically this weight has a very large lever arm about the sacrum. In the example of the 200 lb man the weight lifted is 40 lb and the force on the sacrolumbar disc is about 815 lb. The anatomical consequences of this great compressive force is clear. Again the disc is compressed. If it is weakened by age or injury the disc can herniate, or it can extrude to the point that the nerves in the posterior longitudinal ligaments, nerve root, or the articular facets are irritated. This leads to pain and muscle spasm. The muscle spasm in turn can cause even more compression of the disc and complicate the problem.

It is clear that to reduce the force in the erector spinae muscles it is very advantageous to reduce as much as possible the lever arm of the weight being lifted. We

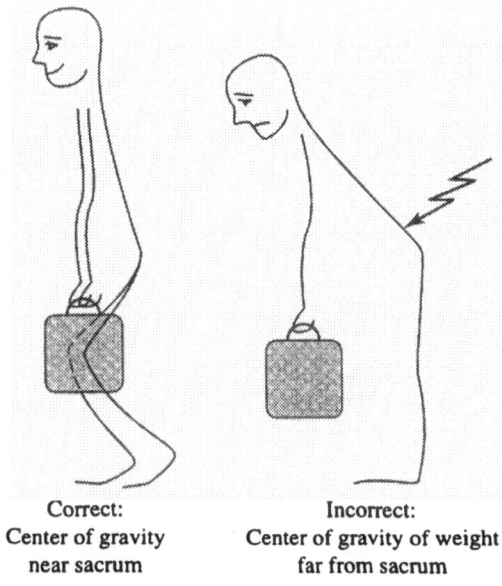

Correct:
Center of gravity
near sacrum

Incorrect:
Center of gravity of weight
far from sacrum

Figure 3.33. Correct and incorrect ways of carrying a load. (Redrawn from Cailliet [8].)

therefore can understand why the correct and incorrect positions for lifting weights are those shown in Figure 3.33.

A very fine account of the basic anatomy, pathology, diagnosis, and treatment of low back pain is to be found in the paperback book by Cailliet [8] which we have used as a reference in parts (i) and (iii) of this section.

(iii)   Shear Stress in the Lumbo-Sacral Disc with "Sway Back"

We can conclude our treatment of the forces acting on the vertebra by briefly considering the forces acting on the lumbo-sacral disc in the standing position. It is a common complaint that standing erect for extended periods can produce fatigue or even pain in the lower back. We will give a brief discussion of this problem, and will be able to see how this fatigue or pain is related to the posture of the body.

We show in Figure 3.34 the free-body diagram of the vertebral column in the erect position, but now we more accurately indicate the lumbar lordosis. In the erect position the forces acting on the vertebral column are all vertical. The weight of the trunk, head, and arms is $0.6W$. The erector spinae muscle may not be fully relaxed, but we will neglect the force that exists and set $F_e = 0$. The balance of forces thus requires that the reaction force R be equal to the superincumbent

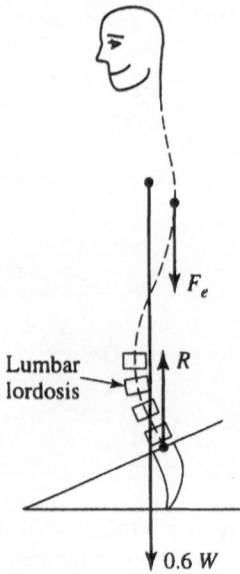

Figure 3.34. Free-body diagram of the vertebral column in the erect position. (Redrawn from Cailliet [8].)

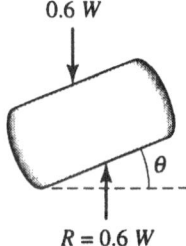

Figure 3.35. Forces acting on the lumbo-sacral disc.

weight $0.6W$. Note now that while this force acts in the vertical direction the plane of the lumbo-sacral disc is inclined from the horizontal. As a result the reaction force produces not only a compressive stress, but also a shear stress on the disc. This is seen in Figure 3.35. Here we make a free-body diagram of the lumbo-sacral disc.

We may decompose the forces on the disc into components normal to the cartilage plates of the disc and tangential to these plates, as shown in Figure 3.36. From the anatomy of the disc it is clear that it is much less capable of withstanding shearing stress than the compressive stress. The shearing stress is proportional to $R \sin \theta$. As we saw in the discussion of the anatomy of the vertebra, the lumbo-sacral angle is normally 30°. In this case the shearing force is one-half the supercumbent body weight. If the lumbar-sacro angle increases to say 40°, as it does in the case of "sway back," then $\sin 40° = 0.64$ and the shearing force becomes equal to 64% of the supercumbent weight. As a result of this shearing action the fourth lumbar vertebra will tend to slip forward on top of the fifth lumbar vertebra and the synovial capsules in the posterior portion of the vertebrae will be squeezed and irritation and pain can develop.

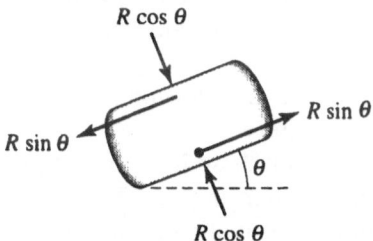

Figure 3.36. Tangential and normal components of forces on the lumbo-sacral disc.

The great majority of cases of low back pain associated with posture is connected with this "sway back" condition. The pelvic tilt and increase in lumbosacral angle generally lead to pain through irritation of the facet synovial tissue [8]. The painful lower back pain associated with pregnancy is also due to the increase in the lumbar lordosis. In pregnancy, of course, the weighty uterus contributes to the tilting forward of the pelvis.

Clearly, the direction for alleviation of this pain is the development of a posture in which the lumbar lordosis is flattened and the lumbo-sacral angle is reduced. There are many exercises which will help develop correct pelvic orientation [8]. Nevertheless, constant attention to flattening this lordosis is required throughout the day during the training period.

3.3.D.   Further Applications. Comments on the Torque Condition for Static Equilibrium

In the sections above we have seen a few examples in which detailed anatomical measurements have been combined with the equations of static equilibrium to compute the magnitude and direction of the force acting in muscle groups at certain joints in the body. In addition to the examples above, careful studies of the shoulder joint have been made by Inman and his co-workers [15]. In the book by Williams and Lissner [4] one can find interesting computations of the forces at work at the knee and the ankle joint in various body positions.

The variation of the forces in muscles during locomotion has also been studied. For those interested in this subject a recent work is the paper by J. B. Morrison [16] which has a good bibliography. A discussion of the mechanical properties of muscle can be found in the article by Inman and Ralston [17].

In concluding our discussion of the static equilibrium of rigid bodies, it is appropriate that we look briefly, but critically, at our derivation of the torque equation, i.e., that the sums of the torques taken about any origin must be zero for static equilibrium. As the reader will recall, this fundamental result was derived using intuitive symmetry arguments following the line of argument first given by Archimedes. In view of the fact that Newton's laws of motion are the basis of mechanics, it is natural to inquire whether the torque condition ($\sum_i T_i = 0$ for equilibrium) can be obtained from these laws of motion, without an appeal to intuition. The answer to this question is yes. The motion of a rigid body can be looked upon as being composed of motion of the center of mass of the body, plus rotary motion of the rigid body about its center of mass. By applying Newton's laws of motion to each particle inside a rigid body, it can be shown that the body will have no "rotary acceleration" (or, more precisely, that the rate of change of its "angular momentum" will be zero) if the total torque acting on the body is equal to zero. Normally, and logically, a consideration of angular momentum appears in mechanics texts toward the end of the book, following the development of the conservation laws; see, for example, D. Kleppner and R. Kolenkow [18] or Resnick and Halliday [19]. At this point in our development it is sufficient to recognize that while our derivation of the torque condition for equilibrium seemed separate from Newton's laws of motion, this equilibrium condition, in fact, can be shown to be rigorously deducible from Newton's laws of motion.

3.4   Static Equilibrium of Deformable Bodies: Stress, Strain, and Fracture

3.4.A.   Introduction

In our previous analyses we assumed that the bodies we studied were perfectly rigid and would not deform or break under the action of external forces. Real materials, on the other hand, do not behave in this way. In actual fact, external forces acting on a real, deformable body set up internal forces which stretch, bend, and if large enough, can break the body. In the design of houses, bridges, or in the study of orthopedics the deformation and fracture of structural members, be they beams or bones, is of central concern. In this section we present an elementary analysis of the forces inside a deformable body, and the resulting deformation including fracture. As an illustration of the questions which face us, let us consider the distribution of forces *inside* a rectangular beam supported at the ends, and loaded in the middle as shown in Figure 3.37.

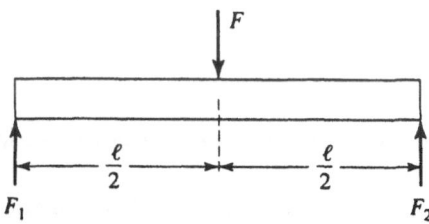

Figure 3.37. Free-body diagram of a rectangular beam.

The condition for static equilibrium of the external forces acting on the body is

$$F_1 = F_2 = \frac{F}{2}.$$

With this choice of F_1 and F_2 the total external force and external torque on the body is zero.

Let us now apply the conditions of static equilibrium in a more subtle way. Not only is the beam as a whole in equilibrium, but also each part of it must also be in equilibrium. Put in other words, let us make a free-body diagram (see Figure 3.38) of a part of the beam and examine the forces that act within the beam to permit that part to be in static equilibrium.

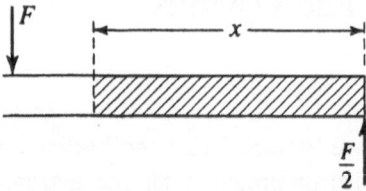

Figure 3.38. Free-body diagram of a section of the beam.

The free body now under consideration is the shaded section of the beam of length x. In order for this section to be in equilibrium, we see, first of all, that the part of the beam to the left of the position x must exert a force downward equal to $F/2$ so that the net force on the body is zero; this is illustrated in Figure (3.39) below:

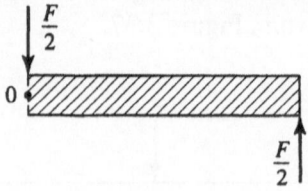

Figure 3.39. Free-body diagram for the shaded section of the beam.

Now, however, we note that this free body is not in equilibrium insofar as the torque condition is concerned! There is a torque about the point 0 of magnitude $(xF/2)$. In order then for the beam to be in equilibrium, we see that the material

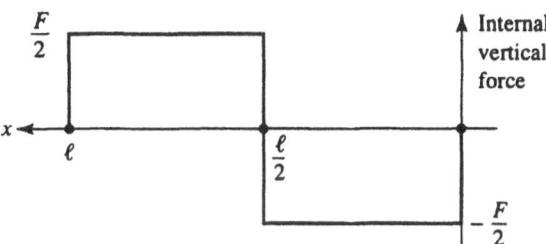

Figure 3.40. Internal vertical force on the right-hand section of the beam.

to the left of the shaded part must also exert a torque on the section at x whose magnitude is equal to $(xF/2)$! Thus within the beam there is at each point an internal force and a torque.

In Figures 3.40 and 3.41 we plot as a function of x the internal force and torque which must exist at each position so that the corresponding shaded part of the beam be in equilibrium. The left-hand section of the beam exerts a torque internally on the right-hand part. This torque increases linearly as the length x of the right-hand section increases until the center of the beam, then it decreases linearly until the internal torque is zero at the far left-hand end.

Thus we see that there must be internal forces and torques. We now ask, in particular: How do these internal torques arise? A little reflection shows that the deformation of the beam under the external loading provides the needed torques. This is seen in Figure 3.42. Due to the load $\vec{F}$, the beam is slightly bent. Figure 3.43 shows an exaggeration of this situation. As a result of this bending, the upper end of the beam is slightly compressed and the lower end slightly stretched. In particular, we have now a situation where

$$\text{distance}(AB) < x < \text{distance}(CD).$$

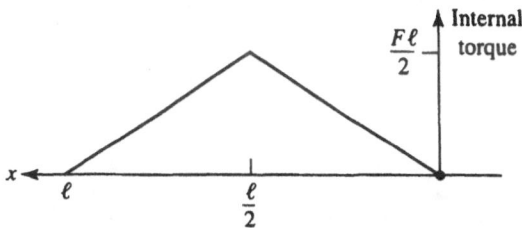

Figure 3.41. Internal torque on the right-hand section of the beam.

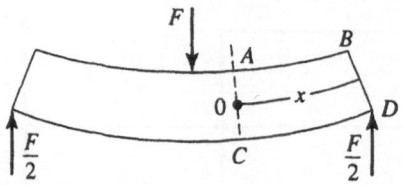

Figure 3.42. Bending of a loaded beam.

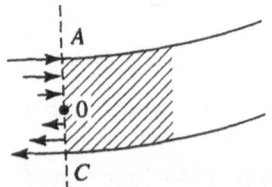

Figure 3.43. Internal stresses in a bent beam.

This elastic deformation is accompanied by forces called stresses which act on the cut. At point A, the part to the left of the cut exerts a *push* to the right, at point C it exerts a *pull*. In Figure 3.44 we see this situation qualitatively: It is easily seen that these stresses have a nonzero moment about the point 0. The torque they produce is clockwise. Its magnitude and direction is equal and opposite to the torque of $F_2 = F/2$ about point 0:

$$\text{Torque} + \frac{Fx}{2} = 0.$$

We can see then, in a qualitative way, what is happening. Under the action of the external forces the body deforms to the point that the consequent internal

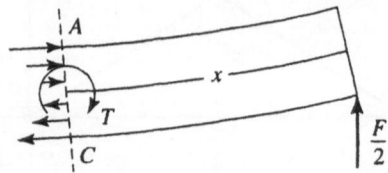

Figure 3.44. Torque within a bent beam.

stresses permit static equilibrium to be achieved everywhere inside the body. We may use this qualitative understanding now to examine quantitatively the deformations of real bodies and the conditions under which they will fail or fracture. To provide this quantitative treatment, we must first examine the rudimentary connection between the stress and strain in deformable bodies.

3.4.B.   Stress and Strain in Tension and Compression, Hooke's Law, Young's Modulus, Failure or Fracture

The simplest and most important example of the relationship between stress and strain is to be found in the stretching or compression of a rod of uniform cross sectional area. Such a rod is shown in Figure 3.45. The rod of length ℓ, and cross sectional area A, is subjected to a stretching or compressive force F at the ends. Experimentally it is found for a wide variety of elastic materials such as metals, crystalline solids, wood, bone, and rubber that for small forces the change in length of the rod ($\delta\ell$) is proportional to the original length ℓ and to the applied force per unit area:

$$\delta\ell = \left(\frac{1}{E}\right)\ell\left(\frac{F}{A}\right). \tag{3-53}$$

The quantity (F/A), the force per unit area is called the stress (σ):

$$\sigma = F/A. \tag{3-54}$$

The fractional change in length $\delta\ell/\ell$ is the strain (ϵ):

$$\epsilon = (\delta\ell/\ell). \tag{3-55}$$

The experimental finding then is that for small stresses the stress and strain are directly proportional to one another. This is called Hooke's law. The stress–strain

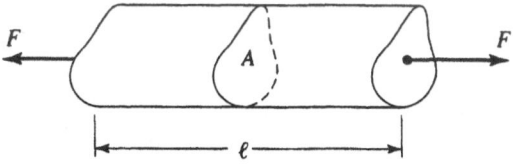

Figure 3.45. Stretching of a uniform rod.

regime in which this law applies is called the "elastic" region. An elastic body is one in which the strain is proportional to the stress, and removal of the stress leads to a return of the body to its initial shape.

The coefficient of proportionality between stress and strain (E) is Young's modulus

$$E = \frac{\sigma}{(\delta\ell/\ell)}. \tag{3-56}$$

The dimensions of E are N/m^2 or dyn/cm^2, or lb/in^2. Sometimes one also finds the unit kg/mm^2. In this case one must multiply E by $9.8 \ m/s^2$ to express E in units of force per unit area—N/mm^2. Another unit which appears and which we will sometimes use is the *bar*:

$$1 \ bar = 10^5 \ N/m^2.$$

This is very closely identical to the pressure of one atmosphere (mean air pressure at sea level). It is related to the pound per square inch (psi) by

$$1 \ psi = 6.9 \times 10^{-2} \ bar,$$

$$1 \ bar = 14.5 \ psi. \tag{3-57}$$

In Table 3.1 we give some examples of values of Young's modulus. We see from this table 3.1 that given the same stress, that bone will deform in tension or compression about the same as hardwood—provided, of course, that the stress remains in the elastic region. It is important to examine the stress–strain relation-

Table 3.1: Young's Modulus.

Material	E (in Bars)
Steel	2×10^6
Bone (limbs, along axis)	2×10^5
Hardwood	$\sim 10^5$
Tendon	200
Rib cartilage	120
Rubber	10
Blood vessels	~ 2

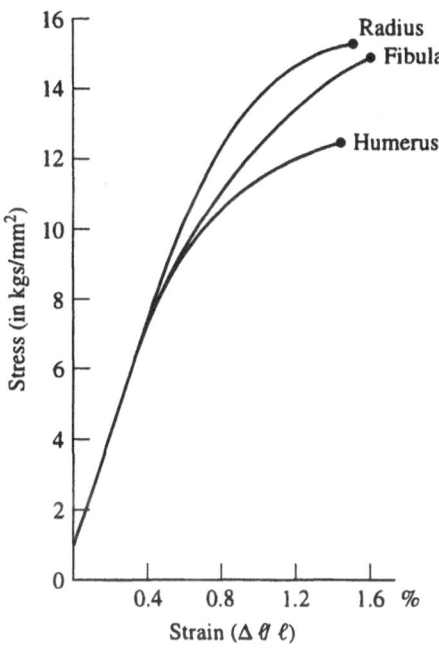

Figure 3.46. Stress–strain curves for wet, compact limb bones of persons between 20 and 39 years. (Redrawn from Yamada [11].)

ship outside the elastic region if we are to understand plastic deformation and fracture. In Figure 3.46 we show graphs of stress versus strain for wet, compact human bones taken from various limbs [11]. We observe that for strains less than 0.5% (i.e., $\delta\ell/\ell < 0.005$) these bones are elastic. Above about 0.5% increasing the stress produces much greater strain than would be obtained from an elastic medium. As the stress continues to increase the curves simply stop. This corresponds to fracture of the bone. All the failure occurs at about the same strain value; i.e., about 1.5%. The stress which corresponds to this point is called the ultimate tensile strength [11]. In Table 3.2 we list the ultimate tensile strength for these various bones along with their Young's modulus (also sometimes called modulus of elasticity) [11].

In the case of compressive stress the ultimate compressive strength of wet femoral bone of persons aged 20 to 39 years is $\sim 1.5 \times 10^3$ bar. Young's modulus in compression should be the same as that in tension. It is interesting to observe that the ultimate tensile strength of steel forgings or fairly strong aluminum alloys is only two or three times larger than that of bone.

Table 3.2: Properties of Wet Human Bone in Tension.

| Bone | (Age of Person: 20–39 Years) | |
	Ultimate tensile strength (Bars)	Young's modulus (Bars)
Femur	1.21×10^3	1.72×10^5
Tibia	1.40×10^3	1.80×10^5
Fibula	1.46×10^3	1.85×10^5
Humerus	1.22×10^3	1.71×10^5
Radius	1.49×10^3	1.85×10^5
Ulna	1.48×10^3	1.84×10^5

From Yamada [11].

3.4.C.   Bending Moment and Curvature of a Beam

Now that we know the relationship between stress and strain for an elastic material in tension or compression, we may return to the problem of relating the torques within a body to the bending or deformation of that body. Our analysis will proceed along the following lines: We will show quantitatively the relation between the radius of curvature of a beam and the internal torques. By relating the radius of curvature to the deflection of the beam, we can obtain an equation for the deflection of the beam as a function of position along its length. This equation then can be solved (at least in principle) for any distribution of external stress and thus one can compute quantitatively the deflection of a beam in terms of its Young's modulus, its cross sectional shape, and the external applied forces.

In Figure 3.47 we show a section of a beam which has undergone bending. For a short enough section we can characterize the bending by the radius of curvature R of the bend as indicated in the figure.

If we examine the bent section of the beam we observe that the extension of the lower part of the beam and the compression of the upper part of the beam can be related very simply to the distance of these parts from the neutral axis and the radius of curvature of the bend. The neutral axis of the bent section is that surface within the beam whose length is unchanged by the bending. We see in Figure 3.48 in detail a diagram which enables us to compute the strain at any distance x from the neutral axis. The lines $A'B'$ and AB represent mathematical cuts of the bent section.

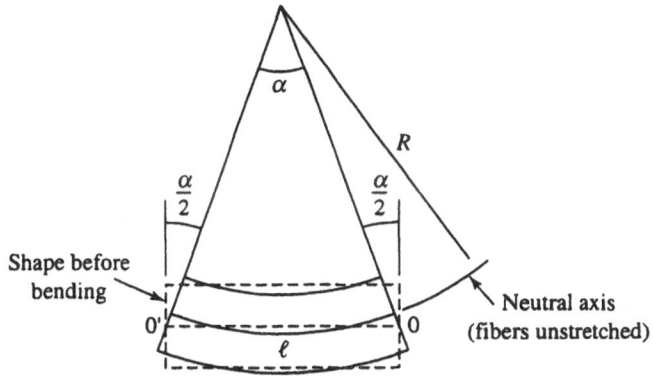

Figure 3.47. Geometry of a bent beam.

The neutral axis is the curve OO'. The length of this section is ℓ. The length of the section CC', which is a distance x from the neutral axis, is

$$CC' = OO' + 2x \tan \alpha/2. \tag{3-58}$$

Since α is very small, we can replace $\tan(\alpha/2)$ by $\alpha/2$ and also since $OO' = \ell$ we can express the change in length $\delta\ell$ of the section CC' by

$$CC' = \ell + \delta\ell. \tag{3-59}$$

Thus we may write

$$\ell + \delta\ell \cong \ell + x\alpha. \tag{3-60}$$

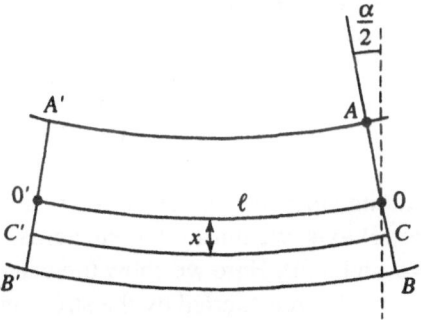

Figure 3.48. Detail of a bent section.

But α itself is related to the radius of curvature R by

$$\alpha = \frac{\ell}{R}, \tag{3-61}$$

thus we see that the stretching (or compression) of a fiber in the beam a distance x from the neutral axis is related to x and the radius of curvature R by the equation

$$\delta\ell = \ell\left(\frac{x}{R}\right) \tag{3-62}$$

or the strain $\epsilon = (\delta\ell/\ell)$ in the beam a distance x from the neutral axis is

$$\epsilon = \left(\frac{\delta\ell}{\ell}\right) = \frac{x}{R}. \tag{3-63}$$

We now have expressed quantitatively the strain inside the beam in terms of the radius of curvature of the bending at any point along the length of the beam. Equation (3-63) is valid for small deflections regardless of the shape of the cross section of the beam.

The strain has associated with it a stress and so there is a stress distribution inside the beam. At each mathematical cut in the beam this stress distribution amounts to a torque about the neutral axis. This torque is called the bending moment, and the bending moment within the beam must be exactly equal to that needed to produce the static equilibrium of each section of the beam. Thus we see that the actual deflection of the beam will be that required to produce static equilibrium in each section of the beam under the action of the external forces. Let us then compute the internal torque on the bending moment along the section $A'B'$.

At each point on the section the stress, i.e., the force per unit area, is for an elastic body

$$\sigma = E\left(\frac{\delta\ell}{\ell}\right) = E\left(\frac{x}{R}\right). \tag{3-64}$$

This force is the same along surfaces parallel to the neutral axis. To compute the net torque about the neutral axis, we must integrate across the area of the beam as indicated in Figures 3.49 and 3.50. Here we show the beam as having an arbitrary cross sectional area. The total force exerted by the stress in the shaded strip is

$$dF = \sigma \, dA, \tag{3-65}$$

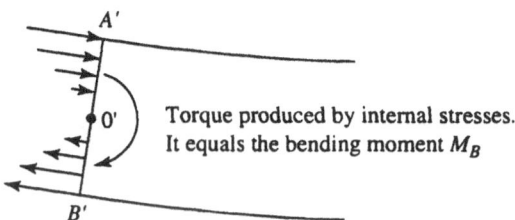

Figure 3.49. Internal torque in a bent section.

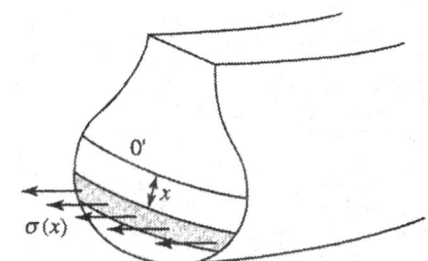

Figure 3.50. Diagram for the computation of total torque in a bent beam.

where dA is the area of the strip. The torque dT about O' by the force on this strip is

$$dT = \sigma x\, dA. \qquad (3\text{-}66)$$

Note now that the stress σ is related to the distance x by the elastic stress–strain relation (3-64). Thus (3-66) becomes

$$dT = \frac{Ex^2}{R}\, dA. \qquad (3\text{-}67)$$

The *total torque* about the neutral axis OO' is called the *bending moment* M_{B}

$$M_{\mathrm{B}} = \int dT = \frac{E}{R} \int x^2\, dA. \qquad (3\text{-}68)$$

The integral of the differential area elements, weighted by the square of their distance from the neutral axis, should properly be called the area second moment.

The conventional, but not really appropriate, nomenclature is to call this integral the "area moment of inertia." In any case, the integral above will be denoted by the symbol I_A. It has units of $(length)^4$ and its magnitude can be computed directly once the actual shape of the beam cross section is specified. Writing then

$$\int x^2 \, dA \equiv I_A, \tag{3-69}$$

we see that the bending moment acting at each cross section of the beam is related to the radius of curvature and Young's modulus by the general result

$$M_B = \frac{E I_A}{R}. \tag{3-70}$$

The degree of bending is measured by $(1/R)$, the "curvature." A sharp bend has a small R, and a gentle bend has a large R. Writing then

$$\left(\frac{1}{R}\right) = \frac{M_B}{E I_A}, \tag{3-71}$$

we see that the curvature will be large in direct proportion to the bending moment M_B. The bending moment within the beam can, of course, be computed as we showed in the introduction to this section by applying the condition of static equilibrium to arbitrary segments of the entire body. The curvature is inversely proportional to Young's modulus. A large Young's modulus corresponds to large stiffness. Finally, the curvature is also inversely proportional to the second moment of the area I_A. This has important consequences in the design and use of structural members, for the appropriate second area moment can be very different for different directions of stress on the beam. Consider, for example, a beam of rectangular cross section which is bending as shown in Figure 3.51. In this case the area second moment is computed from the definition of I_A as

$$I_A = \int_{-h/2}^{+h/2} x^2 W \, dx = \left.\frac{W}{3}x^3\right|_{-h/2}^{+h/2} = \frac{Wh^3}{12}. \tag{3-72}$$

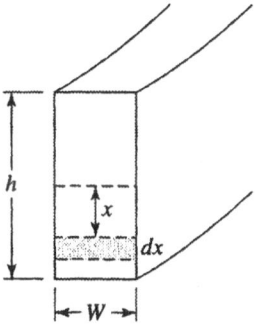

Figure 3.51. Diagram for the computation of I_A.

Thus for a 2 in × 6 in wood beam $I_A = \frac{[2 \text{ in} \times (6 \text{ in})^3]}{12} = 36 \text{ in}^4$. On the other hand, if the 2 in × 6 in wood beam is used in such a way as shown in Figure 3.52, then the area second moment has the value

$$I_A = \frac{Wh^3}{12} = \frac{6 \times 2^3}{12} = 4 \text{ (in)}^4.$$

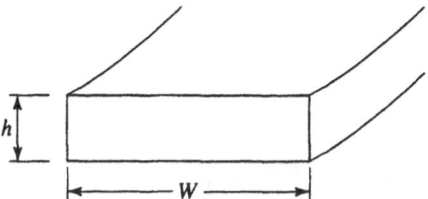

Figure 3.52. Alternate form of bending of a 2 in × 6 in beam.

Thus the curvature of the board will be nine times greater if used in this way rather than with its flat side normal to the bending moment. The crucial point here, of course, is that the second area moment varies as the cube of the dimension of the beam in the plane of bending, and only linearly with the dimension perpendicular to that plane. Thus for stiffness it is desirable that the beam be broad in the plane of bending.

Considerations such as these, along with the need to minimize the amount of material used, results in the famous I beam shape of structural members used in steel building frames:

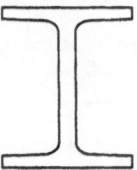

It is sometimes useful to know the second area moment of a solid or hollow rod. The results for such beams are shown below:

Solid circular beam radius a:

$$I_A = \frac{\pi a^4}{4}$$

(3-73)

Hollow circular beam, radii a_2 and a_1:

$$I_A = \frac{\pi}{4}(a_2{}^4 - a_1{}^4)$$

(3-74)

Equation 3-71 tells us the curvature of the beam at each point in terms of M_B, and the material and geometrical quantities E and I_A. To compute the actual deflection of the beam along its length, one would have to laboriously compute, section by section, curved pieces and fit them together mathematically. It is, in fact, possible to do this much more simply by obtaining an equation for the actual beam displacement in terms of the bending moment. To do this, we must relate the radius of curvature to the deflection at each point. The desired equation for the deflection will be a "differential equation" whose solution gives the actual shape

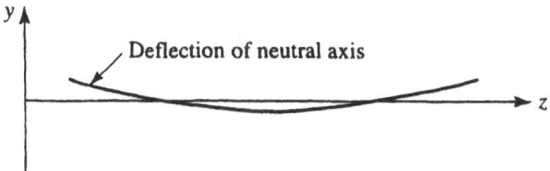

Figure 3.53. Deflection as a function of position along the beam.

3.4.D.   Differential Equations for Deflection of a Loaded Beam

In Figure 3.53 the z-axis represents the axis of an undeformed beam. Let us imagine that some stress distribution deflects the beam so that its neutral axis describes some curve $y(z)$. We shall now show that the equation, obeyed by the deflection $y(z)$ obeys, is

$$\frac{d^2y}{dz^2} = \frac{M_B}{EI_A}. \tag{3-75}$$

This equation is a direct consequence of (3-71) since, for small deflections, the curvature $(1/R)$, of any function $y(z)$, is given simply by

$$\frac{d^2y}{dz^2} = \frac{1}{R}. \tag{3-76}$$

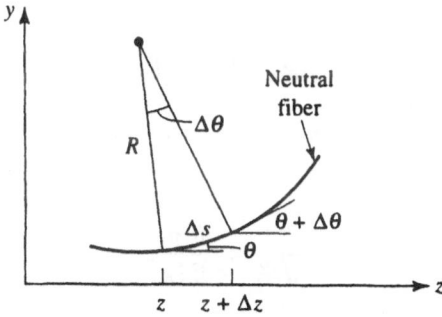

Figure 3.54. Curvature of a neutral fiber and the associated radius of curvature R.

This mathematical result can be proven by reference to Figure 3.54. From Figure 3.54, one sees that

$$\frac{1}{R} = \left(\frac{\Delta\theta}{\Delta S}\right). \tag{3-77}$$

Now if the deflection is always small we have the following equations:

$$\left(\frac{dy}{dz}\right)_z = \tan\theta \cong \theta, \tag{3-78}$$

$$\left(\frac{dy}{dz}\right)_{z+\Delta z} = \tan(\theta + \Delta\theta) = \theta + \Delta\theta, \tag{3-79}$$

$$\Delta\theta \cong \left(\frac{dy}{dz}\right)_{z+\Delta z} - \left(\frac{dy}{dz}\right)_z. \tag{3-80}$$

Also, when the deflection is small,

$$\Delta S \cong \Delta z.$$

Thus

$$\frac{1}{R} = \left(\frac{\Delta\theta}{\Delta S}\right) \cong \left(\frac{\Delta\theta}{\Delta z}\right) \cong \frac{(dy/dz)_{z+\Delta z} - (dy/dz)_z}{\Delta z}. \tag{3-81}$$

The expression on the rhs of (3-82) is, in the limit, $\Delta z \to 0$, nothing but the second derivative (d^2y/dz^2). Thus we have proved

$$\frac{1}{R} = \left(\frac{d^2y}{dz^2}\right) \tag{3-82}$$

and on using this in (3-71) we obtain the differential equation for the deflection $y(z)$ as

$$\frac{d^2y}{dz^2} = \frac{M_B}{EI_A}. \tag{3-83}$$

To apply this equation in the calculation of $y(z)$, we must remember that the bending moment M_B in general is a function of z, the position along the axis. We can

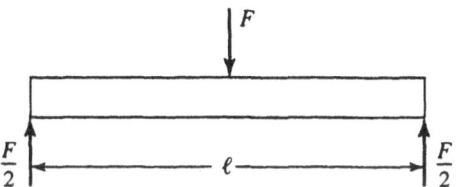

Figure 3.55. Free-body diagram of a loaded beam.

always determine the bending moment from a knowledge of the external forces on the beam. For example, if a beam is loaded as discussed at the beginning of this section (see Figure 3.55), the bending moment at each point z along the beam is calculated as was discussed earlier and has the values shown in Figure 3.56. To obtain the deflection of a beam so loaded, one need only solve the differential equation (3-75) with this variation for $M_B(z)$ subject to the conditions that $y = 0$ at $z = 0$ and at $z = \ell$. We leave this as an exercise for the reader.

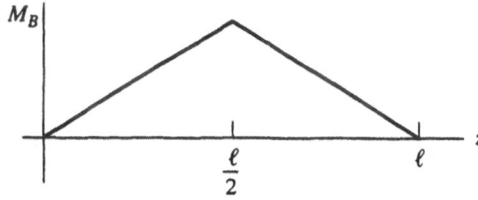

Figure 3.56. The bending moment as a function of position along the beam.

A simpler exercise is to compute the deflection of a beam subject to a single force F acting at the end of the beam, such as illustrated in Figure 3.57. We may compute $M_B(z)$ by applying, as before, the condition of static equilibrium to a section of the beam to the right of a mathematical cut at z.

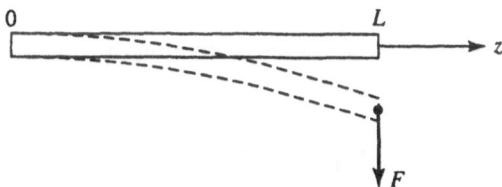

Figure 3.57. Bending of a beam loaded at its end.

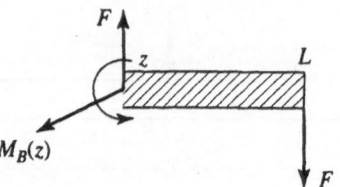

Figure 3.58. The bending moment at each point z along the beam.

Clearly at z there must be a force F exerted by the left-hand part of the beam, and a torque M_B whose magnitude exactly balances the torque about the cut produced by the applied force, see Figure 3.58. Thus one has

$$M_B(z) = F(L - z).\tag{3-84}$$

This relation is illustrated in Figure 3.59. Before inserting this into (3-75) we must take note of the proper sign of M_B. Clearly the effect of M_B is to deflect the beam downward. Such a deflection requires a negative sign for $(d^2 y/dz^2)$. Thus, in Equation (3-83), we must associate a negative sign with $M_B(z)$.

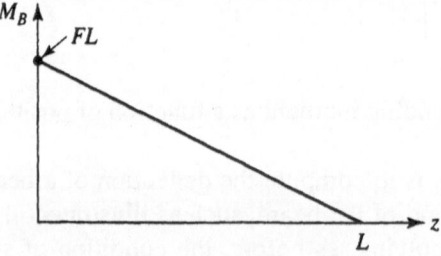

Figure 3.59. M_B versus position z.

The differential equation then, for the deflection of a beam subjected to a force F at the end of the beam, is

$$\frac{d^2 y}{dz^2} = -\frac{F(L - z)}{E I_A}.\tag{3-85}$$

One integration gives

$$\left(\frac{dy}{dz}\right) = +\frac{F}{2EI_A}(L - z)^2 + c_1. \tag{3-86}$$

The second integration gives

$$y = -\frac{F}{6EI_A}(L - z)^3 + c_1 z + c_2. \tag{3-87}$$

The constants of integration c_1 and c_2 are determined by the conditions that, at the left-hand end of the beam, $y = 0$ and $(dy/dz) = 0$; i.e., the beam is held fast in a clamp in such a way that the beam is parallel to the horizontal axis. The condition $(dy/dz) = 0$ at $z = 0$ fixes c_1 from (3-86)

$$c_1 = -\frac{FL^2}{2EI_A}. \tag{3-88}$$

The condition $y = 0$ at $z = 0$ gives c_2 from (3-87) as

$$c_2 = \frac{FL^3}{6EI_A}. \tag{3-89}$$

On inserting these into (3-87) we obtain

$$y = -\frac{F}{6EI_A}\{(L - z)^3 + L^2(3z - L)\}. \tag{3-90}$$

This gives the deflection at each point in the beam. In particular, the deflection at the end of the beam ($z = L$) is

$$y(L) = \frac{FL^3}{3EI_A}. \tag{3-91}$$

This deflection is proportional to the cube of the length of the beam. It is interesting to compute the deflection numerically for a wood plank. We carry the calculation out for a 2 in × 4 in beam in the two orientations shown in Figure 3.60.

Let us assume

$$F = 50 \text{ lb},$$

$$E = 1 \times 10^5 \text{ bar} \cong 15 \times 10^5 \text{ lb/in}^4,$$

$$L = 8 \text{ ft} = 96 \text{ in}.$$

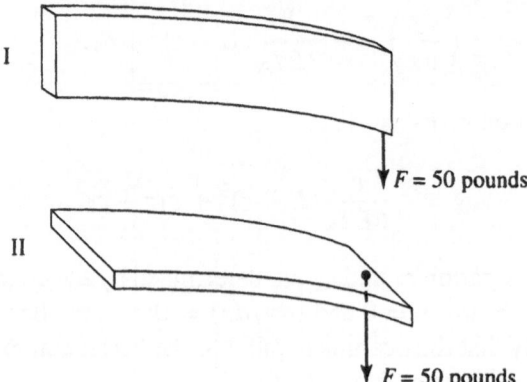

Figure 3.60. Deflection of a loaded board, for two different orientations.

The values for the area second moment in each orientation of the beam are, using (3-72):

Configuration I:

$$I_A = \frac{Wh^3}{12} = \frac{2 \times 4^3}{12} = 10.6 \text{ in}^4.$$

Configuration II:

$$I_A = \frac{4(2)^3}{12} = 2.67 \text{ in}^4.$$

Thus, the deflection in configuration I is

$$y(L) = -\frac{(50)(96)^3}{3(15 \times 10^5)(10.6)} \cong 0.93 \text{ in}$$

and the deflection in configuration II is

$$y(L) = -\frac{(50)(96)^3}{3(15 \times 10^5)2.67} = 3.7 \text{ in}.$$

The deflection in configuration I is about 1 in, while that in configuration II is four times greater.

3.4.E.   Bending and Breaking of a Beam or a Bone: Application to Fracture of the Tibia

As the external stresses on a beam are increased, the beam bends more and more. At some point it will break. The criterion for yielding or fracturing can be obtained using the following line of reasoning. As the beam deforms, its radius of curvature will decrease. In accordance with (3-64) this means an increase in the internal stresses at each point a distance x from the neutral axis. Quantitatively, the internal stress varies with distance from the neutral axis as

$$\sigma = E\left(\frac{x}{R}\right). \tag{3-92}$$

The maximum stress in tension is at the bottom of the stressed beam, and the maximum stress in compression is at the top of the beam. Thus, from Figure 3.61,

$$(\sigma_{\text{max compress}}) = \frac{E\ell'}{R} \tag{3-93}$$

and

$$(\sigma_{\text{max tension}}) = \frac{E\ell}{R}. \tag{3-94}$$

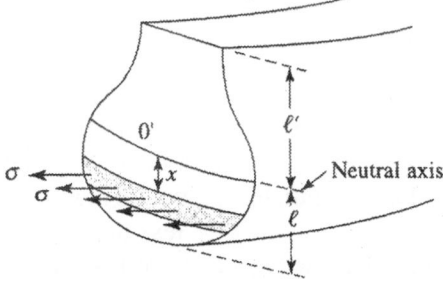

Figure 3.61. Stress σ in a bent beam varies with position x.

We can expect that yielding or fracture will occur when the maximum stress in the bent beam exceeds, for example, the ultimate tensile strength (Y_T) or ultimate compressive stress (Y_C) whichever is smaller. With this very simple criterion, the condition for yielding would be when

$$(\sigma_{max})_{compress} > Y_C$$

or

$$(\sigma_{max})_{tension} > Y_T.$$

More careful consideration shows that this is not quite the precise criterion. In the case of yielding in simple tension the member under tension is not supported in any way along the surface. In the present case of yielding under bending, a small section of maximally stressed fibers on the lower surface of the beam are being supported by neighboring fibers closer to the neutral axis. This support implies that the ultimate bending stress Y_B can be expected to be larger than that in tension or compression. We list in Table 3.3 the ultimate bending stress Y_B for human wet, long bones for persons aged 20–39 years [11].

Table 3.3: Ultimate Bending Stress Y_B for Wet Human Long Bones; Persons Aged 20–39 Years.

Bone	Ultimate bending stress Y_B in bars
Femur	2.08×10^3
Tibia	2.13×10^3
Fibula	2.16×10^3
Humerus	2.11×10^3

From Yamada [11].

Comparing these entries with those in Table 3.2 we see that while Y_B and Y_T are of the same general size, the ultimate stress for bending is about 70% or 80% greater than that for tension. It is also worth observing that on increasing age the ultimate bending stress Y_B does fall, but not by too much. Typically, Y_B for persons in the age group 70–89 is only about 25% less than those in the 20–39 year age group. We may speculate that the frequency of fractures associated with falls in elderly persons is connected with their inability to cushion the fall or distribute their weight effectively upon falling.

The quantity which is most easily available to calculate in a stressed bone or beam is its bending moment M_B. We may now relate the failure criterion

$$\sigma_{max} \geq Y_B \tag{3-95}$$

to the bending moment quite simply. According to (3-71), the bending moment M_B is related to the radius of curvature of the beam by

$$\frac{1}{R} = \frac{M_B}{EI_A}, \tag{3-96}$$

where E is Young's modulus and I_A is the second area moment of the beam. This equation is valid regardless of the form of the beam cross section. Since the maximum stress in the bent beam is related to the radius of curvature by

$$\sigma_{max} = \frac{E\ell}{R}, \tag{3-97}$$

where ℓ is the maximum distance from the neutral plane to the outer or inner surface of the beam. In a rectangular beam, ℓ, is clearly equal to one-half of its width h: $\ell = \ell' = h/2$. For a cylindrical beam $\ell = \ell' = a$ where a is the radius of the cylinder.

Combining (3-95) and (3-97), we find that the criterion for fracture can be expressed in terms of the bending moment as follows:

$$\sigma_{max} = \frac{E\ell}{R} = \frac{M_B\ell}{I_A} \geq Y_B. \tag{3-98}$$

Thus, for fracture or yielding, the bending moment must be large enough to satisfy the following inequality

$$M_B \geq \frac{Y_B I_A}{\ell}. \tag{3-99}$$

Let us now apply this criterion to an examination of a common fracture situation. If a person's foot and ankle are fixed in position, and the support of the other foot is lost, the tibia and fibula can easily be fractured. Consider, for example, the case where the right foot and ankle are in a hole in frozen snow and the left foot slips. If the person then falls to the side as shown in Figure 3.62, the bending mo-

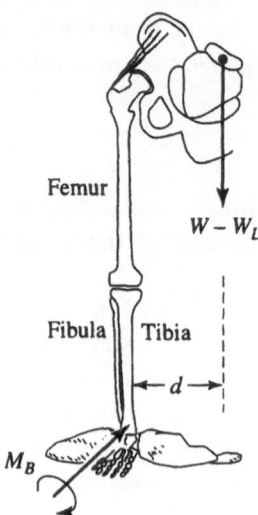

Figure 3.62. Bending moment at the pinned ankle, as a person falls.

ment is equal to the weight of the body (minus the weight of the leg W_L) times the distance between the center of gravity and the pinned leg (d).

A similar situation can occur in athletics where the base of the leg is pinned by a tackler or a fallen player, and the body falls over to the side. The pelvis and trunk, of course, can pivot about the head of the femur to a limited degree until the muscles there prevent further rotation. Beyond this though the full moment is applied to the tibia and fibula. Another oft occurring situation is that in which the foot and ankle are encased in a rigid ski boot fixed to a ski that is firmly stuck to the ground. If the body then falls to the side, the tibia and fibula will break. These bones will also break if the body falls forward or backward.

We can compute how far over the body must fall until the bone breaks, quite simply. From the geometry of the free body-diagram in Figure 3.62 in which the hip and trunk are regarded as rigidly attached to the leg, we see that the condition for fracture is that

$$M_B = d[W - W_L] \geq \frac{Y_B I_A}{\ell}. \tag{3-100}$$

The tibia and fibula are the long bones of the leg. The support provided by the smaller fibula in this fracture process is negligible. Treating the tibia as a bar of circular cross section, we note that the minimum cross section occurs at about one-third of the distance up from the bottom of the tibia. At this point the effective diameter of the bone is 0.8 in. For this circular cross section, (I_A/ℓ) is given using (3-73), which for $a = \ell$ reads

$$\frac{I_A}{\ell} = \frac{\pi a^4}{4a}.$$

With $\ell = a = 0.4$ in, this gives

$$\frac{I_A}{\ell} = \frac{\pi a^3}{4} = 5 \times 10^{-2} \text{ in}^3.$$

Since $Y_B \simeq 2 \times 10^3$ bar $\simeq 29 \times 10^3$ lb/in^2, the bending moment at which fracture occurs is

$$M_B = (W - W_L)d \geq 29 \times 10^3 \times (5 \times 10^{-2}) \text{ lb-in}$$

or

$$(W - W_L)d \geq 1450 \text{ lb-in}.$$

This numerical result shows that a person, whose body weight minus leg weight $(W - W_L)$ is about 145 lb, will suffer a fracture of the tibia when the midline of his body has moved 10 in away from the pinned lower leg.

3.4.F.   Stress and Strain in Shear

In our discussion of the static equilibrium of each part of a loaded beam, we found that acting at each mathematical section of the beam there was a torque or bending moment *and* a force. This is indicated in Figure 3.63 for the shaded section. If

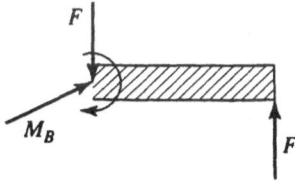

Figure 3.63. Force and bending moment M_B acting on a loaded beam.

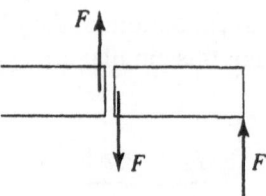

Figure 3.64. Free body diagram for a section of a bent beam.

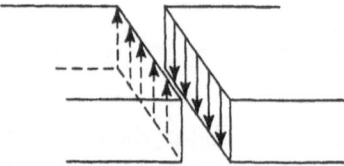

Figure 3.65. Shearing stresses parallel to the plane of the mathematical cut.

we consider the material to the left of the shaded section, we see immediately that this experiences a force equal and opposite to that shown in the figure 3.63. This is shown in Figure 3.64 in fact, then, each cross section of the beam is subject to a "shearing stress" which acts parallel to the plane of the mathematical cut, as indicated in Figure 3.65. The material is in effect being "sheared" just as if a scissors was applied to it as shown in Figure 3.66.

Under such a stress there will be a shearing deformation of the body. We now characterize quantitatively the shear stress, the shear strain, and the relationship between them. The essential geometry of the shearing process is shown in Figure 3.67. We imagine a block fixed at its bottom surface and a force exerted parallel to the top surface. In this geometry the deformation of the block δ is in the direction of F. The magnitude of the deformation δ, when F is sufficiently small,

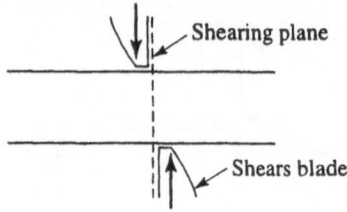

Figure 3.66. Equivalent action of a scissors in producing a shear stress.

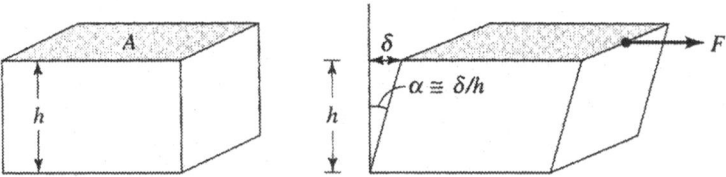

Figure 3.67. The essential geometry of the shearing process.

is found experimentally to be proportional both to the original height of the block h and to the force per unit area (F/A) exerted parallel to the plane as shown. In the region of elastic shear deformation it is found that

$$\delta = \left(\frac{1}{G}\right) h \left(\frac{F}{A}\right)$$

In analogy to the tensile geometry we may define a shear stress $\sigma_s = (F/A)$, a shear strain $(\delta/h) \equiv \epsilon_s$, and a shear modulus G, that fixes the constant of proportionality relating the stress and strain

$$\epsilon_s - \frac{1}{G}\sigma_s. \tag{3-101}$$

Typically, G, which has the dimension of force/area or pressure, has a size about half that of Young's modulus, the modulus of elasticity. Notice that the strain (δ/h) is equal, for small deformations, to the angle α produced by the shear deformation.

We see then that the full description of the deformation of a loaded beam will involve both the bending due to the internal torques, and the shear deformation produced by the internal forces. Generally, in a transversely loaded beam, the bending deformations are dominant and the shear deformations can be neglected. However, when the beam is subjected to a twisting moment, there is no bending at all and the shear stresses fully determine the resulting twisting of the beam. This can be seen in the discussion which follows.

In Figure 3.68 we show a cylinder before and after the application of a torsional stress. We can express the deformation of the cylinder in terms of the position in the cylinder by imagining the cylinder to be made up of fibers running parallel to the axis.

For those fibers on the outer surface of the cylinder, a twist angle of ϕ at the top of the cylinder implies that the shear deformation angle α for the surface fibers

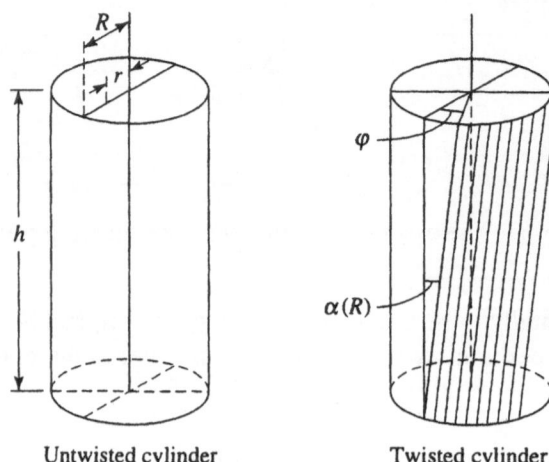

Untwisted cylinder   Twisted cylinder

Figure 3.68. Geometrical representation of the effect of a torsional stress.

is

$$\alpha(R) = \frac{R\phi}{h}. \tag{3-102}$$

The shearing deformation of fibers closer to the cylinder axis is smaller than this. In fact, the shear angle α_r for a fiber a distance r from the axis is

$$\alpha(r) = \frac{r\phi}{h}. \tag{3-103}$$

We now wish to compute, using the shear Hooke's law relation (3-101), the connection between the deformation angle ϕ of a long cylinder and the torque applied to it.

Consider a total torque T applied at the top as shown in Figure 3.69, and imagine an upper piece of the cylinder to be defined by a mathematical horizontal cut. On the lower part of the cut cylinder shear stresses will be set up which set up a countertorque equal to $-T$. $-T$ is the contribution from all surface elements $ds = 2\pi r\, dr$, each multiplied with the moment of the shear stress (= force per unit surface area) proper for that value of r:

$$T = \int ds(r\sigma(r)).$$

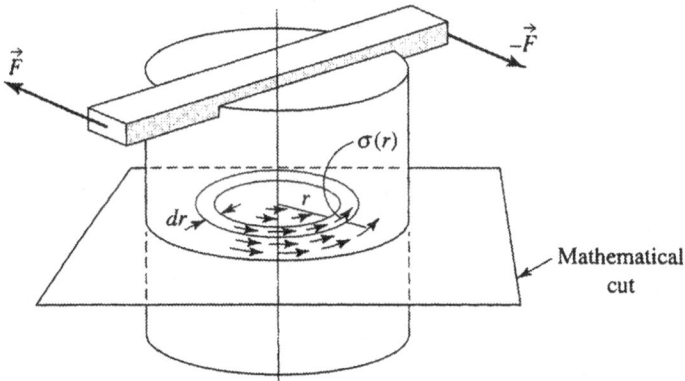

Figure 3.69. Diagram for the calculation of the relation between torque and defor-mation of a twisted cylinder.

The value of the shear stress $\sigma_s(r)$ is related to the angular deformation of the cylinder at the point r by the shear Hooke's law relation

$$\epsilon(r) = \alpha(r) = \frac{r\phi}{h} = \frac{1}{G}\sigma_s(r). \tag{3-104}$$

Thus

$$\sigma_s(r) = \frac{G\phi}{h}r. \tag{3-105}$$

The countertorque exerted on the upper part of the mathematical cut in Figure 3.69 is then

$$T = \int (2\pi r)\,dr\,r\left(\frac{G\phi}{h}\right)r, \tag{3-106}$$

$$T = \frac{2\pi G\phi}{h}\int_{r=0}^{R} r^3\,dr, \tag{3-107}$$

or, finally,

$$T = \frac{G\phi}{h}\frac{\pi R^4}{2}. \tag{3-108}$$

This gives the relation between the twist angle of the rod, its length, radius, and the applied torque.

The result (3-108) can be rearranged in the form

$$\phi = \frac{hT}{G((\pi/2)R^4)}.$$

This emphasizes the fact that the twist angle is inversely proportional to the shear modulus, or shear stiffness (G) and also inversely proportional to the fourth power of the radius of the cylinder. Increasing the radius by a factor 2 makes it 16 times stiffer in its resistance to twisting. We list in Table 3.4 numerical values for the shear modulus G for some materials.

Table 3.4: Table of Shear Moduli.

Material	G (Bars)
Steel	0.8×10^5
Aluminum	0.25×10^5
Long bone	$(0.8–1.5) \times 10^5$

Let us now consider briefly the failure of beams under torsional stress. In the case of bending we saw that the deformation of a beam produced compressive and tensile stresses, and when these exceeded the ultimate tensile or compressive strength the material failed. In the case of torsion, one can analyze the compressive and tensile stresses associated with the twisting deformation. Frankel and Burstein [13] argue that the direction of maximum compression and tension in a cylinder under torsion lies at an angle to the cylinder axis. If the beam or bone then fracture when the tensile or compressive limit is reached the fracture line will be at an angle to the beam axis. This is the situation that does occur in "spiral fracture" associated with the twisting of the tibia in skiing accidents.

We show in Figure 3.70 an x-ray photograph of a spiral fracture of the tibia sustained in a skiing accident. The photography was kindly supplied by Robert Runyon, M. D.

In Table 3.5 we give a table of the actual breaking torque for structural biomaterials like long bones or intervertebral discs. All the data in this table are taken from Yamada ([11]).

In skiing, of course, we are arranging the most favorable circumstances for torsional fracture. The long lever arm of the ski means that only a relatively small

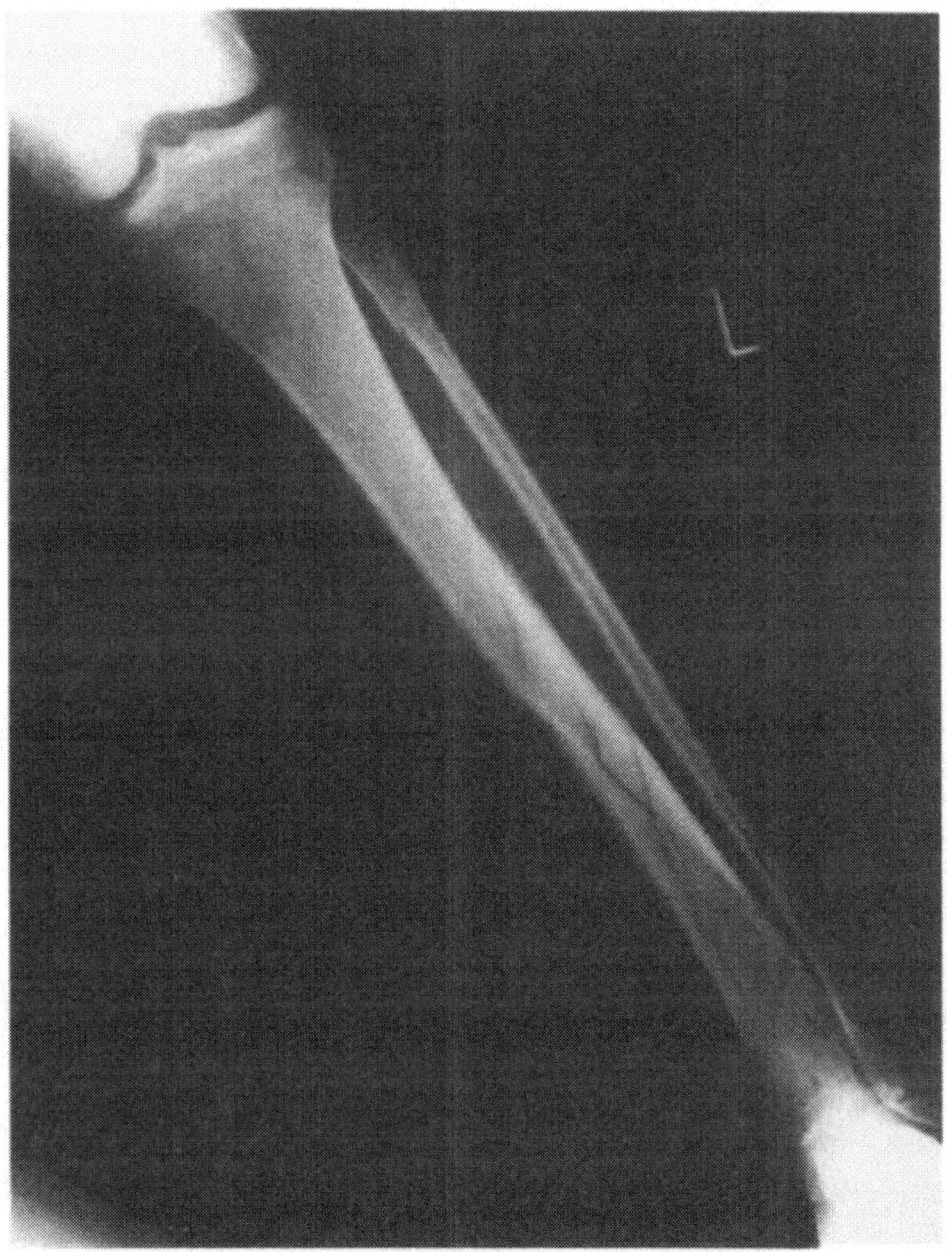

Figure 3.70. Spiral fracture of the tibia.

Table 3.5: Breaking Torque of Various Bones.

	Object	Breaking torque Nm	Breaking angle of twist
Leg	Femur	140	1.5°
	Tibia	100	3.4°
	Fibula	12	35.7°
Arm	Humerus	60	5.9°
	Radius	20	15.4°
	Ulna	20	15.2°
Intervertebral discs	Cervical	5	~ 38°
	Mid. thoracic	17	~ 24°
	Lumbar	44	~ 15°

force need be applied at the tip before fracture of the tibia occurs. We can see this quantitatively from Table 3.5. Since the distance between the foot and tip of the ski is approximately 1 m, a force of about 100–150 N is all that is required to fracture the tibia in torsion. Since 1 N is equal to about 0.224 lb, we see that a force of 25–35 lb at the end of the ski is all that is needed for fracture. Of course, modern skis are equipped with release bindings. We can easily compute the lateral force at the tip of the ski boot at which, or below which, the binding should release. Since the boot tip is about 0.2 m from the heel, the lever arm at this point is about five times smaller than at the ski tip. Thus the fracture will occur when the force at the tip of the boot exceeds, say, 100–150 lb. For safety's sake it is reasonable then to set the binding release at a value, say, of about one third this value, i.e., at about 30–40 lb. In fact, there are machines that permit a measurement of the force setting for lateral release of ski bindings, and one should be sure that the release pressure is at least in the range estimated above, to avoid sprains as well as breaks.

3.5   Static Equilibrium of Fluids

3.5.A.   Introduction and Definition of Hydrostatic Pressure

In the previous sections of this chapter we have examined the static equilibrium of rigid bodies, and of deformable solids. We now examine the application of the conditions of static equilibrium to a fluid. This subject, hydrostatics, has interesting applications in high-altitude physiology, in deep-sea diving, in the determination of protein molecular weights using the ultracentrifuge, and in understanding the effects of blood pressure.

The quantity that plays the role of stress in a fluid is the hydrostatic pressure P. To understand the properties of this quantity, let us consider a container filled with liquid or gas, closed by a movable piston of area A. Let the piston be subject to a force F as shown in Figure 3.71. If the force of gravity and other external forces that act on the fluid are zero, then the state of stress inside the container is one of uniform pressure P. The magnitude of the pressure P is everywhere equal to (F/A)

$$P = \left(\frac{F}{A}\right).$$

(3-109)

While the pressure has the same dimensions as stress in solid bodies, in a fluid the pressure is defined as a scalar quantity. It is a number having only magnitude and no direction. (In the more general case of a fluid subject to external forces the pressure remains a scalar, but its magnitude can vary from point to point in the fluid.)

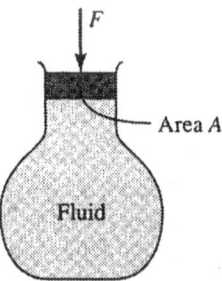

Figure 3.71. Hydrostatic pressure P in a fluid is uniform and is equal to F/A.

Having defined the pressure P inside a fluid, we must now examine the forces associated with this pressure inside and at the surface of the container. The notion of uniform pressure in the container implies that the pressure (or force per unit area) generated by the piston is transmitted uniformly throughout the liquid. In particular, at the walls of the container each small element of area ΔA experiences a force ΔF perpendicular to that area. The magnitude and direction of the force is given by

$$\Delta \vec{F} = -\hat{n}(\Delta A)P.$$

(3-110)

As shown in Figure 3.72, $\hat{n}$ is a *unit vector* which points normal to the plane of ΔA. Thus, at the bounding surfaces, the usefulness of the concept of pressure is quite clear. The liquid transmits the applied force F in such a way that at each point on the surface the force per unit area is equal to that directly under the piston. The net force on any area of the boundary surface can be found by summing vectorially the forces $\Delta \vec{F} = -\hat{n}(\Delta A)P$ on each small area.

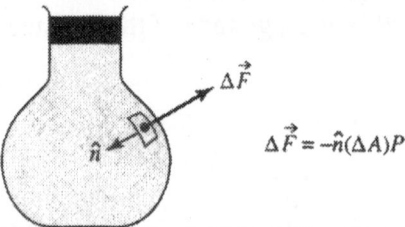

Figure 3.72. Normal force $\Delta \vec{F}$ on a surface element of ΔA of a container is determined by the pressure P and area $\Delta A : \Delta \vec{F} = -\hat{n}(\Delta A)P$.

The effect of the pressure within the interior of the fluid is a bit more subtle. We can see that the forces *inside* the fluid also obey the relation

$$\Delta \vec{F} = -\hat{n}(\Delta A)P, \tag{3-111}$$

provided that the fluid is in static equilibrium. We shall then show that the conditions of mechanical equilibrium enable us to compute the magnitude and direction of the force exerted at any point within the fluid.

Let us examine a part of the liquid in the container. Let this part be a cylindrical wedge whose bounding surface is three planes, as shown in Figure 3.73. These mathematical surfaces each have an inward and outward pointing normal. Let us

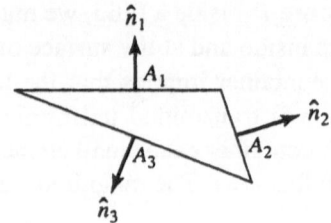

Figure 3.73. Mathematical surfaces in a fluid.

denote as $\hat{n}_1$, $\hat{n}_2$, and $\hat{n}_3$ the unit vectors pointing outward from each of these plane surfaces. (The *caret* ˆ notation denotes a vector of *unit* length.) The area of a unit height of each surface is A_1, A_2, and A_3.

We now demonstrate that if we regard the internal forces acting on these planes as related to the pressure, in just the same way as at the containing boundary that the fluid within the wedge will be in static equilibrium. The forces within the fluid act on the boundaries of the wedge in a direction opposite to the outward normals, with magnitudes shown in Figure 3.74 since we are assuming $\Delta \vec{F} = -\hat{n}(\Delta A)P$. We now examine whether this mathematical wedge is in static equilibrium with this assumption of the nature of the internal forces.

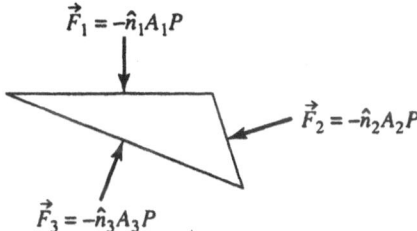

Figure 3.74. Forces acting on plane mathematical surfaces in a fluid under pressure.

The first condition to be examined is whether the net force on the wedge-shaped cylinder is equal to zero. To this end, we observe that

$$\vec{F}_1 + \vec{F}_2 + \vec{F}_3 = -\hat{n}_1 A_1 P - \hat{n}_2 A_2 P - \hat{n}_3 A_3 P$$
$$= -P\left(\hat{n}_1 A_1 + \hat{n}_2 A_2 + \hat{n}_3 A_3\right). \qquad (3\text{-}112)$$

Now observe that the unit normal vectors $\hat{n}_1$, $\hat{n}_2$, and $\hat{n}_3$ can be generated from the unit vectors $\hat{n}'_1$, $\hat{n}'_2$, $\hat{n}'_3$ in the plane of the paper and pointing along the surfaces A_1, A_2, and A_3 as follows:

$$\hat{n}_1 = \hat{n} \times \hat{n}'_1, \qquad \hat{n}_2 = \hat{n} \times \hat{n}'_2, \qquad \hat{n}_3 = \hat{n} \times \hat{n}'_3, \qquad (3\text{-}113)$$

where $\hat{n}$ is a unit vector pointing out of the plane of the paper, and $\hat{n} \times \hat{n}_1$ is the vector cross product of $\hat{n}$ and $\hat{n}'_1$ and is equal to $\hat{n}_1$, the outward pointing normal to the area A_1, etc. This is illustrated in Figure 3.75 using these three equations in

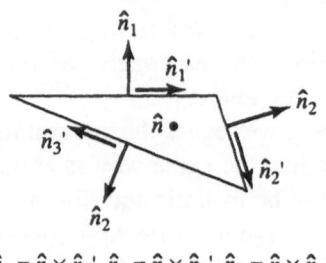

$$\hat{n}_1 = \hat{n} \times \hat{n}_1', \; \hat{n}_2 = \hat{n} \times \hat{n}_2', \; \hat{n}_3 = \hat{n} \times \hat{n}_3'$$

Figure 3.75. Normal vectors $\hat{n}_i$ as cross products.

the expression for $\vec{F}_1 + \vec{F}_2 + \vec{F}_3$ we find

$$\vec{F}_1 + \vec{F}_2 + \vec{F}_3 = -P\hat{n} \times \left(\hat{n}_1' A_1 + \hat{n}_2' A_2 + \hat{n}_3' A_3\right). \qquad (3\text{-}114)$$

But we see at once that the vector sum in parentheses is exactly equal to zero because each vector is along the side of the triangle shown in Figure 3.76. Thus, $\hat{n}_1' A_1 + \hat{n}_2' A_2 + \hat{n}_3' A_3 = 0$ and $\vec{F}_1 + \vec{F}_2 + \vec{F}_3 = 0$, and the triangular wedge-shaped cylinder does satisfy the condition that the net force is equal to zero. The present derivation can readily be extended to any shaped cylinder since we see that for an arbitrary shaped body, like that shown in Figure 3.77, and since $\sum_i F_i$ is proportional to this (under the assumption that $F_i = -\hat{n}_i A_i P$) the total force on any closed mathematical surface will be zero.

The total torque on the wedge-shaped triangle of Figure 3.75 is also equal to zero. This is seen simply as follows. The force on each side acts effectively along the perpendicular bisector of each side. The three perpendicular bisectors of a triangle always intersect at a point. Thus, if one takes torques about the point of intersection, the total torque will be zero. Since the net torque is zero about one

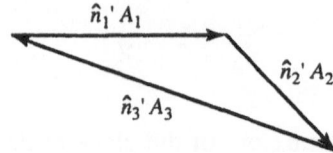

Figure 3.76. Sum rule for tangential forces $\hat{n}_i' A_i$.

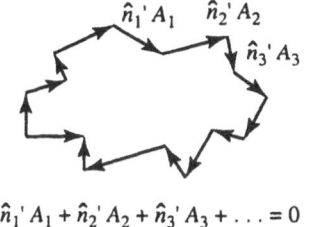

$$\hat{n}_1{}'A_1 + \hat{n}_2{}'A_2 + \hat{n}_3{}'A_3 + \ldots = 0$$

Figure 3.77. The sum of tangential surface forces on any shape object is zero.

point, and since the net force is zero, it follows that the torque taken about any point will be zero.

We conclude our discussion of the definition of hydrostatic pressure by observing that the pressure in a fluid is a scalar quantity. The forces in the fluid or on its bounding surfaces can be computed from the pressure at that point using the formula

$$\Delta \vec{F} = -\hat{n}(\Delta A)P$$

where $\Delta \vec{F}$ is the vector force on a small area ΔA and $\hat{n}$ is the direction of the normal pointing outward from the surface of which ΔA is a part.

If the fluid is subject to no external body forces like gravity, the pressure in the fluid will be uniform throughout. In the following section we consider the effect of gravity in producing a variation of pressure in the fluid. We will examine the implications of this pressure variation for such questions as the variation of air pressure with altitude above the Earth's surface and for the buoyancy of bodies immersed in the fluid.

3.5.B.   Fluid in a Gravitational Field: The Variation of Pressure with Height

Let us consider a fluid at rest in a container. Let each part of the fluid be subject to the force of gravity. We will now compute the variation of fluid pressure beneath the fluid surface using the conditions of static equilibrium; i.e., we shall use the fact that each interior region of the fluid is at rest. Construct with the body of the fluid a mathematical pill box, as shown in Figure 3.78. The upper and lower faces of the pill box are at positions z and $z + \Delta z$. Since the fluid is at rest the total force

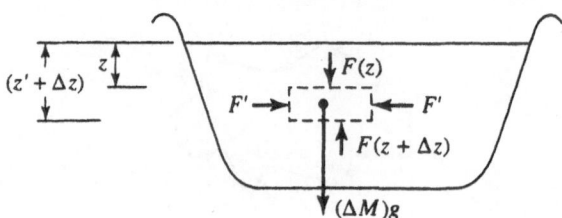

Figure 3.78. Forces acting on a small volume of fluid under the action of gravity.

acting on this mathematical volume must add to zero. We show in Figure 3.79 the free-body diagram of the forces acting on the pill box.

The force acting downward at the top surface is the area A times the pressure at height z. Similarly, the force acting upward at the bottom surface is the area A times the pressure $P(z + \Delta z)$ at that level. The mass of fluid inside the pill box is taken as ΔM, and the force of gravity on this is $(\Delta M)g$ where g is the acceleration of gravity. The force exerted by gravity (F') on the sides of the pill box cancels in the horizontal plane at each height (using the argument given in Section 3.5A) and need not be further considered.

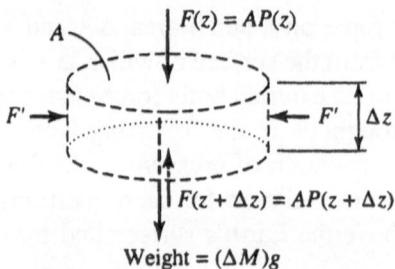

Figure 3.79. Free-body diagram of forces acting on small volume in fluid.

Since the fluid within the mathematical pill box is in static equilibrium, the total force acting upon it is zero. We, therefore, have the equation

$$[AP(z + \Delta z) - AP(z)] - (\Delta M)g = 0, \qquad (3\text{-}115)$$

where the upward direction is taken as positive.

If the distance Δz is small, we can write

$$P(z + \Delta z) = P(z) + \left(\frac{dP}{dz}\right)\Delta z. \tag{3-116}$$

On inserting this into (3-115) we have

$$A\Delta z\left(\frac{dP}{dz}\right) = \Delta Mg.$$

The mass ΔM of the fluid in the box is connected with the volume $A\Delta z$ of fluid through the density ρ

$$\rho = \left(\frac{M}{V}\right) = \frac{\text{mass}}{\text{volume}} = \frac{\Delta M}{(A\Delta z)}. \tag{3-117}$$

Thus we have the result that at each point in the fluid the derivative of the pressure with height satisfies the equation

$$\left(\frac{dP}{dz}\right) = \rho g. \tag{3-118}$$

In general, of course, the density of the fluid can vary with the pressure (or the height z) so that we should properly write this equation including this

$$\left(\frac{dP}{dz}\right) = \rho(P)g = \rho(z)g. \tag{3-119}$$

Clearly, the pressure increases as one goes deeper into the fluid from its upper surface. To compute the pressure at various heights in the fluid, it is necessary only to integrate this equation. We will do this for the two cases of an incompressible liquid and a gas whose density varies with pressure in accordance with the ideal gas law.

(i)   Incompressible Fluid: Pascal's Law: Pressure Units

In an incompressible fluid ρ, the density is a constant. It does not change with the pressure. Any real liquid is, of course, not incompressible. The compressibility, however, is so small as to permit neglecting it. For example, in the case of water the fractional change in density of water is only 20% when it is subjected to a pressure as large as 10,000 atmospheres. For our purposes this density variation can be neglected, and we can integrate (3-120) immediately to obtain (P_0 is the pressure at the surface of the fluid):

$$P = \rho g z + P_0. \tag{3-120}$$

This is Pascal's law. The reader can easily show that this result is equivalent to the statement that the pressure, a distance z below the surface, is equal to that produced by the weight per unit area of a column of the superincumbent fluid. Using this equation, we can compute numerically the increase of pressure with depth. Choosing water as the fluid, we have $\rho \cong 1$ g/cm^3, $g \cong 980$ cm/s^2. Therefore

$$P = 980 \text{ dyn/cm}^2 z \text{ (in cm)}. \tag{3-121}$$

For $z = 1$ m, the pressure increase is

$$\left(\frac{\Delta P}{\Delta z}\right) = 9.8 \times 10^4 \frac{(\text{dyn/cm}^2)}{\text{m}}$$

Since 1 atm $= 14.7$ lb/in$^2 = 1.03 \times 10^6$ dyn/c^2, we see that the increase of pressure with depth in water is

$$\left(\frac{\Delta P}{\Delta z}\right) \cong 0.1 \text{ atm/m}. \tag{3-122}$$

Thus, we see that on descending beneath the surface the pressure of water increases by about 1 atm for each 10 m (33 ft), i.e.,

$$\left(\frac{dP}{dz}\right) \approx 1 \text{ atm/10 m}. \tag{3-123}$$

We will shortly consider the consequences of this in the physiology of divers.

Note on Pressure Units

   The units of pressure are, of course, force per unit area. In MKS units the unit of pressure is the N/m^2. In CGS units the unit is the dyn/cm^2. An official mks unit of convenient size for our purposes is the bar which is defined as follows:

$$1 \text{ bar} = 10^5 \text{ N/m}^2 = 10^6 \text{ dyn/cm}^2.$$

This is a convenient unit as it is very nearly equal to the atmospheric pressure at sea level. In fact, calling this atmospheric pressure "one atmosphere" we have that

$$1 \text{ atm} = 14.7 \text{ lb/in}^2 = 1.013 \text{ bar} = 1.013 \times 10^6 \text{ dyn/cm}^2.$$

Alternatively, one can express pressure in units of the height of a column of fluid with known density. Pascal's law shows us that the pressure can be related to the height of the column z by the formula $P = \rho g z$. In the medical literature one often finds pressures expressed in millimeters of mercury or in centimeters of water. By using Pascal's law and the density of mercury ~ 13.6 g/cm^3, or water 1 g/cm^3, one finds the following conversions between the pressure units and height units

$$1 \text{ atm} = 1.013 \text{ bar} = 760 \text{ mmHg} = 1033 \text{ cmH}_2\text{O}.$$

To give some estimate of the magnitude of pressures which occur physiologically, it is interesting to observe that the pressure exerted upon and within the blood by the heart at the moment of maximum contraction of the heart (systole) is normally 120 mmHg. Alternatively, it is to be noted that in normal breathing the maximum pressure changes associated with the inspiration or expiration of air from the lungs is about 6 cmH$_2$O.

(ii)   Compressible Fluid: Ideal Gas

We can integrate (3-120) in general when the density of the fluid is known as a function of pressure. In the case of the gases in the Earth's atmosphere the dependence of density on pressure can be estimated with good accuracy using the ideal gas law. According to this, the pressure, density, and temperature of a gas are related in accordance with the formula

$$P = nkT, \tag{3-124}$$

where n is equal to the number of molecules per cubic centimeter, k is Boltzmann's constant $= 1.38 \times 10^{-16}$ erg/°K. T is the absolute temperature in degrees Kelvin. The number density n can be related to the mass density ρ using the mass M of each gas molecule, viz.:

$$\rho = Mn = \text{mass of gas/cm}^3. \tag{3-125}$$

Thus for gas consisting of one molecular type, we have, using (3-126) and (3-125), that the relation between density and pressure is

$$P = \left(\frac{\rho}{M}\right) kT, \tag{3-126}$$

where M is the mass of an individual gas molecule.

The Earth's atmosphere is a mixture of gases—primarily Nitrogen (N_2) 78%, Oxygen (O_2) 21%, and Argon (A) 1%. The ideal gas law describes this mixture provided that the mass M chosen in (3-127) is the average mass of the atmospheric constituents, i.e.,

$$M_{ave} = M_{N_2}(0.78) + M_{O_2}(0.21) + M_A(0.01),$$

where M_{N_2}, M_{O_2}, M_A are, respectively, the masses of a nitrogen, oxygen, and argon molecule.

Combining (3-127) and (3-120), we obtain

$$\left(\frac{dP}{dz}\right) = \left(\frac{gM}{kT}\right) P. \tag{3-127}$$

Let us apply this carefully to the problem of calculating the variation of P in the atmosphere. In this computation it is most convenient to choose as our starting point not the top of the atmosphere and then move downward, but instead to compute the pressure variation starting at the bottom of the atmosphere, i.e., the Earth's surface and then move upward. In this case, the direction in which gravity acts and the direction of increase of z are opposite, and so we must account for this by using a negative sign in (3-128), i.e.,

$$\left(\frac{dP}{dz}\right) = -\frac{gM}{kT} P. \tag{3-128}$$

At the surface of the Earth, $z = 0$, $P = 1$ atm $= P_0$. This equation is immediately integrable as follows:

$$\int_{P=P_0}^{P(z)} \frac{dP}{P} = \frac{gM}{k} \int_{z=0}^{z} \frac{dz}{T(z)}. \tag{3-129}$$

In (3-130) we include for the moment the temperature under the integration over z. This is because the temperature of the atmosphere is known to decrease with altitude z. To obtain $P(z)$ accurately, we should, in principle, include this temperature variation. However, once again the effect on the pressure is not great, and we can neglect the effect and assume that the temperature T is constant. Under these conditions we integrate (3-130) to obtain

$$\ln\left(\frac{P(z)}{P(0)}\right) = -\frac{gM}{kT}z \qquad (3\text{-}130)$$

or

$$P(z) = P(0)e^{-(gM/kT)z}. \qquad (3\text{-}131)$$

The quantity (kT/gM) has the dimensions of length. It has an important meaning and can be called the scale height z_0:

$$z_0 = \left(\frac{kT}{gM}\right). \qquad (3\text{-}132)$$

On ascending to a height z_0 the pressure falls to $(1/e)$ of its value at the surface, at $2z_0$ the pressure falls to $(1/e^2) \sim \frac{1}{9}$ of its value at the Earth's surface. Thus z_0 measures the scale distance over which the pressure changes quite substantially. Above a few scale heights the pressure is quite tiny compared to the atmospheric pressure P_0. In Figure 3.80 we show diagrammatically the pressure variation with height above ground.

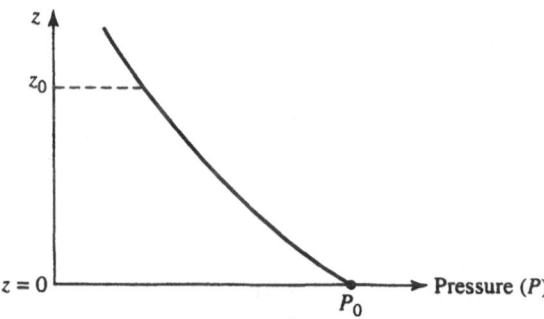

Figure 3.80. Variation of pressure with height in atmosphere z_0 is the "scale height."

The scale height z_0 for air at $T = 32\,°F$ or $0\,°C$ can be computed as follows. The average of the molecular masses of the air molecules is

$$M \cong 47.3 \times 10^{-24}\ \text{g}.$$

Thus

$$z_0 = \left(\frac{kT}{Mg}\right) = \left(\frac{1.38 \times 10^{-16}(273)}{47.3 \times 10^{-24}(980)}\right) = 8.15 \times 10^5\ \text{cm},$$

or $z_0 = 8150$ meters   or $z_0 = 26{,}800$ feet.

Thus we may write numerically that the pressure variation with height above the Earth varies as

$$P = 760e^{-z/26,800}\ \text{mmHg}, \tag{3-133}$$

when P is measured in millimeters of mercury and z is measured in feet. Alternatively, if z is measured in meters we have

$$P = 760e^{z/8,150}\ \text{mmHg}. \tag{3-134}$$

Clearly, as one ascends to heights comparable to 26,000 ft there will be very substantial changes in pressure, with associated physiological effects which we will shortly discuss. In Table 3.6 we list the pressures calculated, using (3-134) at a number of interesting heights. We also list the corresponding partial pressures

Table 3.6: Air Pressure and Oxygen Partial Pressure at Various Heights.

Site	z (Feet)	(Meters)	P mmHg	P_{O_2} mmHg	P/P_0
Sea level	0	0	760	160	1.0
Mt. Washington	6,288	1,917	670	141	0.88
Pikes Peak	14,100	4,300	448	94	0.59
Mt. Blanc	15,780	4,810	421	88	0.55
Mt. McKinley	20,300	6,190	357	75	0.47
Mt. Everest	29,000	8,841	259	54	0.34
Jet travel	35,000	10,670	205	43	0.27
	40,000	12,195	170	36	0.22

of oxygen at each height. It may be of interest to mention that the first observation of the decrease of air pressure with height was made in 1648, 8 years after Toricelli invented the mercury barometer. The measurement was made by Perier at the suggestion of B. Pascal (1623–1662) by ascending a volcanic mountain, Puy de Dome, in central France. The height of this peak is 1365 m. The height of the mercury column was observed to decrease in proportion to the elevation during ascent [20].

(iii) Buoyant Force on Bodies Immersed in a Fluid

In his book *On Floating Bodies* [1], Archimedes studied the forces exerted on a body immersed in a fluid. He discovered that the upward or buoyant force on an immersed body is equal to the weight of the fluid displaced by the body.

We can obtain this result by recognizing that the origin of the buoyancy is found in the increase of fluid pressure with depth. Since the fluid has greater pressure the deeper one goes, an immersed body is pushed upward because the surrounding fluid exerts a greater force on the lower parts of the body than on its upper parts. We can see this simply and numerically by considering the buoyancy of a thin, solid, right cylinder immersed in a fluid, as indicated in Figure 3.81. The height of the cylinder is h and its top is at a distance z beneath the surface. The net upward force on this cylinder is

$$\Delta F_{\text{up}} = F_2 - F_1 = \Delta A P(z+h) - \Delta A P(z)$$

$$= \Delta A (P(z+h) - P(z)), \tag{3-135}$$

$$\Delta F_{\text{up}} = \Delta A \int_{z}^{z+h} \left(\frac{dP}{dz}\right) dz. \tag{3-136}$$

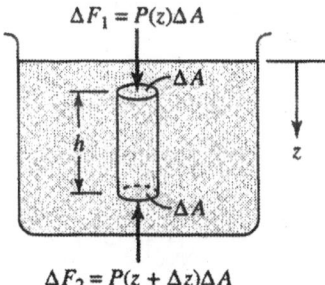

Figure 3.81. Buoyant forces acting on an immersed cylinder.

Now, according to our basic equation (3-119), (dP/dz) is given by

$$\left(\frac{dP}{dz}\right) = \rho(P)g, \tag{3-137}$$

even when the density ρ of the fluid varies with the pressure (or, equivalently, with position (z)). Of course, for the incompressible fluid, ρ is constant, but we may obtain Archimedes' result even for compressible fluids. Regarding the density as a function of position, i.e., $\rho = \rho(z)$, we can insert (3-137) into (3-136) and find

$$(\Delta F)_{\text{up}} = g \int_{z}^{z+h} \Delta A \rho(z)\, dz. \tag{3-138}$$

But since $\Delta A\, dz$ is a small element of volume of the cylinder and $\rho(z)$ is the density at the volume, we see that the integral over the height of the cylinder is equal to exactly the total mass ΔM_{f} of the fluid excluded or displaced by the cylinder, i.e.,

$$\int_{z}^{z+h} \Delta A \rho(z)\, dz = \int_{0}^{\Delta M} dM = \Delta M_{\text{f}}. \tag{3-139}$$

Thus we see that the total force upward on this thin cylinder is exactly equal to

$$\Delta F = g \Delta M_{\text{f}}, \tag{3-140}$$

i.e., the weight of the displaced fluid. We see, therefore, that this thin cylinder is buoyed up by a force equal to the weight of the displaced fluid even in a compressible fluid. The same result applies to a body of arbitrary shapes since it can be looked upon as being made up of many such small cylinders. (See Figure 3.82).

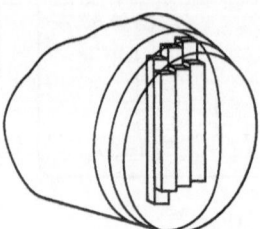

Figure 3.82. Decomposition of a solid body into small cylinders.

In the interest of rigor it should be pointed out that the top and bottom surfaces of the cylindrical sections of a body will, in general, tilt in an arbitrary direction relative to the horizontal, i.e., each cylinder is not a right cylinder. The pressure force is normal to the surface of each cylinder, but on breaking the force on each surface element into its horizontal components, all add to zero at each level z, and the vertical component is related to the horizontal projection of the area. These considerations are illustrated in Figure 3.83. On summing up the total force on the tilted surfaces, we have

$$F_{\text{horizontal}} = \sum (\Delta F)_{xy} = 0$$
$$F_{\text{vertical}} = \sum \Delta F_z = \sum (\Delta A)_z (P(z+h) - P(z))$$
$$= g \sum \Delta M_{\text{f}} = g M_{\text{f}}, \qquad (3\text{-}141)$$

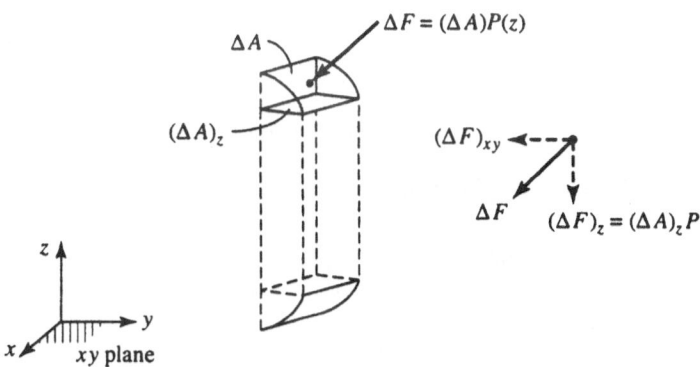

Figure 3.83. The horizontal and vertical components of the force exerted on a cylinder with tilted upper and lower faces.

where M_{f} is the total mass of fluid excluded by each of the cylindrical sections with upper and lower bases in the horizontal planes, which approximate accurately the volume of the arbitrary body. Insofar as the summation of the net pressure force is concerned, the body is effectively made up then of right cylindrical sections whose upper and lower faces are in the horizontal plane, as in the simple example discussed above (see Figure 3.84). Thus, in general, regardless of the shape of the body, and regardless of the compressibility or incompressibility of the medium,

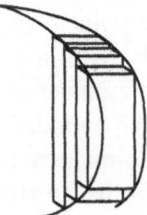

Figure 3.84. Decomposition of a body into right cylinders.

a body immersed in a fluid is buoyed up with a force equal to the weight of the displaced fluid.

We shall shortly discuss the effect of buoyancy in the determination of the molecular weight of proteins. At this point it may be worth computing the size of this buoyancy effect in the simple and, sometimes, useful case of the moving of underwater rocks. Imagine that one ties a rope around a rock under water and lifts the rock to another location. The force required to lift the rock under water will be called W_{eff}, the effective weight of the rock. From the free-body diagram in Figure 3.85 we see that the effective weight is related to the mass of the rock M_R and to the mass of displaced fluid M_F as follows:

$$W_{eff} = M_R g - M_F g. \tag{3-142}$$

The ratio of the effective weight to the actual weight $W = M_R g$:

$$\frac{W_{eff}}{W} = \frac{W_{eff}}{M_R g} = \left(1 - \frac{M_F}{M_R}\right), \tag{3-143}$$

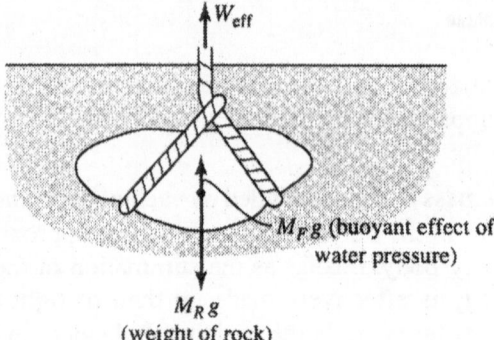

Figure 3.85. Free-body diagram of the forces on underwater rock.

the masses of rock and fluid are related to their densities ρ_R and ρ_F by

$$M_R = \rho_R V_R,$$

$$M_F = \rho_F V_R,$$

since the volume of fluid displaced is equal to the volume of rock. Thus, we see that the effective weight in water is reduced relative to the weight in air by the factor

$$\frac{W_{\text{eff}}}{W} = \left(1 - \frac{\rho_F}{\rho_R}\right). \tag{3-144}$$

A table of minerals shows that the density of rocks is typically two to three times greater than that of water. Thus,

$$\left(\frac{\rho_F}{\rho_R}\right) \approx \tfrac{1}{2} \text{ to } \tfrac{1}{3}.$$

Thus, we expect that the effective weight of a rock in water is approximately one-half to two-thirds its weight in air. This reduction is practically very useful in permitting the movement of rather heavy rocks under water.

3.5.C. Physiologic Effects of Increased Fluid Pressure, Underwater Diving, Postural Effects on Blood Pressure, and Effect of High Acceleration

(i) Underwater Diving

In Section 3.5B(i) we saw that as a diver descends more deeply beneath the surface the surrounding water pressure increases at a rate of one atmosphere for each 33 ft of descent (i.e., $\Delta P = 1$ atm for $\Delta z = 10$ m $\cong$ 33 ft). The pressure of the water is transmitted to all the fluids within the body—to the blood, the tissues, etc. In order to have pressure equilibrium within all parts of the body, the gas inspired into the lungs must have the same pressure as the surrounding fluid. In this way, the pressure throughout the interior of the body is made equal to that of the surrounding water so that no difference in pressure is allowed to develop within the body. If a pressure difference were allowed to develop, there would be a rupture of blood vessels because of the difference in pressure.

On ascent and descent the diver must arrange to have the pressure of gas in his lungs be the same as that of the surrounding water. He can do this either by breathing out on ascent or by adjusting the output pressure of his compressed air tanks. Second to drowning, the most serious underwater diving accident is produced by taking a full breath of air at depth, and holding this breath as the diver rises to the surface quickly. For example, if the diver did this at 99 ft he would have gas as 4 atm in his lungs. This is fine at 99 ft, but if he holds this total volume of gas on ascending, then at the surface the surrounding water is at 1 atm, and his lungs are holding air at 3 atm. This can do two things. His lungs can rupture, thereby allowing gas to flow into the space between lungs and ribs. This is called pneumothorax. Also the great pressure of air in the lungs can force air bubbles into the blood stream. These air embolisms can then occlude blood vessels in the brain or the coronary circulation, and this can lead to death. Of course, the obvious necessity of balancing pressure in the ears, sinuses, and intestines must be realized.

What are the effects of gases at increased pressure on the physiology of the body? If the diver breathes air (79% N and 20% O_2) at higher and higher pressure, it follows that, in effect, he is taking in oxygen at much greater amounts than exist in air at atmospheric pressure. High-pressure oxygen has serious poisoning effects. It is known, for example, that newborn premature infants can be blinded by giving them pure oxygen at atmospheric pressure. Normally, of course, oxygen in the air has a pressure of only $\frac{1}{5}$ atm. At one time, premature infants were put in incubators with an atmosphere of pure oxygen. This had the effect of stopping the development of the blood vessels in the retina of the eye and produced blindness. The condition is known as retrolental fibroplasia because a secondary effect is the growth of fibrous tissue behind the lens of the eye. In the case of the diver it is found that if the partial pressure of oxygen exceeds 2 atm, the diver can go into convulsions which are indistinguishable from those of grand mal epilepsy or electroshock therapy. Two or three minutes of kicking and twitching is followed by unconsciousness. Unless the excess oxygen is removed, death results. If it is removed promptly, the process is reversible.

Suppose that a diver breathes compressed air at the pressure of the surrounding fluid. At a depth of 297 ft the air pressure must be 1 atm, plus (297/33 = 9) 9 atm, i.e., a total of 10 atm. Since the partial pressure of oxygen in air is 20%, it follows that at this depth the diver is, in fact, inhaling oxygen at the pressure at which it causes convulsions and death. If the diver breathes *pure* oxygen as he descends, he is at the dangerous 2 atmosophere level at 33 ft depth. The reasons for the convulsions associated with high oxygen pressure are not understood. However, it is suspected [21] that the high oxygen content in the lungs oxidizes enzymes whose structure is fixed by sulfhydryl bonds. The oxygen breaks these bonds and the enzymes fail to function because their structure is changed.

Despite the danger of oxygen poisoning, breathing pure oxygen during dives is important in military diving, and poisoning can be avoided if the dive is not too long in duration. In Table 3.7 we give a table of safe diving times using pure oxygen [22].

Table 3.7: Diving Times at Various Depths Using Pure Oxygen ([22]).

Depth (feet)	Time (minutes)
10	240
15	150
20	110
25	75
30	45
35	25
40	10

The reason that oxygen is used in military diving is that it permits rapid ascent and descent. If air is used, it would seem that one could go to much greater depths without the danger of oxygen poisoning. This is true, but another danger arises in the presence of nitrogen in the air. The nitrogen under pressure dissolves in the blood at high pressure. As the diver ascends, the rate of ascent must be slow enough so that the nitrogen can diffuse out of the blood and tissues without forming bubbles. If bubbles are formed due to too rapid decompression, the diver will suffer the effects of a painful and crippling disease called the caisson disease or "bends." As might be expected, severe lesions can occur in bone. These occur because gas bubbles, blocked from moving by the surrounding network of bone, can produce infarctions or a stoppage of the circulation of blood through the bone tissue. These infarctions cause necrosis—death—of the bone tissue. The region around the joint surfaces of bones particularly contain a rich blood supply because active metabolism is going on there. Infarctions or blockages of blood flow in these joint areas can cause death of the tissues with joints. Since both divers and caisson workers digging tunnels or bridge footings under water are subject to this condition, detailed investigations have been made of the precise decompression versus time stages that are needed to permit the equilibration of nitrogen between the body and the atmosphere without the formation of bubbles. It is now known [23] that 75% of the dissolved nitrogen diffuses rather readily out of the tissues into the blood and out of the body at the lungs. After a saturation exposure this fraction of the dissolved nitrogen is eliminated after a "routine" decompression lasting about 2 hours. The difficult part of the dissolved nitrogen is that which is associated with fatty tissue. This fraction $\sim 25\%$ takes many hours for elimination depending on the degree of dissolution in the fatty tissue. For example, at a depth of 90 ft, after exposure to the compressed air for 100 min, 57 min is required for decompression. However, should the exposure last for 9 hours, 12 hours will be required for the decompression!

Helium is, on the other hand, much less soluble in fat than is nitrogen. Thus, if the nitrogen in the air is replaced by helium, the time for decompression after long exposure to elevated pressure can be reduced markedly [23].

We see then that the use of compressed air in diving limits the speed with which a diver can ascend safely to the surface. This speed can be a vital consideration in war time, as can be seen by the following considerations presented by Captain O.D. Yarborough, U.S.N. [21]. During World War II:

> German mine-laying activities had drastically restricted shipping within the coastal waters surrounding the British Isles, to an extent that threatened the survival of water traffic and overwhelmed the capabilities of conventional deep-sea diving to cope with these enemy activities.

To counteract this, many divers were used, but still they were insufficient because of the lengthy time involved in decompression after their search of a chosen volume of water. To overcome this difficulty, the idea of using pure oxygen was hit upon—and was used before much was known about the effects of oxygen poisoning. It was soon found that divers became disabled by nausea and convulsions. The military comment on this by Captain Yarborough is

> This led to complete inactivation or loss of the diver with or without subsequent recovery of the bodies as a result of the convulsions.

Only after much subsequent investigation were safe diving times at various depths using pure oxygen established.

(ii)   Postural Effects on Blood Pressure

The human heart is a muscular pump which on maximum contraction (systole) can exert a hydrostatic pressure of about 120 mmHg on the blood within it. On relaxation of the heart there is still sufficient tension in the left ventricle of the heart muscle so that it exerts a minimum pressure of about 80 mmHg during its relaxation period (diastole). Under the action of the pressure exerted by the heart, blood is forced out of the left ventricle into the aorta and then down the branching arterial tree, ultimately to pass through the capillaries. The venous side of the capillary beds collect the blood in coalescing branches of the venous circulation until the blood is returned to the right atrium of the heart at very nearly zero pressure. On average, the mean pressure of the blood is then about 100 mmHg = 135 cmH$_2$O, and the maximum pressure difference between the arterial and venous blood is about 120 mmHg. Since the blood has a density only slightly different from water, it is clear that a person 6 ft tall (180 cm) standing upright has a hydrostatic pressure difference of 180 cmH$_2$O between head and feet. In Figure 3.86 we give a diagram of the *mean* arterial pressure and venous pressure in the large veins at each position in a standing 6 ft person.

From this diagram we see that a change in posture from horizontal to vertical can have a very marked effect on the pressure pattern in the body. One of the important consequences of this is the faintness one can experience on rising. If, in fact, one faints on suddenly rising from a recumbent position, it is described medically as "postural hypotension with syncope" (syncope = fainting).

The cause of this condition is that the veins are very expansile. On rising suddenly the pressure of blood in the veins increases markedly as indicated in Figure 3.86. In the absence of correcting mechanisms the veins expand markedly and the blood "pools" there. The venous return to the heart drops drastically and the blood circulation to the brain consequently falls—resulting in dizziness or even loss of consciousness. Normally the body possesses three mechanisms for the maintenance of good blood return to the heart. These are:

(1) pressure reflexes which induce constriction of the diameter of the arteries and arterioles;

(2) reflex acceleration of the heart rate when receptors in the aorta or the carotid artery sense a drop in pressure;

(3) muscular activity in the limbs which help to maintain the diameter of veins and act to reduce pooling of the blood.

Dysfunction of these reflexes, or too rapid motion, which produces blood pooling before the postural reflexes come into play, can produce dizziness and/or fainting.

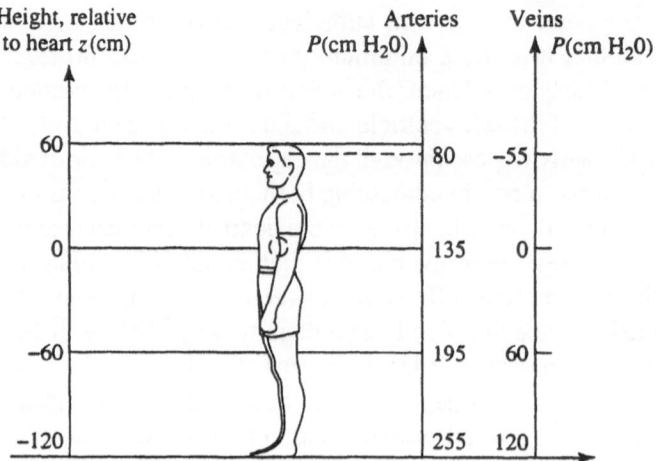

Figure 3.86. Mean arterial and venous pressure at various positions in an erect person 6 ft tall.

Figure 3.86 also shows that a measurement of the arterial pressure on the arm will give quite different results if the arm is at the level of the heart or raised above the head.

(iii)   Effect of High Acceleration

The variation of pressure with depth in a fluid depends upon the weight of the fluid. If the fluid experiences an acceleration, for example, a centripetal acceleration $\vec{a}$, addition to the gravitational acceleration $\vec{g}$, then the effective weight of mass m of the fluid is $m(\vec{g} - \vec{a})$. Thus, the form of Pascal's law in the presence of a centripetal acceleration is

$$\left(\frac{dP}{dz}\right) = \rho(\vec{g} - \vec{a}) \cdot \vec{z}.$$

This effect can markedly change the pressure distribution in a body subject to a centripetal acceleration, and this does, in fact, occur in the case of "black out" experienced by pilots during high "g" acceleration of their aircraft. In Figure 3.87 we show the trajectory of a plane climbing after a dive. The pilot is subject to a centripetal acceleration upward of magnitude

$$a = \frac{V^2}{R}.$$

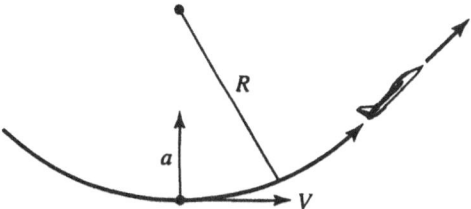

Figure 3.87. Trajectory of a plane coming out of a dive.

The total hydrostatic pressure difference between two points a distance Δz apart in the body is

$$\Delta P = \rho(g + a)\Delta z.$$

If we choose, for example $V = 200$ m/s, $R = 2$ km, one has $a = 20\text{m/s}^2 \cong 2$ g and the hydrostatic pressure differences triple in value. Consulting Figure 3.86, we see that the arterial pressure in the brain is $(135 - 3 \times 55) = -30$ cmH$_2$O. The blood pressure to the brain falls markedly. The additional weight of blood in the veins can distend them and also reduce venous return. Clearly, the blood circulation in the brain will fall markedly. Since a rich blood supply is needed to maintain both the visual response in the retina of the eye and to maintain consciousness, this reduction of blood supply can produce temporary loss of vision and/or unconsciousness.

Studies of the effects of acceleration [24] show that centripetal acceleration toward the head of magnitude between 4 g and 7 g produces blackout and cerebral ischemia, i.e., loss of vision and loss of consciousness due to inadequate blood supply. Acceleration toward the feet of between 3 g and 5 g can result in asystole— the inability of the heart to contract completely, and bleeding from the delicate membranes inside the eyelids and then on the surface of the eyeball, and from the mucous membranes.

3.5.D.   Physiologic Effects of Decrease of Air Pressure. Mountain Sickness; Balloon Ascensions, Physiology of the Storage and Delivery of Oxygen by the Blood

(i)   Mountain Sickness

As man began the ascent of high mountains, he experienced and gave written accounts of a disabling ailment called "mountain sickness." Below about 3000 m travelers will detect only slight changes in pulse rate and breathing rate. In the range 3000 to 4000 m (10,000–13,000 ft) the symptoms increase markedly in intensity. Above 4000 m hiking becomes extremely difficult. In his fascinating book [20] *Barometric Pressure*, P. Bert (1833–1886) has collected in a scrupulously accurate manner detailed accounts of early explorers and contemporary climbers describing the effects of mountain sickness. Bert was the most able early investigator of the various biologic and medical effects of raised or lowered atmospheric pressure. He was the Professor of Physiology at the Faculté des Sciences, in Paris, succeeding to that post in 1869 following the great physiologist Claude Bernard.

A modern textbook of medicine will describe accurately but sparsely the symptoms of mountain sickness as: dyspnea (difficult or labored breathing), tachycardia (heart rate in excess of 100 beats per minute), malaise (vague body discomfort), headache, nausea and vomiting, and insomnia. These words, however, are inadequate to convey adequately the full dimensions of this ailment. A much more satisfactory account is to be found in the description given by M. Lortet who, in August 16, 1869, ascended to the summit of Mt. Blanc, carefully equipped to make quantitative observations of the changes in his breathing, heart rate, and temperature [21] as a function of altitude. We give below P. Bert's discussion [20] and quotation of Lortet's account [25], as translated by M. A. and F. A. Hitchcock:

> I now come to the two ascents of Mont Blanc which were noteworthy from the standpoint that interests us because for the first time the whole combination of physiological phenomena was studied with the precision instruments used in laboratories. Disturbances of circulation and respiration were thus determined in the conditions which the present exactness of physiological research demands. Besides, these observations serve as a basis for an entirely new theory of mountain sickness, which will be discussed in its proper place.
>
> M. Lortet begins with a rapid historical survey of the symptoms felt by the most celebrated travellers. Then, before beginning the account of his journey, he lets escape the precious confession of an incredulity of which I

have often heard alpine travellers boast, even those who had made the most difficult ascents:

'However, in spite of so many data and proofs reported by these distinguished men worthy of credence, I had been a little incredulous and I could not help believing that imagination played a great part in the production of these phenomena. On the main range of Monte Rosa I had often ascended heights of more than 4300 meters without any difficulty and without the least discomfort, and I could not believe that 500 meters more were enough to affect an organism which had stood the test very well up to this altitude. Now I am forced to admit it, I have been convinced de visu, and even a little at my expense, of the very real existence of symptoms which, above this altitude, attack anyone who breathes and particularly anyone who moves in this rarefied air.'

He then comes to the account of his first ascent with Dr. Marcet, August 16, 1869. I copy the important points of his description, which is remarkable for its exactness and moderation:

'Up to the Grands-Mulets (3050 meters), where we arrived at 3 o'clock to pass the night, we were well; no one felt the least discomfort; we all had excellent appetites; but already our instruments announced serious disturbance of circulation, respiration, and especially calorification (heat production).

'The night at the Grands-Mulets was horrible.... At half-past two we set out.'

At daybreak they reached the Grand-Plateau (3932 meters).

'We stopped a moment to breathe....The guides took a little nourishment; but it was completely impossible for me to swallow a single mouthful, although I still felt quite well.

We climbed very slowly; we all felt an inclination to sleep which was very difficult to struggle against and an intense occipital headache, thirst and dryness of the throat, only a few palpitations, but a wretched pulse which varied between 160 to 172 per minute.

When we reached the ridge, we were all tired, and it seemed to me that it would be completely impossible for me to go further. None of us vomited, but almost all of us were nauseated. Like those who are attacked by seasickness, I was completely indifferent about myself and the others, and I wanted only one thing, to remain motionless. The Englishmen who were following us seemed even more affected than we were; one of them was obliged to stop and soon retraced his steps.'

At last they reached the summit of Mont Blanc:

'I no longer felt any kind of illness, but the breathlessness was extreme as soon as I wished to take a few steps rapidly. The least movement caused me disagreeable palpitations. One of my companions, who had felt no ill effect until then, was attacked suddenly, as soon as he had reached the summit, by dizziness and almost constant vomiting which did not cease until he reached the Grand-Plateau on the way down. His stomach was empty, so that he vomited only glairy and bilious matter with very painful efforts. Nothing succeeded in stopping this stomach trouble; only one thing seemed to relieve his condition at all, that was small fragments of pure ice which he managed to swallow from time to time. His pulse was very uneven, very wretched, and the thermometer placed under his tongue hardly went above $+32°$!

The sun was warm, the atmosphere fairly calm, so it was with surprise that I observed that the temperature of the air was $-9°$.

We remained at the summit nearly two hours to make the experiments of which I shall speak later. While I was resting, I felt quite well, although it was impossible for me to take the least nourishment.'

The second ascent went much better. The night at the Grands-Mulets was good; magnificent weather made the walking easy:

'We felt almost no discomfort except a leaden sleepiness while we were climbing the slope which leads to the Dome. I have never felt anything like it, and I am sure that I slept while I was walking. But when I reached the ridge, the cold air and rubbing my forehead with snow removed this congestion.

I felt much better than on the first ascent. I even had an appetite and could eat some morsels with pleasure. However, breathlessness at the slightest movement was still intense. One of our companions experienced great nausea, complete lack of appetite, but did not vomit.'

After this general description, M. Lortet passes to the analysis of the disturbances in the various functions. And at the beginning he is careful to say:

'Hardly noticeable while going from Lyons to Chamounix, that is, passing from a height of 200 meters to an altitude of 1000 meters, their disturbance is, on the contrary, very appreciable from Chamounix to the Grands-Mulets (from 1050 to 3050 meters), still plainer from the Grands-Mulets (3050 meters) to the Grand-Plateau (3932 meters); finally this change becomes very great from the Grand-Plateau to the Bosses-du-Dromadaire (4556 meters), and at the summit of the Calotte of Mont Blanc (4810 meters).

We shall therefore review the variations undergone by the respiration, the circulation, and the inner temperature of the body, taken under the tongue at different altitudes, either while walking, or after a suitable period of rest.

Respiration: from Chamounix to the Grand-Plateau (from 1050 to 3952 meters) disturbances of respiration are slight in those who know how to walk in the mountains, who keep their heads lowered to lessen the laryngeal orifice, who breathe with their mouths closed, being careful to suck an inert object, such as a hazelnut or a little piece of quartz, which considerably increases salivation and prevents the drying out of the air passages. From Chamounix to the Grand-Plateau, the number of respiratory movements is hardly changed; while at rest, we find twenty-four per minute, as in Lyons and in Chamounix; but from the Grand-Plateau to the Bosses-du-Dromadaire and to the summit, we find thirty-six movements per minute. The breathing is very short and very difficult, even when one remains quiet; it seems as if the muscles are stiffened and the ribs are held in a vise. At the summit, the slightest movements brings on panting; but after two hours of rest these discomforts disappear little by little. Respiration drops to twenty-five per minute, but it still remains painful.

Circulation: During the ascent, although progress is excessively slow, the circulation is accelerated extraordinarily. At Lyons, when I am resting and fasting, my average pulse rate is sixty-four per minute. While I was climbing from Chamounix to

Mont Blanc, it increased progressively, following the altitudes, to 80, 108, 116, 128, 136; and finally, while I was climbing the last ridge which leads from the Bosses-du-Dromadaire to the summit, to 160 and sometimes more. These ridges, it is true, are very steep, they have a grade of forty-five to fifty degrees; but slowness of the walking is very great. One generally takes thirty-two steps per minute and often much less when steps have to be cut constantly. The pulse is feverish, hasty, and weak. It is plain that the artery is almost empty. The slightest pressure stops the current in the blood vessel. The blood must pass very rapidly in the lungs, and this rapidity increases still more the insufficient oxygenation which has already results from the rarefaction of the air. It does not have time to receive the oxygen adequately, and neither does it have time to give off its carbonic acid entirely. Above the elevation of 4500 meters, the veins of the hands, the forearms, and the temples are distended. The face is pale with slight cyanosis, and everyone, even the guides acclimated to these lofty regions feel a heaviness in the head and a drowsiness which are often very painful, due probably to a venous stasis in the brain or to a failure of oxygenation of the blood.

Even after two hours of complete rest at the summit and fasting, the pulse always remains between 90 and 108 beats per minute.'

The account of M. Lortet provides a much more vivid and meaningful description of the symptoms of mountain sickness than can be found in the words dyspnea, nausea, malaise, etc. His account shows clearly that a very marked onset of symptoms occurred in his ascent at a threshold height of about 4000 m. Above this height the symptoms grew markedly more severe with the increase in altitude.

(ii) Balloon Ascensions and the Physiological Effects of Decreased Air Pressure

The serious altitude limitations required for the maintenance of life became quite clear as a result of the experiences of balloonists in the latter half of the nineteenth century. The accounts of several balloon pioneers are also recorded in *Barometric Pressure* [20]. The accounts established the heights at which unconsciousness, and then, soon after, death comes for balloonists without the aid of a supply of oxygen. Both the early and more recent scientific studies of fatal accidents associated with decreased air pressure and low oxygen supply (hypoxia) demonstrate time and again the fact that the victim of hypoxia is not warned or frightened by any painful symptoms of the danger of the continued decrease of pressure. We can summarize briefly the functional changes with altitude as experienced by normal, healthy subjects not engaged in physical activity [26]. Below 10,000 ft (3150 m) there is no detectable effect on performance and respiration and heart rates are unaffected. Between 10,000 ft and 15,000 ft (3150 m–4570 m) is a region of so-called "compensated hypoxia" [26]. There is a measurable increase in heartbeat and breathing rate, but only slight loss of efficiency in performing complex tasks. Between 15,000 and 20,000 ft (4570 m–6100 m), however, dramatic changes start to occur.

The respiratory and heartrates increase markedly; there is a loss of critical judgment and muscular control, and a dulling of the senses. Emotional states can vary widely from lethargy to excitation with euphoria and even hallucinations. This is the state of "manifest hypoxia" [26]. The final fatal regime is in the altitude region from 20,000 ft to 25,000 ft (6100 m–7620 m). This is the region of "critical hypoxia." The symptoms are rapid loss of neuromuscular control and consciousness followed by cessation of respiration and finally death.

These various symptoms are shown very clearly in the tragic balloon ascent of the "Zenith" carrying the balloon pioneers Tissandier, Sivel, and Croce-Spinelli on April 15, 1875. During this ascent Sivel and Croce-Spinelli died. The balloon's maximum elevation as recorded on their instruments was 8600 m. Though gas bags containing 70% oxygen were carried by the balloonists, the rapid and insidious effect of hypoxia reduced their judgment and muscular control and prevented their use of the oxygen when it was most needed. Though these balloonists were indeed trying to establish an altitude record, their account shows clearly that their judgment was severely impaired during critical moments near the maximum tolerable altitudes. As they were on the verge of losing consciousness at 7450 m, they decided to throw out the ballast and rise even higher. They lost consciousness above this altitude, but by good fortune the balloon descended rapidly after reaching 8600 m. On falling to about 6500 m the balloonists revived and—under the influence of the hypoxia did exactly the wrong thing once again—they threw out ballast! The second rise to high elevation killed Croce-Spinelli and Sivel. We give below a partial account of this flight written by the survivor Gaston Tissandier [27] as quoted in Bert's book [20, p. 965], and translated by M. A. and F. A. Hitchcock:

> I come to the fatal hour when we were about to be seized by the terrible influence of the atmospheric decompression. At 7000 meters we are all standing in the basket; Sivel, numbed for a moment, has revived; Croce-

Spinelli is motionless in front of me. 'Look,' he says to me, 'how beautiful these cirrus clouds are!' The sublime spectacle before our eyes was indeed beautiful. Cirrus clouds, in different forms, some long, others rounded, formed a circle of silvery white around us. And leaning out of the basket one could see, as if at the bottom of a well, whose walls were formed by the cirrus clouds and the vapor below, the surface of the Earth which appeared in the abysses of the atmosphere. The sky, far from being dark or black, was a clear and limpid blue; the glowing sun burned our faces. However, the cold had already begun to be felt, and we had already wrapped ourselves up. Numbness had seized me; my hands were cold, icy. I wanted to put on my fur gloves; but without my realizing it, the action of taking them from my pocket demanded an effort which I could no longer make.

At this height of 7000 meters, however, I was writing in my notebook almost mechanically; I copy verbatim the following lines, which were written without my having a clear memory of them at present; they are nearly illegible, written by a hand which the cold caused to tremble strangely:

'My hands are icy. I am well. We are well. Vapor on the horizon with little rounded cirrus clouds. We are rising. Croce is panting. We breathe oxygen. Sivel closes his eyes. Croce also closes his eyes. I empty the aspirator. Temp. $-10°$. 1:20. $H = 320$ mm. Sivel is drowsy ... 1:25. Temp. $-11°$, $H = 300$ mm. Sivel throws out ballast. Sivel throws out ballast.' The last words are hardly legible.

Sivel, in fact, who had remained for some instants thoughtful and motionless, sometimes closing his eyes, had no doubt just remembered that he wanted to pass above the limits where the Zenith was then soaring. He drew himself up, his energetic face lighted up suddenly with unusual animation; he turned toward me and said to me: 'What is the pressure?'—'30 centimeters (about 7450 meters altitude).' 'We have plenty of ballast, shall I throw some out?'—I answered, 'Do as you please.'—He turns to Croce and asks him the same question. Croce nods his head with a very energetic sign of affirmation.

Sivel seized his knife and cut three cords successively; the three bags emptied and we rose rapidly....

... Towards 7500 meters, the numbness one experiences is extraordinary. The body and the mind weaken little by little, gradually, unconsciously, without one's knowledge. One does not suffer at all; on the contrary. One experiences inner joy, as if it were an effect of the inundating flood of light.

One becomes indifferent; one no longer thinks of the perilous situation or of the danger; one rises and is happy to rise. Vertigo of lofty regions is not a vain word. But as far as I can judge by my personal impressions, this vertigo appears at the last moment; it immediately precedes annihilation, sudden, unexpected, irresistible.

I wanted to cry out, 'We are at 8000 meters!' But my tongue was paralyzed. Suddenly I closed my eyes and fell inert, entirely losing consciousness. It was about 1:30.

At 2:08, I awoke for a moment. The balloon was descending rapidly, I succeeded in cutting a bag of ballast to check the speed, and in writing in my notebook the following lines, which I copy:

> 'We are descending; temperature $-8°$; I am throwing out ballast, $H = 315$ mm. We are descending. Sivel and Croce still unconscious at the bottom of the basket. Descending very rapidly.'

Hardly had I written these lines when a sort of trembling seized me and I fell inanimate again. The wind was blowing violently upward, and indicated a very rapid descent. Some moments after, I felt myself shaken by the arm, and recognized Croce, who had revived. 'throw out some ballast,' he said to me, 'we are descending.' But I could hardly open my eyes, and did not see whether Sivel had awakened.

At this point the confused balloonist Croce threw out ballast. This act soon results in the ascent of the balloon and the consequent death of Croce and Sivel.

At about 3:30, I opened my eyes again, I felt numb, weak, but my mind was active. The balloon was descending with terrifying speed; the basket was swinging violently and describing great oscillations. I dragged myself on my knees and pulled both Sivel and Croce by the arm. 'Sivel! Croce!' I cried, 'Wake up!'

My two companions were crouched in the basket, their heads hidden under their traveling rugs. I assembled my strength and tried to raise them. Sivel's face was black, his eyes dull, his mouth open and full of blood. Croce's eyes were half shut and his mouth bloody.

To tell in detail what happened then is impossible. I felt a terrible wind rushing upward. We were still at an altitude of 6000 meters. There were in the basket two bags of ballast which I threw out. Soon the Earth drew near, I wanted my knife to cut the rope of the anchor: impossible to find it. I was frantic, I kept crying: 'Sivel! Sivel!'

Luckily, I succeeded in finding my knife and unfastening the anchor at the right moment. The shock as we struck the ground was extremely violent. The balloon seemed to be flattened and I thought that it was going to remain where it was, but the wind was strong and carried it away. The anchor did not hold and the basket slid flat over the fields; the bodies of my unhappy friends were jostled this way and that, and at every moment I thought that they would fall out. However, I got hold of the valve cord, and the balloon soon emptied, then ripped against a tree. It was four o'clock.

As I set foot on the ground, I was seized by a feverish excitement, and fainted, growing livid. I thought I was going to join my friends in the other world.

However, I recovered little by little. I went to my unhappy companions, who were already cold and rigid. I had their bodies sheltered in a neighboring barn. Sobs choked me!

Tissandier then goes on to discuss the evidence for the maximum elevation achieved in their ascent, and the reason for the death of his companions.

As at the moment of my unconsciousness, at 8000 meters, the needle of the barometer was passing rapidly over the pressure number of 28 (8002 meters) and indicating thus an ascent of great speed, I am convinced that we reached this altitude of 8600 meters in the first ascent. After the first descent, Croce-Spinelli and very certainly Sivel were still alive; they were struck by death when the balloon reached for a second time the high levels which it had just left, but which it was not to pass beyond, its weight and volume certainly not permitting it to mount higher.

I am convinced that Croce-Spinelli and Sivel would still be living, in spite of their prolonged sojourn in the higher strata, if they had been able to breathe oxygen. Like me, they must have suddenly lost power of movement. The tubes conducting the vital air must have slipped from their paralyzed hands! But these noble victims have opened new horizons to scientific investigation; these soldiers of science in death have pointed out the dangers of the way, so that their successors may know how to foresee and avoid them.

In Figure 3.88 is Tissandier's graph of the course of the balloon on that day. It described in space a sort of gigantic "M," 8600 m high. Paul Bert was the first person to recognize and demonstrate that the effects of high-altitude flight were produced primarily by the reduction in partial pressure of oxygen. In a series of experiments using low-pressure chambers, he demonstrated that the effects

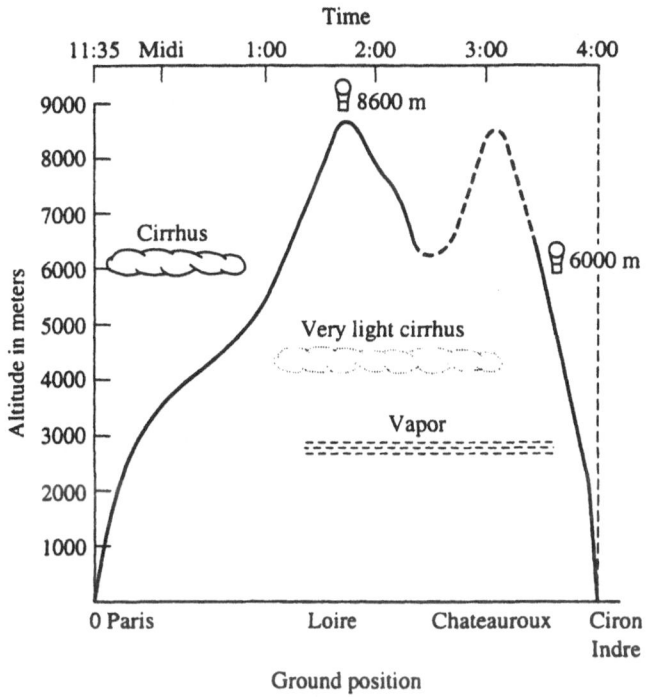

Figure 3.88. Graph of the flight of the balloon "Zenith" April 15, 1875. (After Tissandier [20] and [27].)

of high-altitude sickness could be completely removed provided that the pressure of oxygen is maintained at 160 mmHg (i.e., the partial pressure of oxygen at the Earth's surface) even as the total pressure of the inspired gas mixture falls. The actual quantitative analysis of the role of oxygen in altitude and mountain sickness played a vital role in the development of pulmonary physiology. Perhaps the most important feature of that advance was the growth of understanding of the vital role of hemoglobin in the transport of oxygen and its release in the tissues.

It is, in fact, possible to understand quantitatively why the effects of hypoxia start to occur in the range of altitude of about 15,700 ft (or 4710 m), and why a relatively small increase in elevation above this results in marked enfeeblement. What is needed to explain these features is an elementary understanding of the mechanism of oxygen storage and delivery by the blood. In the following section we shall discuss in detail how the blood picks up oxygen in the lungs and releases it in the body tissues. We shall see that in mountain sickness and in balloon ascents the amount of oxygen released from the blood into the tissues decreases very markedly as the oxygen pressure in the lungs falls below a certain threshold value determined by the shape of the so-called "oxygen dissociation curve of hemoglobin." We shall see that the quantitative properties of oxygen storage and release by hemoglobin explains all the detailed symptoms of mountain sickness and the high-altitude symptoms of balloonists.

(iii) Oxygen Storage and Delivery by the Blood, [28] Mountain Sickness, and High-Altitude Anoxia

In the absence of red blood cells containing hemoglobin the only mechanism for the capture of oxygen would be the solubility of oxygen in the blood plasma. In fact, the amount of oxygen carried by blood in this way represents only $\sim 1\frac{1}{2}\%$ of the carrying capacity of oxygen saturated blood containing normal amounts of hemoglobin.

The solubility of a gas in a fluid such as water or blood plasma is described by Henry's law. If V_{O_2} is the volume of gas at $0\,^{\circ}C$ which dissolves in a volume V_P of liquid (plasma), then Henry's law states that V_{O_2} is proportional to V_P, and the partial pressure of the gaseous oxygen P_{O_2}, i.e.,

$$V_{O_2} = \alpha V_P P_{O_2}.$$

The fractional volume of gaseous oxygen dissolved in the plasma is (V_{O_2}/V_P), and if we express P_{O_2} as a fraction of atmospheric pressure, then one finds experimentally that the fraction (V_{O_2}/V_P) is equal to [28]

$$\left(\frac{V_{O_2}}{V_P}\right) = 0.023\frac{P_{O_2}\ (\text{in mmHg})}{760\ (\text{mmHg})}.$$

Though the partial pressure of oxygen in the inspired air is 160 mmHg, the process of warming and mixing the air with saturated water vapor in the bronchial tree reduces the partial pressure by 47 mmHg by the time it is in the alveolar sacs of the lung [26, p. 1103]. In addition, the increased fraction of CO_2 in the alveoli and the process of transfer of gas between the blood and air reduces the partial pressure of oxygen another 9 mmHg so that at atmospheric pressure the partial pressure of oxygen in the alveoli is actually about 104 mmHg. Thus, the fractional volume of oxygen dissolved in blood in the alveolar capillaries is

$$\left(\frac{V_{O_2}}{V_P}\right) = 0.023\left(\frac{104}{760}\right) = 0.0032,$$

or as it is often expressed, the solubility is 0.32 vol. %.

Blood containing a normal density of red blood cells, on the other hand, has a much greater oxygen storing capacity than this. If fully oxygenated arterial blood in equilibrium with alveolar oxygen at 100 mmHg is shaken in vacuo, or washed with an inert gas like nitrogen or argon, the blood will become fully deoxygenated and will give up about 20 cm³ of oxygen per 100 cm³ of blood. Thus, if the volume fraction of oxygen held by hemoglobin in the red cells is denoted by $(V_{O_2}/V_B)_H$, we have that

$$\left(\frac{V_{O_2}}{V_B}\right)_H \bigg/ \left(\frac{V_{O_2}}{V_P}\right) = \frac{0.20}{0.0032} = 62.5.$$

Thus the red blood cells are 62.5 times more effective in storing oxygen than is the process of direct solubility of oxygen in the plasma.

The attachment of oxygen to the hemoglobin molecules in the red cells is governed by the laws of statistical mechanics. Each hemoglobin molecule when fully saturated can hold four oxygen molecules. The hemoglobin molecules release these molecules or reattach them reversibly. The actual fractional degree of saturation depends upon the partial pressure (P_{O_2}) of the oxygen surrounding the blood. The shape of the curve relating the degree of saturation S (S = number of oxygen molecules held on average by one hemoglobin divided by four) to the partial pressure of oxygen can be calculated by means of statistical mechanics, using various models for the interaction between the oxygen molecule and the hemoglobin molecule. We will calculate the predicted shape of this curve in Volume II of this series. At present let us content ourselves with the experimentally measured relationship between degree of saturation and the partial pressure of oxygen in the gas surrounding the blood. The experimental results for this oxygen dissociation curve are shown in Figure 3.89.

This figure shows the percentage saturation of each hemoglobin molecule in normal blood (pH = 7.40) at body temperature $T = 38°$. The general shape of this curve is described as "sigmoidal," i.e., like an S. The pressure at which the descending limb occurs in the dissociation curve is a marked function of: the partial pressure of CO_2, the pH of the blood, and the temperature. As the CO_2 partial pressure in the surrounding gas increases, the descending limb of the curve shifts to the right. This effect is called the Bohr effect after its discoverer. As the pH

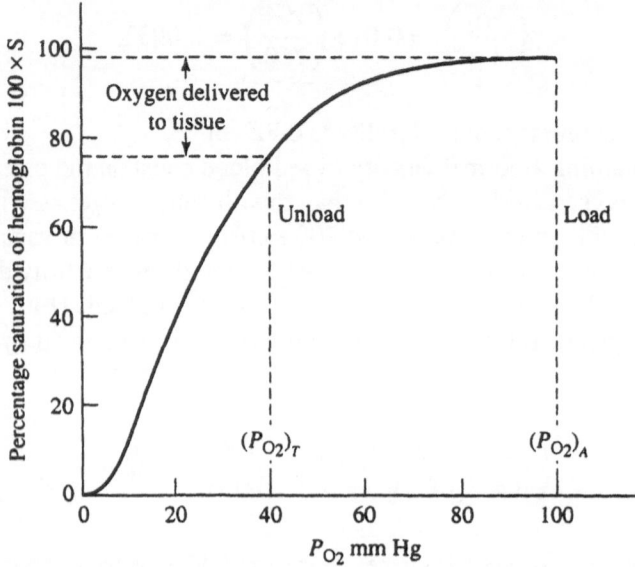

Figure 3.89. Oxygen dissociation curve for hemoglobin at $T = 38\,°C$ and pH = 7.40: normal human blood (From Roughton, [28]). The hemoglobin loads up at $(P_{O_2})_A$ and unloads at the partial pressure of oxygen $(P_{O_2})_T$ in the tissues, delivering the fraction of carried oxygen indicated.

of the blood decreases, the dissociation curve also shifts to the right. The "normal" range of pH in blood is between pH = 7.6 and 7.2. Increase in temperature also shifts the dissociation curve to the right. Physiologically, of course, human blood is normally kept at accurately controlled temperature.

We can now see from Figure 3.89 how the blood picks up oxygen in the alveolar capillaries in the lung, and delivers it as needed in muscle and brain tissue throughout the body. In the alveoli the partial pressure of the oxygen is ~ 104 mmHg. Under these conditions the hemoglobin becomes saturated to about 98% of its full capacity. As the blood circulates around the body, it passes through tissues in which the partial pressure of oxygen is on average about 40 mmHg. This pressure occurs directly on the sharply descending portion of the dissociation curve. The hemoglobin, being in equilibrium with the tissue gases, then gives up oxygen until its saturation is reduced to about 75%. Thus on average each hemoglobin molecule gives up about one oxygen molecule. Note that if the dissociation curve shifts to the left, there will be a considerable reduction of the oxygen delivered, while shifts to the right, associated with increased CO_2 in the tissue, will increase the release of oxygen. On average then the blood returning to the heart is only about 75% saturated. This corresponds to a change of the volume fraction of oxygen in blood from $(V_{O_2}/V_{blood}) = 0.2$ to $(V_{O_2}/V_{blood}) = 0.15$ which is what is found normally.

To see the reason for altitude sickness, we now need to know only one more fact: that is, the partial pressure of oxygen in the lung alveoli as a function of the altitude. We already know (see Table 3.6) the partial pressure of oxygen in the inspired air. The partial pressure $(P_{O_2})_A$ of oxygen in the alveoli is different from that in the inspired air by about 56 mmHg for the reasons previously mentioned (a mixture of saturated water vapor, increased CO_2 partial pressure, and gas exchange between air and blood). In addition, physiological compensatory mechanisms, such as improved pulmonary ventilation and better gas exchange as the altitude increases, act in such a way as to increase the $(P_{O_2})_A$ to values higher than one would find by simply subtracting 56 mmHg from the partial pressure of the inspired gas at each altitude. Detailed experimental measurements of $(P_{O_2})_A$ as a function of altitude have been made and are presented in Figure 3.90 [26].

The straight-line portion of the relation between $(P_{O_2})_A$ and P corresponds to the region where $(P_{O_2})_A = (P_{O_2})_I - 56$, i.e., where the P_{O_2} is 56 mmHg less than the inspired P_{O_2}. The departure of the line from linearity above about 11,000 ft represents the effects of physiologic compensation: increased gas exchange and pulmonary ventilation. The dotted line at 30 mmHg for $(P_{O_2})_A$ corresponds to the pressure beneath which life cannot be sustained. It occurs at above $\sim 25,000$ ft.

Using this data and the information contained in the oxygen dissociation curve of Figure 3.89, we can understand the nature of the onset of altitude sickness. As the altitude increases, the loading pressure of the hemoglobin in the alveoli decreases. This reduction of $(P_{O_2})_A$ at first has no effect on the loading of oxygen because the dissociation curve is essentially flat above $(P_{O_2})_A \sim 70$ mmHg. As the atmospheric or barometric pressure falls further, $(P_{O_2})_A$ approaches the knee of the oxygen dissociation curve. This point is at about $(P_{O_2})_A \approx 60$ mmHg. This pressure corresponds to an altitude of about 10,000 ft (from Figure 3.90). Up to this altitude we expect relatively little effect on the body because the hemoglobin continues to deliver about the same amount of oxygen to the tissues. However, as the $(P_{O_2})_A$ now starts to approach more closely the unloading pressure $(P_{O_2})_T$, the loading of oxygen in the alveoli starts decreasing markedly and the amount of oxygen delivered to the tissues falls. At about 15,000 ft $(P_{O_2})_A$ has fallen to about 45 mmHg. From the oxygen dissociation curve we see that if the tissue pressure of oxygen $(P_{O_2})_T$ remains on average at 40 mmHg, the amount of oxygen delivered to the tissues falls at this altitude to between one-third and one-fourth of the normal amount. At this level the symptoms of altitude sickness clearly assert

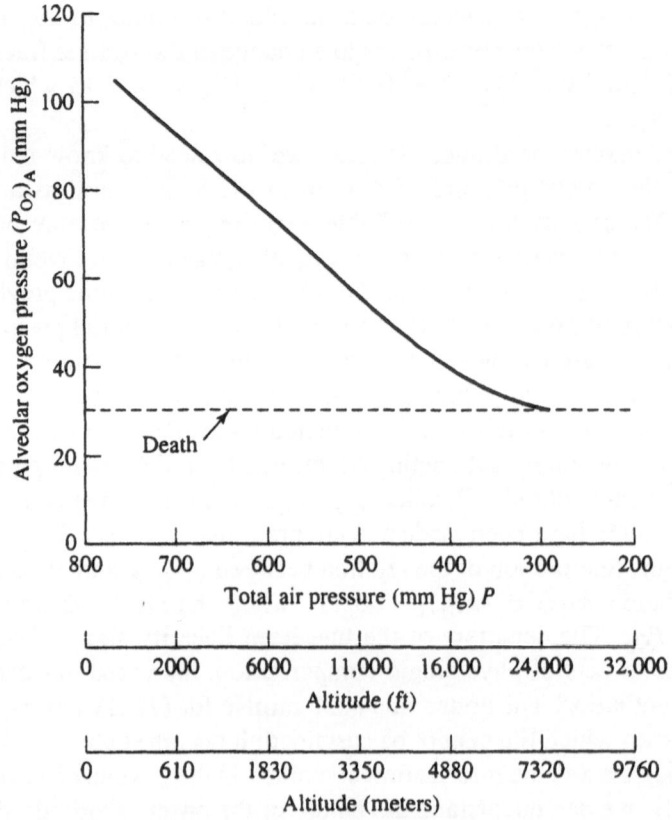

Figure 3.90. Alveolar oxygen pressure as a function of air pressure and height (after Luft [26]).

themselves. As the height increases so that the alveolar pressure falls even less, the difference $(P_{O_2})_A - (P_{O_2})_T$ gets quickly very small and the blood cells simply do not give up any of their oxygen to the tissue. Apparently what can occur is that the tissue oxygen pressure can fall as low as a level of about 30 mmHg as part of the body's effort to extract more oxygen from the circulating hemoglobin. However, if the altitude increases to about 25,000 ft, $(P_{O_2})_A$ is equal to 30 mmHg and again the blood simply will not unload any oxygen whatsoever and death follows. In essence then as the altitude increases, the difference between the loading pressure $(P_{O_2})_A$ and unloading pressure $(P_{O_2})_T$ becomes quite small, and this effect is accentuated greatly by the fact that the oxygen dissociation curve is steep near the unloading pressure. This produces a very marked reduction in oxygen delivery for relatively small changes (e.g., from 45 to 40 mmHg) in the alveolar pressure. These results are in excellent agreement with the observations of the level at which altitude sickness is felt and their marked increase with further elevation.

(iv)   High Altitude Air Travel: Cabin Pressurization

To function satisfactorily at high altitudes, it is clear that oxygen must be supplied at a pressure above say $\sim$ 115 mmHg. At this value of the inspired oxygen pressure the alveolar oxygen pressure is about 60 mmHg and the hemoglobin loads oxygen at a pressure just above the knee of the oxygen dissociation curve. When the partial pressure of oxygen is 115 mmHg, the total pressure of air is about 550 mmHg and the equivalent altitude is about 8500 ft. In fact, to assure the health of air travelers in jet aircraft, which fly at altitudes as high as 40,000 ft, the aircraft interior is provided with air pressurized to about 560 mmHg, i.e., an altitude of about 8600 ft. A modern airliner generally provides about 1 lb (in mass) of air per passenger per minute for good ventilation. This requires a substantial and sophisticated air processing plant when one considers the fact that at 40,000 ft the outside air pressure is about 140 mmHg, and the air temperature can be as low as $-30°$ below 0 °F. The air processing plant in a plane carrying, say, 100 passengers, must take in air, pressurize it to about 560 mmHg and warm it to an accurately controlled temperature near 70 °F at a rate corresponding to 100 lb of air per minute.

It is important to realize that the difference in air pressure between the pressurized cabin and the outside atmosphere represents a very large net internal force tending to burst the fuselage. This internal stress increases to a maximum and decreases each time the airplane goes up to altitude and then lands. The constant repetition of this very large stress can fatigue the structural members of the aircraft in a serious way. Thus, the airframe must be designed to withstand this continued pressurization and depressurization of the cabin.

Loss of cabin pressure at high altitudes can constitute a serious hazard to passengers. Thus, jet aircraft are equipped with oxygen masks, one for each passenger. On take-off the cabin crew instruct passengers on the use of these masks that are designed to drop automatically from the console containing overhead lights and air vents above each seat.

It is interesting to observe that if one breathes pure oxygen at 40,000 ft, i.e., if the pressure of the pure oxygen is the same as that of the surrounding air at 40,000 ft, the oxygen pressure would be $\sim$ 140 mmHg. This is clearly above the level of 115 mmHg needed to load oxygen into the hemoglobin. If one uses oxygen masks to go higher in altitude while breathing oxygen at the ambient air pressure, we observe that there is an altitude limit. At 44,000 ft the total air pressure is about 115 mmHg. Thus, if the inspired air is 100% oxygen, one cannot expect to go much higher unless the inspired oxygen is pressurized above the level of the surrounding air.

There are also mechanical effects associated with the reduction of pressure as the airplane cabin falls from atmospheric pressure 760 mmHg to 560 mmHg on ascent and the subsequent increase of pressure on descent. Many of us have experienced pain in the ears or sinus passages under such circumstances. The middle ear and the sinuses are spaces in which the enclosed air must adjust itself to the pressure of the surrounding air. This equalization of air pressure in the middle ear is accomplished by means of a tube called the eustachian tube which connects the middle ear to the nasopharynx. This tube contains flap-like valves which permits easy escape of air from the middle ear if its pressure is greater than the surroundings. Thus, on ascent, one normally does not feel ear pain. However, on decreasing altitude, when the middle ear is at lower pressure, then atmospheric air must flow in from the nasopharynx. To open the eustachian tube, one must use small muscles at the back of the nose and throat. This is done by yawning, chewing, and swallowing. In combat aircraft marked changes in altitude can lead to painful pressure on the ear drum or its rupture. Damage to the ear by this mechanism is called otitic barotrauma, while damage to the sinuses is called sinus barotrauma.

3.5.E.   Buoyancy and the Measurement of Molecular Weight of Macromolecules

Archimedes' law of buoyancy plays an important role in the measurement of the molecular weight of proteins and other biological macromolecules using the ultracentrifuge. In this method in essence the molecular weight is found by measuring the speed with which a molecule falls through water when the molecules and the water are subjected to very large centrifugal accelerations (g'). g' in fact represents an artificial gravity which is very much larger than the terrestrial 9.8 m/sec^2. Figure 3.91 describes the essential elements.

Now we shall describe Newton's law of motion in a frame of reference, within which the test tube is at rest. That frame clearly rotates in the laboratory with an angular velocity ω. Let $\hat{n}$ be a unit vector perpendicular to the plane in which the test tube rotates, and introduce the angular velocity vector by $\vec{\omega} = \hat{n}\omega$. Also let m stand for the mass of a macromolecule in the test tube, and $\vec{v}_T$ its velocity relative to the tube. The velocity $\vec{v}_L$ of the molecule as observed in the laboratory frame of

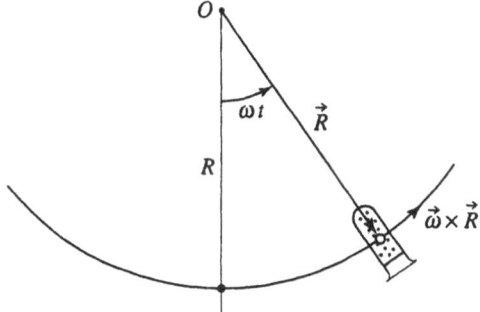

Figure 3.91. Test tube, containing a solution of macromolecules, rotating about 0 with angular velocity ω.

reference is then

$$\vec{v}_L = \vec{v}_T + \vec{\omega} \times \vec{R},$$

and the associated acceleration will be

$$\vec{a}_L = \frac{d}{dt}\vec{v}_L + \vec{\omega} \times \frac{d}{dt}\vec{R} = \frac{d}{dt}\vec{v}_L + \vec{\omega} \times (\vec{\omega} \times \vec{R}) = \frac{d}{dt}\vec{v}_L - \vec{\omega}^2\vec{R}$$

[Here we used the algebraic relation $(\vec{a} \times (\vec{b} \times \vec{c})) = \vec{b}(\vec{a} \cdot \vec{c}) - \vec{c}(\vec{a} \cdot \vec{b})$]

Now let $\{\sum \vec{F}\}_L$ stand for the sum of the forces on the macromolecule, as seen from the lab frame. We may then write Newton's law for the molecule in the form

$$\left\{\sum \vec{F}\right\}_L = m\vec{a}_L = m\left\{\frac{d}{dt}\vec{v}_T - \omega^2\vec{R}\right\},$$

giving the effective equation of motion for the test particle in the tube

$$m\frac{dv_T}{dt} = \left\{\sum \vec{F}\right\}_L + m\omega^2\vec{R}$$

The force m $\omega^2 R$ is called the *centrifugal force* acting on the macromolecule.

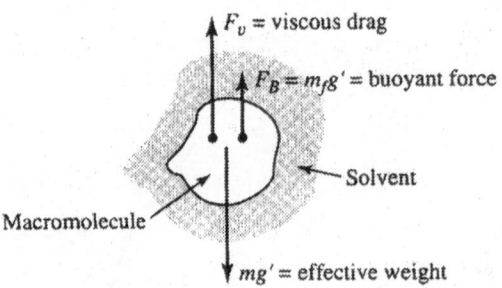

Figure 3.92. Free body diagram of the forces acting on a macromolecule in the ultracentrifuge.

Fig. 3.92 shows the effective forces on a suspended protein molecule. The forces acting on each molecule are

1. The effective weight of the molecule mg'.

2. The buoyant force of the surrounding fluid $m_f g'$. Here m_f is the mass of the fluid displaced by the macromolecule.

3. The viscous friction force F_V with which the fluid resists the fall of the molecule.

The effective weight force is produced by the fluid surrounding the molecule. Since the molecule is accelerating toward the center with acceleration g', Newton's law says that it must be experiencing a force equal to mg'. In effect we may say that the molecule is in a "centrifugal field" rather than a gravitational field. Its "weight" in the centrifuge can be as large as 10^5 times larger than in the gravitational field, i.e., $g' \sim 10^5 g$.

The buoyant force $F_B = m_f g'$ follows the law of Archimedes, i.e., the immersed body is buoyed up with a force equal to the total weight of the displaced fluid. Now the displaced fluid has a weight $m_f g'$ where m_f is the mass of displaced fluid and g' is the centripetal acceleration.

In addition to these two forces, the fluid exerts a frictional or viscous resistance to the motion of the molecule through it. This viscous force is proportional to the velocity of the molecule so we may write F_V as

$$F_V = fV,$$

where V is the speed of the molecule and f is a so-called friction factor whose magnitude is determined by the shape and size of the molecule and by the viscosity of the fluid. In Volume II of this work we shall see that f can be measured in an independent experimental determination of the diffusion coefficient of the molecule. For our purposes we will assume that f is a known constant for our molecule.

According to Newton's second law of motion, the net force on the molecule is equal to its mass times its acceleration. Taking the downward direction as positive, we have

$$mg' = m_f g - fV = m \left(\frac{dV}{dt} \right).$$

This equation tells us that under the action of the "effective gravity" and buoyancy force, the body accelerates. As its velocity increases, the friction force increases. The viscous friction force (fV) increases until it is equal to the net of the gravity and buoyancy forces. The total force then is zero, and no further acceleration occurs. The body reaches terminal velocity and continues to fall at constant speed. This constant speed is called the "sedimentation velocity" V_S. Precisely the same factors are at work in balloon flight or in the slowing effect of a parachute

$$m \left(1 - \frac{m_f}{m} \right) g' = fV_S.$$

Writing the mass of displaced fluid as

$$m_f = \rho_f \Omega_d$$

and writing the mass of the molecule as

$$m = \rho_m \Omega_d,$$

where Ω_d is the volume of fluid displaced by the molecule; and

$$\overline{\Omega} \equiv \frac{1}{\rho_m} = \text{volume of fluid displaced per unit mass of the molecule}$$

$$= \text{"partial specific volume" of the molecule,}$$

we find that the relation between molecular mass and the terminal velocity V_S is

$$m(1 - \overline{\Omega}\rho_f)g' = f V_S$$

or

$$m = \left(\frac{V_S}{g'}\right)\frac{f}{(1 - \overline{\Omega}\rho_F)}.$$

This is called the Svedberg equation after the inventor of this method for molecular weight determination [29].

The effect of buoyancy is very important in this equation. It appears in the term $(1 - \overline{\Omega}\rho_f)$. In the absence of buoyancy this quantity is equal to unity. For a wide range of proteins and viruses $\overline{\Omega}$ ranges between 0.61 and 0.75. Since ρ_f is about 1 g/cm^3, we see that the buoyancy term actually is a number between 0.40 and 0.25. The accurate determination of $\overline{\Omega}$, the so-called partial specific volume, is thus very important in measuring with precision the molecular weight of biological macromolecules.

Table 3.8: Sedimentation Coefficient, Partial Specific Volume and Molecular Weight of Various Proteins (from [30] and [31]).

Macromolecule	$S = (V_S/g')$	$\overline{\Omega}$	m
in units of	10^{-13} s	(g/cm^3)	(g/mol)
Hemoglobin	4.31	0.749	60,000
Lysozyme	1.87	0.688	14,100
Myosin	6.4	0.728	570,000
Bacteriophage virus T_4	890	0.617	$192.5 \pm 6.6 \times 10^6$

In Table 3.8 we give the values of the molecular weight and partial specific volumes of a number of proteins and viruses along with measured values of the sedimentation coefficient (V_S/g').

3.6 References and Supplementary Reading

[1] *The Works of Archimedes*, edited by T. L. Heath. Dover, New York (1953).

[2] *Episodes from the Early History of Mathematics* by A. Aaboe. Random House, New Mathematical Library, New York (1964).

[3] "Functional Aspects of the Abductor Muscles of the Hip," by V. T. Inman. *Journal of Bone and Joint Surgery (Amer.)* **29**, 607–619 (1947).

[4] *Biomechanics of Human Motion* by M. Williams and H. R. Lissner. W. B. Saunders, Philadelphia (1962).

[5] *Cunningham's Manual of Practical Anatomy*, Vol. 1—*Upper and Lower Limbs*, 13th ed.; revised by G. T. Romanes. Oxford University Press, London (1966).

[6] *Grant's Method of Anatomy*, 8th ed. (Chap. 24) by J. V. Basmijian. Williams & Wilkins, Baltimore (1971).

[7] "Don't Throw Away the Cane," by W. Blount. *Journal of Bone and Joint Surgery* **38**, 695 (1956).

[8] *Low Back Pain Syndrome* by R. Cailliet, F. A. Davis, Philadelphia (1962).

[9] "Simple Illustrations of Physical Principles Selected from Physiology and Medicine," by L. A. Strait, V. T. Inman, H. J. Ralston. *Am. J. Phys.* **15**, 375–382 (1947).

[10] *Cunningham's Manual of Practical Anatomy*, Vol. II—*Thorax and Abdomen*, 13th ed. (pp. 86–91); revised by G. J. Romanes. Oxford University Press, London.

[11] *Strength of Biological Materials* by H. Yamada, edited by F. G. Evans (p. 90 ff). Williams and Wilkins, Baltimore (1970).

[12] "Studies on the Strength for Compression, Tension and Torsion of the Human Vertebral Column," by J. Kyoto, *Pref. Med. Univ.* **71**, 659–702 (1962).

[13] *Orthopedic Biomechanics* by V. H. Frankel and A. H. Burstein. Lee and Febiger, Philadelphia (1970).

[14] *Anatomy* by W. Foster, Foster Art Service, Tustin, CA.

[15] "Observations on the Function of the Shoulder Joint," by V. T. Inman, M. Saunders, and L. C. Abbott. *Journal of Bone and Joint Surgery* **36**, 1–30 (1944).

[16] "The Mechanics of Muscle Function in Locomotion," by J. B. Morrison, *J. Biomechanics* **3**, 431–451 (1970).

[17] *Human Limbs and Their Substitutes* by V. T. Inman and H. J. Ralston (296 pp.), edited by P. E. Klopsteg and P. D. Wilson, McGraw-Hill, New York (1954).

[18] *An Introduction to Mechanics* by D. Kleppner and R. Kolenkow. McGraw-Hill, New York (1973).

[19] *Physics*—Part I, by D. Halliday and R. Resnick. Wiley, New York (1966).

[20] *Barometric Pressure* (Researchers in Experimental Physiology) by P. Bert; translated by M. A. Hitchcock and F. A. Hitchcock. College Book, Columbus OH (1943).

[21] "Proceedings of the Underwater Physiology Symposium," edited by L. G. Goff, Publication #377. National Academy of Sciences and National Research Council, Washington, DC (1955).

[22] "Diving Physiology," by E. H. Lanphier, *New England Journal of Medicine* **256**, 120 (1957).

[23] "Underwater Physiology" by A. H. Behnke, Jr. and E. H. Lanphier, in *Handbook of Physiology: Respiration*, Vol. II (pp. 1159–1193). Williams and Wilkins, Baltimore (1965).

[24] "Problems of Air and Space Travel," 6th ed. by S. Bondurant in *Harrison's Principles of Internal Medicine* (pp. 710–714). McGraw-Hill, New York (1970).

[25] Deux Ascensions au mont Blanc en 1869: Recherches Physiologique sur le mal des Montagnes (Lyon, Medical 1869) by M. Lortet.

[26] "Aviation Physiology—Effects of Altitude," by U. C. Luft (Chap. 44, pp. 1099–1145), in *Handbook of Physiology: Respiration*, Vol. II; edited by W. O. Fenn and H. Rahn. Williams and Wilkins, Baltimore (1965).

[27] G. Tissandier in *La Nature*, number of May 1, 1875, third year, first semester, pp. 337–344.

[28] "Transport of Oxygen and Carbon Dioxide," by F. J. W. Roughton (Chap. 31, pp. 767–825) in *Handbook of Physiology: Respiration*, Vol. I; edited by W. O. Fenn and H. Rahn. Williams and Wilkins, Baltimore (1964).

[29] *The Ultracentrifuge* by T. Svedberg and K. O. Pederson. Oxford University Press, Oxford (1940).

[30] *Physical Chemistry of Macromolecules* by C. Tanford (381 pp.). Wiley, New York (1961).

[31] "Molecular Weights of Coliphages and Coliphage DNA: II Measurement of Diffusion Coefficients Using Optical Mixing Spectroscopy and Measurement of Sedimentation Coefficients," by S. B. Dubin, F. C. Bancroft, D. Freifelder, and G. B. Benedek. *J. Mol. Biol.* **54**, 547–556 (1970).

[32] *The Science of Mechanics* by E. Mach, translated by T. J. McCormack from German, 6th ed., Open Court Publishing, La Salle, IL. 1960 (paperback).

3.7 Problems

(The problems are grouped in sections, and in each section ordered roughly by increasing difficulty.)

Understanding the Torque Condition

The problems in this section should help to clarify the concept of torque, and the application of the torque conditions for equilibrium. We will deal mainly with cases where all forces lie in a plane; in this case, the torque, defined with reference to a point P in that plane, is a vector normal to the plane, and can be described by a positive or negative number τ. for the sign of τ due to a force F, we establish the sign convention shown below:

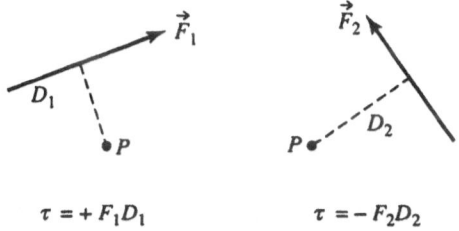

$$\tau = + F_1 D_1 \qquad\qquad \tau = - F_2 D_2$$

In discussing the equilibrium of an extended body, the point of attachment of the forces acting on the body must also be specified. Let us call the *line of action* a line drawn through the point of attachment of a force and parallel to that force.

A body acted on by *two* forces $\vec{F}_1$ and $\vec{F}_2$ is in static equilibrium if and only if $\vec{F}_2 = -\vec{F}_1$, *and* the lines of action of the two forces coincide.

On the other hand, as shown in the left part of the figure, the two forces form a *couple*, whose torque is $\tau = -DF_1$, and the body is not in equilibrium, despite $\vec{F}_1 + \vec{F}_2 = 0$ being satisfied.

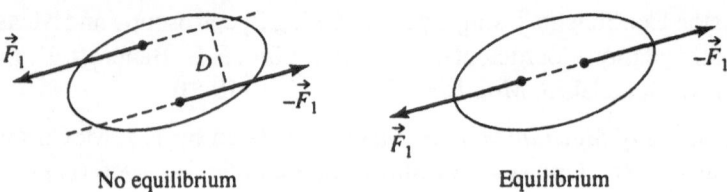

No equilibrium   Equilibrium

It is also useful to define precisely the concept of the *resultant* of two forces in a plane. The resultant of the two forces $\vec{F}_1$ and $\vec{F}_2$ is a force vector

$$\vec{F} = \vec{F}_1 + \vec{F}_2,$$

whose line of action passes through the *intersection* of the lines of action of $\vec{F}_1$ and $\vec{F}_2$. The importance of this construction of the resultant lies in the fact that the resultant $\vec{F}$ is dynamically equivalent to the pair $\vec{F}_1$ and $\vec{F}_2$. In the two following problems you will show that the torque of $\vec{F}$ about a point P is equal to the sum of torques of $\vec{F}_1$ and $\vec{F}_2$ about the same point.

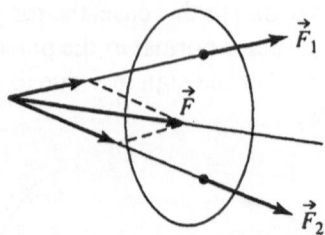

1. A force $\vec{F}$ is the resultant of two perpendicular forces $\vec{F}_1$ and $\vec{F}_2$ in the x–y plane, which act on the same point 0. Show that the torque τ of $\vec{F}$ about any point P in the plane is equal to the sum of the torques of $\vec{F}_1$ and $\vec{F}_2$: $\tau = \tau_1 + \tau_2$, or FD: $F_1 D_1 + F_2 D_2$. *Hint:* Write $F_1 = F\cos\alpha$, $F_2 = F\sin\alpha$, and show from the figure that

$$D = D_1 \cos\alpha + D_2 \sin\alpha.$$

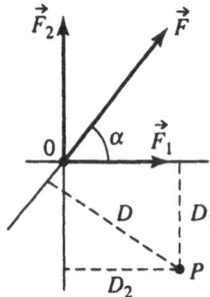

2. Generalize the result of Problem 1 and show that the sum of the torques of any two forces $\vec{F}_1$ and $\vec{F}_2$ lying in a plane is equal to the torque of the *resultant* of the two forces

$$\vec{F} = \vec{F}_1 + \vec{F}_2.$$

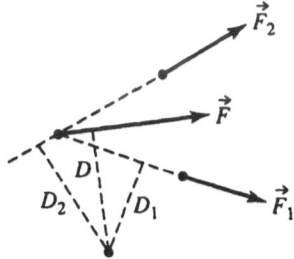

    Hint: Choose an x-axis along $\vec{F}_1$, decompose $\vec{F}_2$ into a part parallel to $\vec{F}_1$ and a part perpendicular to $\vec{F}_1$.

3. Show that the torque of a couple $(\vec{F}, -\vec{F})$ is *independent* of the reference point P.

4. Consider an extended body with three forces $\vec{F}_1$, $\vec{F}_2$, $\vec{F}_3$ acting on it. All force vectors lie in the same plane. Show that, provided

$$\vec{F}_1 + \vec{F}_2 + \vec{F}_3 = 0,$$

the sum of the torques acting on the body has a value independent of the location of the reference point P. *Hint:* Construct the resultants of $\vec{F}_2$ and $\vec{F}_3$; this resultant and $\vec{F}_1$ form a couple. Use the results of Problems 2 and 3.

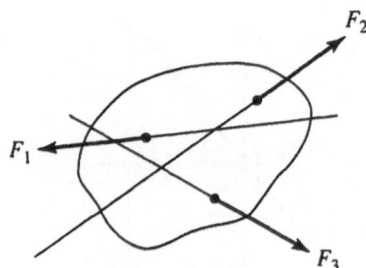

5. Torque meter. Automobile mechanics have a special key to fasten the bolts of the headplate of an engine. It is equipped with a meter which gives them a reading of the torque they apply to the bolts. Imagine for yourself a credible design for such a key.

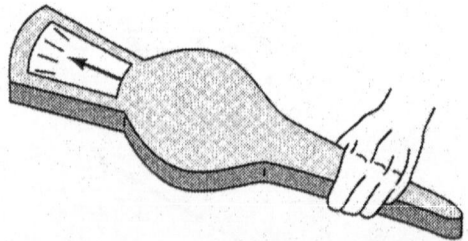

Applications of the torque condition

6. For many applications we need a preliminary understanding of the center of mass, C.

   (a) Consider a uniform bar of length L. Subdivide it into mass elements m_i, so that $\sum m_i = M$ is the total mass of the bar. The total gravitational force is

   $$M\vec{g} = \sum_i (m_i \vec{g}).$$

   The location of c is determined by the torque condition

   $$\sum_i (m_i g) x_i = M g x_c.$$

Let Δx_i be the length of the "slice" m_i, so that $m_i = M(\Delta x_i/L)$. Show that $x_c - \frac{1}{2}L$.

(b) Extending this to a nonuniform object, consider a narrow wedge of length L and mass M. for such a wedge, the mass element m_i is

$$m_i = \frac{2M}{L^2} x_i \Delta x_i.$$

Show that indeed $\sum m_i - M$. Determine the location x_c of the center of mass by using the torque condition

$$\sum m_i x_i = M x_c$$

Show that $x_c = \frac{2}{3}L$.

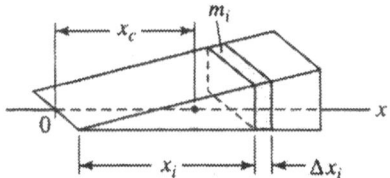

7. Center of mass of your body: You have available:

   (i) A rigid board, 6 ft long, 1 ft wide;

   (ii) A bathroom scale.

   Describe a complete program of measurements (lengths and weights), sufficient to determine the location of your body's center of mass, for the posture where you lie flat on the board.

8. An artist wants to build a Calder-Mobile as shown in the sketch. He has *chosen* the lengths L_1, L_2, ℓ_1, ℓ_2, and the mass M_1. How must he choose the masses m_1 and m_2 to ensure that the mobile is in equilibrium?

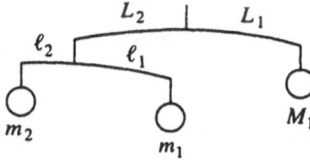

9. A person holds a 10 lb steel ball in his hand, with the forearm horizontally extended. Determine the force $\vec{F}$ exerted on the forearm at the elbow joint, and the tension T in the biceps muscle:

$$\text{Data:} \quad Mg = 10\,\text{lb},$$
$$mg = 3\,\text{lb}.$$

Since the forearm is in equilbrium, you must satisfy the two conditions that the sum of all forces is zero, and the sum of all torques about any reference point is also zero. You can reduce the amount of algebra by a judicious choice of the reference point P.

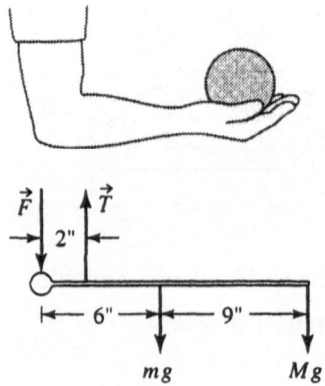

10. You are able to hold out your arm in an outstretched, horizontal position thanks to the action of the deltoid muscle.

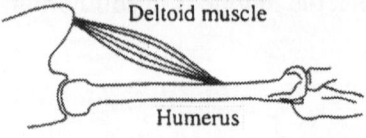

A schematic representation of this situation looks as follows:

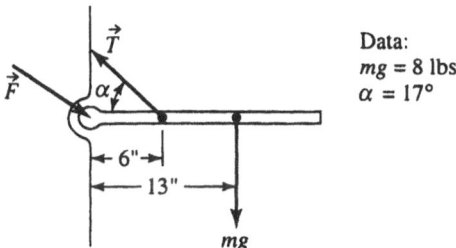

Data:
$mg = 8$ lbs.
$\alpha = 17°$

Using the two equilibrium conditions, determine the tension T in the deltoid muscle, and the vertical and horizontal components of the force exerted by the scapula (shoulder blade) on the humerus.

11. Two tower antenna. Two steel towers support an antenna cable as shown:

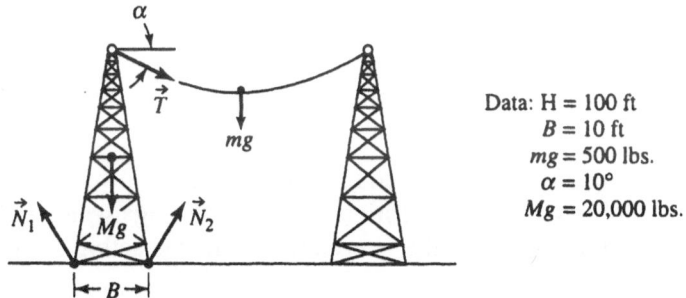

Data: H = 100 ft
B = 10 ft
$mg = 500$ lbs.
$\alpha = 10°$
Mg = 20,000 lbs.

(a) Make a force diagram for the cable. Determine T in terms of mg and α.

(b) Draw a diagram with all the forces acting on the tower (consider only one tower; the setup is symmetric).

(c) State the equilibrium conditions. Compare the number of equations with the number of unknown force components. Explain your findings.

(d) What must be the minimum weight of the left tower so that it does not have to be bolted down at base point 1?

12. Force on an Achilles tendon in a crouching position. In this position, on tiptoe, the Achilles tendon is under considerable tension. A schematic presentation of the geometry is given below. To simplify matters, we neglect the weight of the foot below the ankle joint:

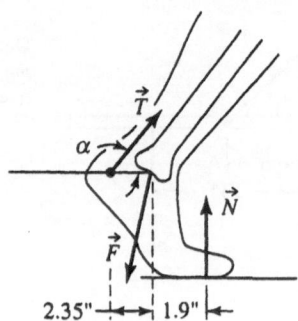

$$\text{Data:} \quad N = \tfrac{1}{2}Mg \text{ (why?)}$$
$$= 100 \text{ lb},$$
$$\alpha = 38°.$$

Find T, as well as the magnitude and direction of $\vec{F}$, from the equilibrium conditions.

13. In the crouching position described in Problem 12, the lower leg as a whole is held in equilibrium through the action of the patellar ligament (attached to the upper tibia and running over the kneecap). The forces acting on the lower leg as a whole are $\vec{N}$, $\vec{R}$, and $\vec{T}$:

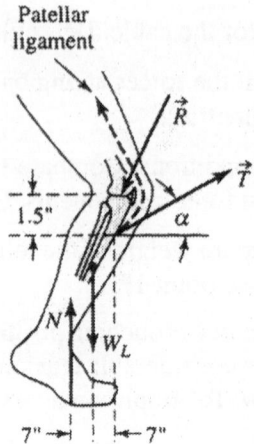

Data: $N = \dfrac{W}{2} = 100$ lb,

$W_L = 20$ lb,

$\alpha = 40°,$ (leg at 45° angle).

From the equilibrium conditions for the lower leg, determine the tension T in the patellar ligament, and the direction and magnitude of $\vec{R}$.

14. Calculate the force on the hip abductor muscle, when a person stands on both feet, as a function of the spacing between the feet. Assume that the body weight W is symmetrically distributed. In what position is the force on the hip abductor muscle equal to zero? (Use the data given in Section 3.3.C(ii).)

15. (a) Suppose a person holds in his right hand a suitcase whose weight is one-third of the body weight W. Compute the force $\vec{F}$ in the hip abductor muscle, and the reaction force $\vec{R}$ on the head of the femur when the body is supported by the left foot. Assume that the suitcase is located 15 ft off the midline of the sacrum. (Use the data given in Section 3.3.C(ii).)

    (b) Repeat the calculations of $\vec{F}$ and $\vec{R}$ for the case that the person carries two suitcases, of weight $W/6$, one in each hand.

16. The stability of a ladder leaning against a wall depends on friction. It is important to understand that the margin of safety decreases as a person moves up the ladder. In this problem you will analyze this point quantitatively.

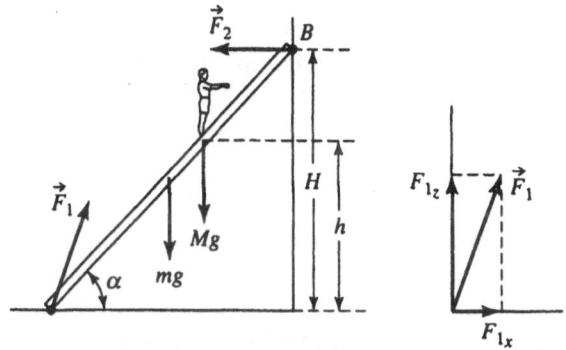

The ladder is of length L and has weight $m\vec{g}$. It leans against a wall at an angle α. Assume that at point B there is *no* friction, so $\vec{F}_2$ is normal to the wall.

At A there is friction, F_{1x}, whose value may be up to μF_{1z} (μ a constant, say 0.5). If the equilibrium conditions call for more than that, *no* equilibrium is possible, and the ladder slips.

A person of weight $M\vec{g}$ stands on the ladder, part way up, at height h:

(a) Find F_{1x}, F_{1z}, and F_2 from the equilibrium conditions.

(b) Plot the increase of the frictional force F_{1x} as the person moves up the ladder.

(c) For $\mu = 0.5$ and $M = 5m$, determine the minimum angle α which allows the person to step up to the top rung of the ladder.

17. Did it ever intrigue you why, in checking your weight on a scale in the doctor's office, or on bathroom scales, it does *not* matter exactly where you stand on the platform? Yet scales are based on the torque condition for equilibrium! They are built as follows:

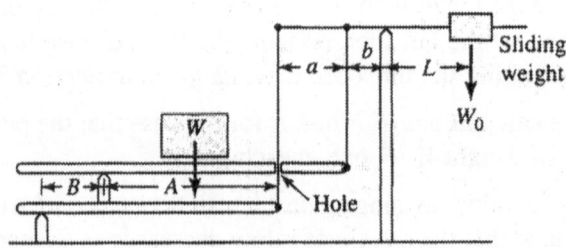

Show that provided $B/A = b/a$, one has the relation $Wb = W_0L$ at equilibriumm *irrespective* of *where* you put the load W on the platform. [To attack this problem, identify first the three components of the system, for which separate equilibrium conditions must be written!]

Stress–Strain. Bending and Shear

18. A long elastic wire can serve as a spring. Suppose a 300 m long piano wire is suspended from the top of the Eiffel Tower, and a mass $M = 100$ kg attached to its lower end. The wire has a radius $r = 1$ mm (10^{-3} m), and its Young's modulus is $E = 2 \times 10^{11}$ N/m^2. Determine the spring constant of the wire and the period T of vertical (up and down) oscillation of the mass M.

19. Calculate the area moment of inertia of a beam with a circular profile of radius R. Show that its value is

$$I_A = \frac{\pi}{4} R^4.$$

20. Suppose a beam of length L is supported as shown and bends under the effect of its *own weight*. Let w be the weight per unit length of the beam; that is, a piece of length $(L - z)$ has a weight $mg = w(L - z)$:

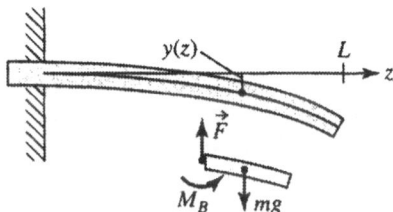

   (a) Determine the bending moment M_B at a distance z from the point of support by considering the torques on the part of the beam beyond z, as shown.

   (b) Using (3-83), write the equation for the deflection $y(z)$ of the beam. Find $y(z)$ and show that the deflection $y(L)$ of the end of the beam is $y(L) = -MgL^3/8EI_A$, M being the total mass of the beam. Compare this with (3-91) and explain the difference.

21. Consider a rectangular beam of length L, supported at both ends, and loaded in the middle with a force P, which is increased to a point where the beam yields at $P = P_B$:

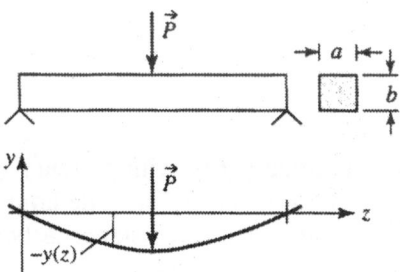

(a) Determine and plot the bending moment $M_B(z)$.

(b) Determine the deflection $y(z)$ of the beam.

(c) Show that at the yielding point, where $P = P_B$, the maximum tensile or compressive stress σ_B in the beam is given by

$$\sigma_B = \frac{3}{2}\left(\frac{L}{ab^2}\right)P_B.$$

22. A car, with a mass M of 1500 kg, is capable of an acceleration a of 2 m/s². Its wheels have a radius R of 30 cm:

(a) Draw a diagram showing all the forces acting on a rear wheel, while the car accelerates. (Since the whole car is much more massive than the wheel, this problem can be treated as an equilibrium problem for the wheel as a reasonable approximation.)

(b) Using the above values of a and M, calculate the torque exerted by the axle on the wheel.

(c) Show that in an axle of radius r_0 the maximum shear stress σ_m in the axle is related to the applied (or exerted) torque τ by

$$\tau = \frac{\pi}{2}\sigma_m r_0^3.$$

(d) The yield stress σ_y of steel is 5×10^8 N/m². The designers do not want σ_m to be in excess of $1/10\sigma_y$. What is the minimum radius r_0 of the axle in this case?

Hydrostatics

23. A U-shaped glass tube is filled partly with mercury ($\rho = 13.6$ g/cm³) and partly with a liquid of unknown density ρ_L. The liquid column has length L, and the level of mercury at the right is a distance Δ below that of the liquid at the left. What is the density of the liquid?

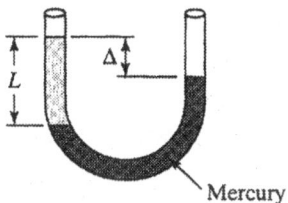

Mercury

24. An imprudent diver straps on his back an air-filled bag containing 5 l of air (at atmospheric pressure, p_0) to give him added buoyancy (against the weight of the equipment). Compare the buoyancy of this air-bag in water, near the surface with its value at a depth of 10 m. Propose a better way to get extra buoyancy. Use: $\rho_{water} = 1$ kg/l. $g = 10$ m/s^2. 1 atm $\simeq$ N/m^2.

25. Consider a balloon of the kind used in sportive or scientific endeavors. The densities of air, hydrogen, and helium are:

$$\rho_{air} = 1.29 \text{ g/l} = 1.29 \text{ kg/m}^3,$$
$$\rho_{H_2} = 0.09 \text{ g/l} = 0.09 \text{ kg/m}^3,$$
$$\rho_{He} = 0.18 \text{ g/l} = 0.18 \text{ kg/m}^3.$$

(a) Explain why helium, although *twice as heavy* as hydrogen, is almost as good a filling gas for balloons as the latter?

(b) Assume the mass of a balloon hull, basket, ballast and men is 1000 kg. What is the minimum radius of a helium-filled balloon needed to get it off the ground?

(c) Suppose the bag containing the helium is made of very strong plastic (e.g., Mylar) and will not stretch once fully inflated. If you wish to design a balloon that will rise to a maximum height of 10 km, what fraction should you inflate the bag to with helium at sea level?

26. Consider the cylindrical fuselage (hull) of a commercial airplane. Assume it flies at an altitude h of 10 km and that the atmospheric *scale height H* is 8 km. Assume that the cabin is kept at ground zero pressure

$$p_0 = 760 \text{ mmHg} = 14.7 \text{ lb/in}^2.$$

(a) What is the outside air pressure at $h = 10$ km?

(b) Assuming the fuselage to be a cylinder of radius $R = 6$ ft, what is the tension in the hull in pounds per foot?

(c) Consider a cabin window of area $A = 10^2$ in^2. What is the total force on the window (and hence on the window frame) under the conditions stated above?

To answer part (b), cut up a cylindrical segment of the hull, and state the equilibrium condition on one of the pieces. Tension is force per unit length along the cut.

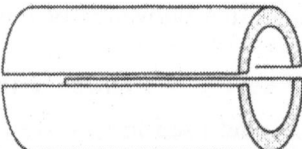

27. Flat tire. You begin to have tire trouble if the pressure inside your tire is less than about 15 psi *above* the outside pressure. Suppose your car weighs 3200 lb, and $\Delta P = 15$ psi. What will be the area of contact between the tire and the road?

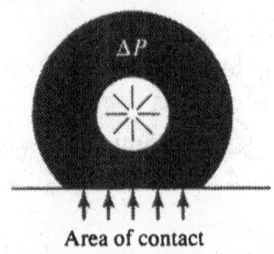

Area of contact

28. The human lung can expel air with a maximum expiratory pressure Δp_{exp} of about 120 mmHg. The Indians of the Amazon make use of this pressure to accelerate a poisoned dart in a blowgun.

(a) Calculate the muzzle velocity of a dart whose mass is 1 g, for a blowgun of length 3 m, and an inner cross sectional area A of 1 cm^2.

(b) In part (a), resistive forces on the dart were neglected. The main resistive force is due to the compression of the air ahead of the dart, and is given by

$$F_{res} \cong \rho_{air} A v^2.$$

Determine the terminal velocity v_∞ of the dart in a very long blowpipe, and the critical distance Z_c (defined in Section 2.7) which the dart must travel to reach a speed close to v_∞. How long would you make a blowgun?

29. The inspirational pressure difference p_{in} which the lung is capable of generating (by maximum expansion of the diaphragm and rib cage) is about 86 mmHg.

   Using this information, and considering the increase of water pressure with depth below surface, show that a person standing upright under water is able to breathe by means of a "snorkel," provided the depth of submersion of the person's mouth is less than 33 in. (A safe depth is only about half this value, since at 33 in, proper ventilation of the lung cannot be maintained.)

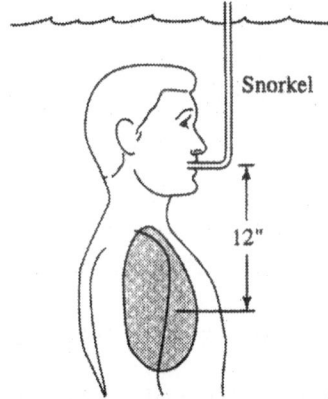

30. In an ultracentrifuge a liquid-filled chamber is spun with high angular velocity about a fixed axis. The problem is to find the hydrostatic pressure $p(R)$ in the fluid at a point a distance R away from the axis of rotation. Let Ω be the angular velocity of the centrifuge, ρ_f the density of the fluid, and consider a volume element of fluid of area A and thickness dR:

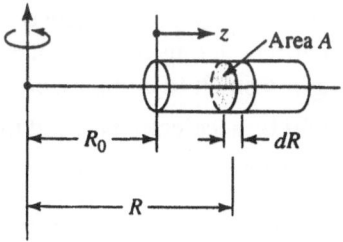

(a) Show that, in equilibrium, the net force on the above-mentioned volume element must be

$$F = \rho_f A \, dR \Omega^2 R$$

and directed radially inward.

(b) This force is the result of the pressure difference between the front and back sides of the volume element. Show that the resulting equation for the pressure $p(R)$ is

$$\frac{dp}{dR} = \rho_f \Omega^2 R.$$

(c) Let the fluid surface be at a distance R_0 from the axis of rotation, and let the pressure be $p_0 = $ atmospheric pressure at this point. Show that the pressure in the fluid is given by

$$p(R) = p_0 + \tfrac{1}{2}\rho_f(R^2 - R_0^2)\Omega^2.$$

(d) Write R as $R = R_0 + z$, so that z is the depth in the fluid. Show that for small z ($z \ll R$), the pressure is approximately given by

$$p(z) = p_0 + \rho_f(\Omega^2 R_0)z.$$

(e) This last expression for $p(z)$ is identical in *form* to the formula for the pressure due to gravity in a stationary fluid, at a distance z below the surface

$$p(z) = p_0 + \rho_f g z.$$

Find the value of $(\Omega^2 R_0)$ for $R_0 = 5$ cm, $(\Omega/2\pi) = 500$ rev/s, and express it as a multiple of g.

31. Suppose a small body of density ρ is submerged in the fluid chamber of an ultracentrifuge. The only forces acting on it are due to the pressure of the fluid (gravity effects are negligibly small in comparison):

(a) Show that the body will be accelerated toward the axis of rotation, or away from it, depending on whether

$$\rho < \rho_f \quad \text{or} \quad \rho > \rho_f.$$

(b) Give an expression for the net buoyancy force acting on the body.

(c) Suppose that the fluid in the chamber has a density which increases linearly with z (salt solution of increasing concentration):

$$\rho_f = \rho_0 \left(1 + \frac{z}{z_0} \right).$$

If macromolecules (proteins), or virus particles, of density $\rho > \rho_0$, are uniformly distributed in the fluid initially, where will they be located after extended centrifugation?

CHAPTER 4

Momentum

4.1 Introduction

The basic laws of mechanics are Newton's three laws which we discussed in Chapter 2. These laws are formulated in terms of the motion of point bodies acting under the influence of applied forces. In Chapter 3 we applied Newton's laws to *stationary, extended* bodies. We showed how to compute the external forces acting on stationary rigid bodies. We also showed how to compute the internal stresses and strains (deformations) of deformable, stationary extended bodies.

In the present chapter we continue our application of Newton's laws to extended bodies. Here, however, we shall consider the *motion* of extended bodies subject to external forces. In general, an extended body can be thought of as an assembly of point masses. Each point mass experiences the effect of both externally applied forces and the internal forces of interaction between the parts of the body.

Under the action of both the internal and external forces the motion of an extended body can be quite complex. In the present chapter we shall demonstrate that though the total motion of the entire body may be quite involved, the motion of a single point, the "center of mass" of the body can be relatively simple. In fact, we shall see that the motion of an extended body consists of the motion of its center of mass, plus orientational motion around the center of mass. The motion of the center of mass is exactly the same as the motion of a point, whose mass is equal to the total mass of the extended body, under the influence of the net external force acting on the body.

By a careful choice of what constitutes "the body" we can often arrange that the net external force is in fact zero. Under these circumstances we will see directly that regardless of the nature of the interactions between parts of the body, the "momentum" of the body as a whole is conserved, i.e., it is constant throughout the motion. This fact permits the formulation of the law of "conservation of momentum." With this law it is possible to give a quantitative description of the essential

250

aspects of the motion of several interacting bodies without knowing the details of the interaction between them. It permits us to describe the result of collisions between bodies, or explosions, or the motion of rockets or the recoil of a boat when we leave it for a pier. We shall apply the conservation of momentum law to the medical effects of collisions on the human body. The conservation of momentum permits us to understand the technique of "ballistocardiography" which is a means of studying, mechanically, the pumping action of the heart using measurements, external to the body, of the changing position of the center of mass of the blood during the cardiac cycle. We shall also apply the ideas of momentum to compute the forces exerted by flowing fluids as they bounce off a surface or move in a curve through a containing tube or pipe.

4.2   Momentum and the Dynamics of a System of Many Particles

Let us begin by examining the application of Newton's laws of motion to each particle in a system of many particles. Such a system is shown diagrammatically in Figure 4.1. Each particle is labeled with a subscript. The jth particle has mass m_j and is located at a position $\vec{r}_j$ relative to some origin O.

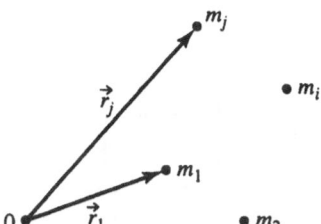

Figure 4.1. A system of many interacting particles.

In general, each particle will experience a force upon it made up of two parts. One part, the external force, is the force exerted on the particle from outside the body. For the jth particle we denote this as $(\vec{F}_j)_{\text{ext}}$. In addition, the jth particle will experience a force due to each of the other particles in the system. We denote by $(\vec{F}_{ij})$ the force that the ith particle exerts on the jth particle. The total force on the jth particle due to all the other particles inside the system is $(\vec{F}_j)_{\text{int}} = \sum_{i \neq j}^{i} (\vec{F}_{ij})$. The term $i = j$ is not included in the sum because the jth particle exerts no force on itself.

We may now write the equations of motion for each of the particles in the system. Particle 1 obeys the equation

$$(\vec{F}_1)_{\text{ext}} + \left(\sum_i \vec{F}_{i1} \right) = m_1 \frac{d\vec{V}_1}{dt}, \tag{4-1}$$

where $\vec{V}_1 = d\vec{r}_1/dt$ is the velocity of particle number 1. In fact, Newton expressed this law of motion by denoting the quantity $m\vec{V}$ as the "momentum" of a particle. If we denote the momentum with the letter p we have that the momentum p of a point body is

$$\vec{p} = m\vec{V}. \tag{4-2}$$

In terms of the momentum then the total force on each point body is equal to the rate of change of the momentum. Thus for body number 1 we have

$$(\vec{F}_1)_{\text{ext}} + \sum_{i \neq 1}^{i} \vec{F}_{i1} = \frac{d\vec{p}_1}{dt}.$$

We can similarly write the equations of motion of each particle in a series of coupled equations

$$(\vec{F}_1)_{\text{ext}} + \left(\sum_{\substack{i \\ i \neq 1}} \vec{F}_{i1} \right) = \frac{d\vec{p}_1}{dt},$$

$$(\vec{F}_2)_{\text{ext}} + \left(\sum_{\substack{i \\ i \neq 2}} \vec{F}_{i2} \right) = \frac{d\vec{p}_2}{dt}, \tag{4-3}$$

$$(\vec{F}_j)_{\text{ext}} + \left(\sum_{\substack{i \\ i \neq j}} \vec{F}_{ij} \right) = \frac{d\vec{p}_j}{dt}.$$

If N is the total number of particles in the system, then, since each equation above is a vector equation, they represent $3N$, second-order, coupled, differential equations in $3N$ unknowns. In general, the motion of each of the particles cannot be calculated. Now we come to the essence of the matter. Though we cannot describe the full motion in detail, we can describe part—an important part—of the motion quite readily. Suppose we add together all the equations in (4-3). This gives

$$\sum_{j} (\vec{F}_j)_{\text{ext}} + \sum_{\substack{j \\ i \neq j}} \left(\sum_{i} \vec{F}_{ij} \right)_{\text{int}} = \frac{d(\vec{p}_1 + \vec{p}_2 + \vec{p}_3 + \cdots)}{dt}. \tag{4-4}$$

Now, let us look carefully at the sum over all the internal forces. We assume that nothing is known in detail as to the nature of these forces. We do not know, for example, how the force of interaction between particles depends upon the distance between them, or how they depend on the speeds of the particles. The only thing we can know about these forces is that they obey Newton's third law. Namely, that the force of action of i upon j is equal and opposite to the force of reaction of j upon i, i.e.,

$$\vec{F}_{ij} = -\vec{F}_{ji}. \tag{4-5}$$

This apparently innocuous statement has vital and very important consequences. It, in fact, permits a vast simplification of (4-4) and will permit the formulation of the law of conservation of momentum. Let us look carefully at a few of the terms in the double sum $\sum_i \sum_j F_{ij}$. These can be written out as

$$\sum_{\substack{i j \\ i \neq j}} \vec{F}_{ij} = 0 + F_{12} + F_{13} + F_{14} + \cdots,$$
$$F_{21} + 0 + F_{23} + F_{24} + \cdots, \tag{4-6}$$
$$F_{31} + F_{32} + 0 + F_{34} + \cdots$$

By examining the terms in this sum, we see that they all cancel in pairs because $\vec{F}_{ij} = -\vec{F}_{ji}$. The net of all the forces of interaction thus gives zero. Thus (4-4), for the rate change of the total momentum of all the particles in the body, can be written as

$$(\vec{F}_{\text{tot}})_{\text{ext}} = \frac{d}{dt} \left(\sum \vec{p}_j \right) = \frac{d\vec{P}}{dt}. \tag{4-7}$$

The total momentum $\vec{P}$ is given by

$$\vec{P} = m_1 \vec{V}_1 + m_2 \vec{V}_2 + m_3 \vec{V}_3 + \cdots. \tag{4-8}$$

Equation (4-7) says that the rate of change of the total momentum of all the particles in the body is equal to the net external force acting upon the body.

We can sharpen even further the interpretation of (4-7). In the following section
we shall show that the total momentum of a system of many particles is equal to
the total mass (M) of the system times the velocity of a single point, the "center of
mass," in the assembly. Thus we shall see that regardless of the complex details of
the motion of each point in the body, the center of mass of the body moves just as
would a point of mass M under the influence of the net external force $(F_{tot})_{ext}$.

4.3   Motion of the Center of Mass of an Extended Body

The total momentum of an extended body is

$$\vec{P} = m_1\vec{V}_1 + m_2\vec{V}_2 + \cdots = \sum m_j\vec{V}_j \qquad (4\text{-}9)$$

$$= m_1\frac{d\vec{r}_1}{dt} + m_2\frac{d\vec{r}_2}{dt} + \cdots$$

$$= \frac{d}{dt}(m_1\vec{r}_1 + m_2\vec{r}_2 + \cdots) = \frac{d}{dt}\left(\sum m_j\vec{r}_j\right). \qquad (4\text{-}10)$$

We may now define the position $\vec{R}$ of the center of mass of the assembly as the
mass weighted average of the positions of each of the constituent particles in the
following way:

$$\vec{R} = \frac{\sum_j m_j\vec{r}_j}{M}, \qquad (4\text{-}11)$$

where $M = \sum_j m_j$ = total mass of the body. The position of the center of mass
then is the vector sum of the position vectors of each particle weighted by a factor
equal to its mass divided by the total mass of the system.

If the body is made up of a continuous distribution of matter, it is more convenient to obtain the center of mass using an integral form for (4-11). The generalization of the discrete sum in (4-11) to an integral expression appropriate to a
continuous distribution of matter, is simply

$$\vec{R} = \frac{1}{M}\int \vec{r}\,dm, \qquad (4\text{-}12)$$

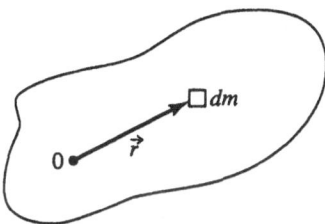

Figure 4.2. Identification of symbols which enter the integral definition for the center of mass.

where dm is a differential element of mass of the body and r is the position of dm relative to some arbitrarily chosen origin as indicated in Figure 4.2. The position of the center of mass in space is independent of the initial choice of origin. We can demonstrate this as follows. In Figure 4.3 we show two origins O and O' separated by a distance $\vec{\ell}$. If we compute the position of the center of mass using origin O, we have

$$\vec{R} = \frac{1}{M} \int \vec{r} \, dm. \tag{4-13}$$

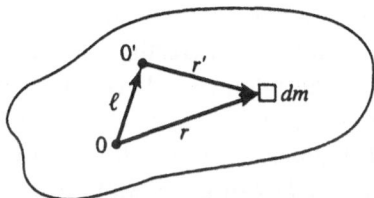

Figure 4.3. The center of mass of a body is located using either origin O, or O'.

On the other hand, the position of the center of mass R' using O' as origin is

$$\vec{R}' = \frac{1}{M} \int \vec{r}' \, dm. \tag{4-14}$$

However, since

$$\vec{r} = \vec{\ell} + \vec{r}' \tag{4-15}$$

we may relate $\vec{R}$ to $\vec{R}'$ by putting (4-15) into Eqn. 4-13 to obtain

$$\vec{R} = \vec{R}' + \vec{\ell}. \tag{4-16}$$

Thus $\vec{R}$ and $\vec{R}'$ are at the same place in the body (see Figure 4.4). Thus we are quite free to choose any arbitrary origin in computing the location of the center of mass.

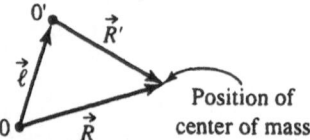

Figure 4.4. Position of center of mass is the same, regardless of the choice of origin.

(i)   Center of Mass of Two Bodies

Let us now compute the location of the center of mass for two simple systems. The first is a system of two bodies (for example, the Earth and its moon) separated some distance r_0 apart. Let the two bodies have masses m_1 and m_2. The location of the center of mass of this example is interesting and useful. If the forces on these particles arise entirely from the interaction between them, the center of mass of the two will move uniformly, i.e., with constant velocity, regardless of the motion of each of the constituent particles. This will be true of the particles and planets which interact according to the law of gravitation, two blocks tied to one another with a spring, or two nuclei being scattered in a high-energy encounter. Let us then find the position of the center of mass for a system of two bodies (see Figure 4.5). According to the definition of the center of mass

$$\vec{R} = \frac{m_1 \vec{r}_1 + m_2 \vec{r}_2}{m_1 + m_2}. \tag{4-17}$$

R is, in fact, located along the line joining the two particles a fractional amount ε along r_0. That is, $\vec{R}$ is at a position

$$\vec{R} = \vec{r}_1 + \varepsilon \vec{r}_0. \tag{4-18}$$

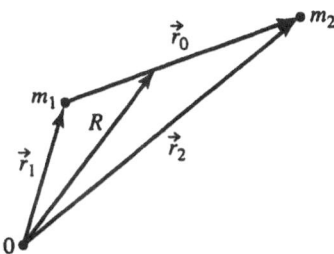

Figure 4.5. Position of two particles and their center of mass $\vec{R}$.

We can obtain the fraction ε from (4-17) as follows. Use the equation $\vec{r}_2 = \vec{r}_1 + \vec{r}_0$ to eliminate $\vec{r}_2$ from (4-17). This then gives

$$\vec{R} = \frac{m_1 \vec{r}_1}{(m_1 + m_2)} + \frac{m_2(\vec{r}_1 + \vec{r}_0)}{(m_1 + m_2)} \tag{4-19}$$

or

$$\vec{R} = \vec{r}_1 + \left(\frac{m_2}{m_1 + m_2}\right)\vec{r}_0. \tag{4-20}$$

Thus we see that the center of mass is along the line $\vec{r}_0$ and is a fractional distance

$$\varepsilon = \left(\frac{m_2}{m_1 + m_2}\right) \tag{4-21}$$

along r_0. If $m_2 \gg m_1$, $\varepsilon \approx 1$, and the center of mass is quite close to mass m_2. Similarly, if $m_1 \gg m_2$, $\varepsilon \approx 0$, and the center of mass is located very near r_1. When $m_1 = m_2$, the center of mass is located half-way between m_1 and m_2.

(ii)   Center of Mass of a Right Triangle

As a second example, we compute the position of the center of mass of a right triangle having constant density ρ. This example will show us how to use the integral formulas for the center of mass. We can carry out the computation with the aid of Figure 4.6. According to the general formula for the position of the center of mass, the x- and y-components are

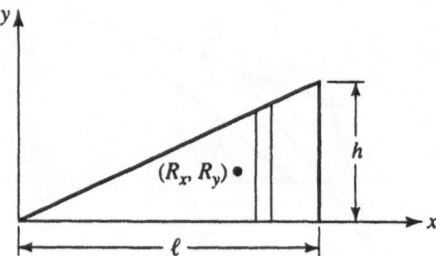

Figure 4.6. Diagram for the computation of the location of the center of mass of a right triangle.

$$R_X = \frac{1}{M} \int_{x=0}^{\ell} x\,dm, \qquad R_y = \frac{1}{M} \int_{y=0}^{h} y\,dm. \qquad (4\text{-}22)$$

All the mass elements in a vertical strip of height y and width dx have the same distance from the origin. Thus, if ρ is the mass per unit area of the triangle we have that

$$dm = \rho y\,dx \qquad (4\text{-}23)$$

and

$$M = \rho A = \tfrac{1}{2}\rho h \ell, \qquad (4\text{-}24)$$

where the area of the triangle A is one-half the base ℓ times the height h of the triangle. Also, since along the hypotenuse y and x are related according to

$$y = \left(\frac{h}{\ell}\right) x, \qquad (4\text{-}25)$$

we see that

$$R_x = \frac{\int \rho y x\,dx}{\tfrac{1}{2}\rho h \ell} = \frac{\rho(h/\ell)\int_0^{\ell} x^2\,dx}{\tfrac{1}{2}\rho h \ell} \qquad (4\text{-}26)$$

or

$$R_x = \frac{2}{\ell^2} \left[\frac{x^3}{3} \right]_0^\ell = \tfrac{2}{3}\ell. \tag{4-27}$$

By carrying out a perfectly similar computation for R_y, one would find that the center of mass is located two-thirds of the distance down from the upper corner of the triangle or one-third of the distance up from the line $y = 0$. Thus

$$R_y = \tfrac{1}{3}h. \tag{4-28}$$

The location then of the center of mass of the triangle is at the point

$$R = (R_x, R_y) = (\tfrac{2}{3}\ell, \tfrac{1}{3}h) \tag{4-29}$$

Now that we see how to locate geometrically the position of the center of mass, let us express (4-7) in terms of the motion of this point. According to (4-7), the rate of change of the total momentum is equal to the net applied external force

$$(\vec{F}_{\text{tot}})_{\text{ext}} = \left(\frac{d\vec{P}}{dt} \right). \tag{4-30}$$

However, from the definition of the center of mass we see that

$$\vec{P} = \frac{d}{dt} (m_1 \vec{r}_1 + m_2 \vec{r}_2 + \cdots) = \frac{d}{dt}(MR).$$

Thus we have that

$$(\vec{F}_{\text{tot}})_{\text{ext}} = \frac{d}{dt} \left(M \frac{d\vec{R}}{dt} \right) \tag{4-31}$$

and if the mass of the system under consideration is constant, this can be written as

$$(F_{\text{tot}})_{\text{ext}} = M \frac{d^2 R}{dt^2}. \tag{4-32}$$

Both (4-31) and (4-32) express the fact that under the action of both external and internal forces, one point on the body, the center of mass, moves as if the total mass of the body was located there; and as if the total external force acted at that point.

Thus, for example, if one throws an extended body like a triangle upward in a gravitational field, the triangle as a whole may tumble and rotate in a complex way. Its center of mass, however, which we located above, travels in a perfect parabola, providing that friction is neglected.

Another important property of the center of mass is that it is the "point of balance" of the body. To put this more rigorously, we may say that if a force $\vec{F}$, equal in magnitude and opposite in direction to the weight ($\vec{W} = M\vec{g}$) of the body, is applied at the position $\vec{R}$ of the center of mass, then the body is in equilibrium; there will be no net force or torque on the body. We can see that this is true using Figure 4.7. Let A be some arbitrary origin of coordinates in the body. Let $\vec{R}$ be the position of the center of mass. Let us compute the total torque exerted by $\vec{F}$ and by each small element of the weight of the body ($\Delta m_j \vec{g}$) about the arbitrary origin A.

$$\vec{T}_A = \sum_j \left(\vec{r}_j \times \Delta m_j \vec{g} \right) + \vec{R} \times \vec{F}.$$

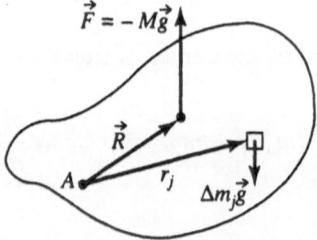

Figure 4.7. Body forces $\Delta m_j g$, balanced by the force $\vec{F} = -\left(\sum m_i \right) g = -Mg$ acting on the center of mass.

Rearranging, we have

$$\vec{T}_A = \left(\sum_j \Delta m_j \vec{r}_j \right) \times \vec{g} + \vec{R} \times \vec{F}$$

But from the definition of the center of mass

$$M\vec{R} = \sum_j \Delta m_j \vec{r}_j,$$

where M is the total mass. Thus

$$\vec{T}_A = \vec{R} \times M\vec{g} + \vec{R} \times \vec{F}.$$

The first term on the right-hand side of this equation is the total torque produced by each element of weight of the body about A, but this term is exactly equal and opposite to $R \times F$ because

$$\vec{F} = -M\vec{g}.$$

Thus $\vec{T}_A = \vec{R} \times M\vec{g} - \vec{R} \times M\vec{g} = 0$. Thus the total torque is zero if $\vec{F} = -M\vec{g}$ acts at the center of mass. Since clearly the net *force* on the body $\vec{F} - M\vec{g}$ is also zero, the entire body is in equilibrium or "in balance" if a force $\vec{F} = -M\vec{g}$ acts through the center of mass.

(iii)   Experimental Determination of the Center of Mass

The fact that the force of gravity, in effect, acts through the center of mass provides a method for determining the location of the center of mass. This method is particularly useful when the shape of the body is irregular and the center of mass is difficult to compute theoretically. In effect, this method consists in finding the point in the body around which the force of gravity exerts no torque. This point can be found experimentally with reference to Figure 4.8. First balance the figure on a knife-edge pivot along one side of the body. Inscribe a vertical line on the figure passing through the pivot. Since the figure is in balance, the center of mass is somewhere along this line AA'. Next balance the figure starting at some other point on the perimeter of the body and again draw a vertical line BB' on the rebalanced body through the pivot. The center of mass lies at the point of intersection of AA' and BB' because the center of mass must be along both AA' and BB'.

It is interesting to observe that Borelli [1], a student of Galileo, used an extension of this procedure in three dimensions to locate the center of mass of the human body in 1679.

The fact that the weight of an extended body acts in effect at the center of mass also provides a means of measuring the weight of parts of the body without

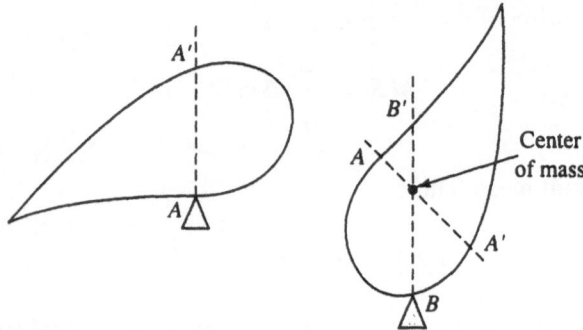

Figure 4.8. Balancing scheme for the experimental determination of the location of the center of mass.

removing them from the patient [2]. For example, one can measure the weight of the leg from knee to foot as follows. Let the subject lie prone on a board supported at one end by a weight scale as shown in Figure 4.9. The center of mass of the entire body is located at some position R_B. Let the scale read some force F_0 with the patient in this position. Next let the subject lift his leg as shown in Figure 4.10.

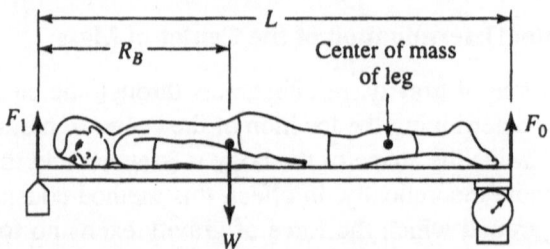

Figure 4.9. Determination of the location of the center of mass of the leg: Position I.

The free-body diagram of the forces acting in the new situation is shown in Figure 4.11.

Here we represent the raising of the leg by first replacing the leg with a "negative leg"; i.e., we imagine that a force W_L equal to the weight of the leg is applied at the center of gravity of the prone leg, but in the direction opposite to the force of gravity. Next we apply a force equal to the weight at the leg at the position just above the knee at center of gravity II. The scale in situation II reads F_0' as the force it exerts. If we take torques about the left hand support in each of Cases I and II,

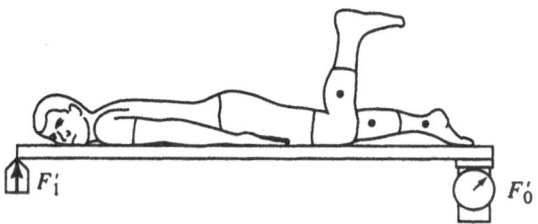

Figure 4.10. Determination of the center of mass of the leg: Position II.

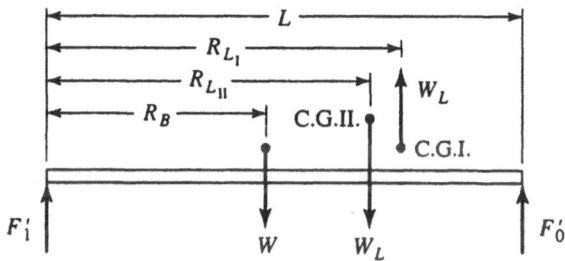

Figure 4.11. Free-body diagram of forces acting in Position II.

we have the two equations for static equilibrium.

$$W R_\mathrm{B} - F_0 L = 0, \quad \text{Case I,} \tag{4-33}$$

$$W R_\mathrm{B} + W_\mathrm{L} R_{\mathrm{L\,II}} - W_\mathrm{L} R_{\mathrm{L\,I}} - F_0' L = 0, \quad \text{Case II.} \tag{4-34}$$

On subtracting these equations, we find

$$W_\mathrm{L}(R_{\mathrm{L\,II}} - R_{\mathrm{L\,I}}) - (F_0' - F_0)L = 0.$$

Thus, the weight of the leg is equal to

$$W_\mathrm{L} = \left(F_0 - F_0'\right) \frac{L}{(R_{\mathrm{L\,I}} - R_{\mathrm{L\,II}})}. \tag{4-35}$$

The weight of the leg is equal to the difference between the two readings of the weight scales times the distance L between the scale and the other end support, divided by the change in position of the center of mass of the leg relative to the supporting board as the leg is rotated upward. Anatomical measurements [2] show that the center of gravity of the leg is located at a point 43% of the distance from the knee joint to the ankle joint. Thus, if ℓ is the measured length from knee to ankle joint $R_{\text{L I}} - R_{\text{L II}} = 0.43\ell$, and the weight of the leg can be found in this method from the equation

$$W_{\text{L}} = \left(F_0 - F_0' \right) \frac{L}{(0.43\ell)}. \tag{4-36}$$

Clearly the same technique can be used to measure the weight of the leg from hip to foot, the trunk of the body, the arms, etc.—provided that the location of the center of gravity of these members is known. (See [2, Appendix A].)

4.4   Conservation of Momentum

Equation 4-30 states that the rate of change of the total momentum $\vec{P}$ of a system of particles is equal to the net external force acting on the system. In many cases it is possible to choose the "system" in such a way that the net external force acting on the system is equal to zero. In this case, regardless of the motion of the various parts of the system, the momentum of the system as a whole remains constant. The momentum is conserved (or constant) when the net force acting on the system is zero. This is the law of conservation of momentum. Algebraically, in terms of (4-30), this is expressed as follows. Since

$$(\vec{F}_{\text{tot}})_{\text{ext}} = \frac{d\vec{P}}{dt},$$

$$\vec{P} = \text{constant} = \vec{p}_1 + \vec{p}_2 + \cdots + \vec{p}_N,$$

when $(\vec{F}_{\text{tot}})_{\text{ext}} = 0$ and where $\vec{p}_1 \ldots \vec{p}_N$ are the momenta of each part of the total system. The conservation of momentum law is very useful in that it permits a determination of the motion of one part of a system if the motion of a second part is known. One need not know the forces acting between the two parts. Let us give some examples.

4.4.A.   The Boy, the Ball, and the Boat

Suppose that a boy is in a rowboat that is stationary in the water. The boy then throws a baseball horizontally from the boat with velocity V_0. If the frictional resistance of the water is neglected, what is the velocity of the boat after the baseball is thrown? See Figure 4.12. If we were to look at this problem from the point of view of Newton's second law, we would have to determine the force that the ball exerted on the boat and boy, as he threw the ball. The use of the conservation of momentum law permits us to solve the problem without knowing this force—provided that we take as our system the ball, the boat, and the boy. There is no net horizontal force on this total system. Thus, in the horizontal plane, the momentum of the system is the same both before and after the ball is thrown. Since the momentum is originally zero, and since the momentum of the ball is $m V_0$ and that of the boat and boy is $M V$, we have that

$$m V_0 - M V = 0$$

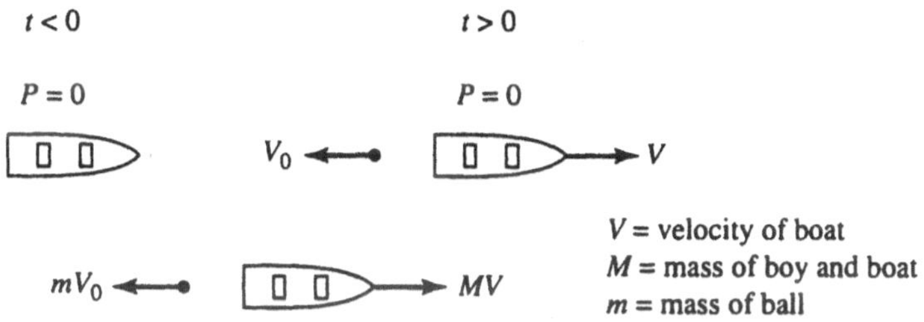

Figure 4.12. Conservation of momentum after throwing a ball from a boat.

That is, the velocity of the boat is less than the velocity of the ball by the fraction (m/M), and the ratio of mass of the ball to that of the boat and boy

$$V = \left(\frac{m}{M}\right) V_0.$$

Thus, we have found the recoil velocity of the boy and boat without any knowledge of the force that the ball exerted upon them.

An equivalent expression of the law of the conservation is to be found in (4-31). Since the total momentum is equal to the total mass of a system times the velocity of the center of mass (dR/dt), we have from (4-31) that

$$(F_{\text{tot}})_{\text{ext}} = \frac{d}{dt}\left(M\frac{dR}{dt}\right).$$

Thus, when the external force is zero, the quantity $M(dR/dt)$ is constant; i.e., the center of mass of the body moves with constant velocity. (Of course, since $\vec{F}$ and

$\vec{R}$ are both vectors, this statement also applies to the components of $\vec{F}$ and $\vec{R}$.) We may use this form of the law of conservation of momentum to solve the problem of the oscillator in a box.

4.4.B.   The Oscillator in a Box

Suppose inside a box we have a mass m attached by a spring to a post in the box as indicated in Figure 4.13. Let there be no friction between the box and table top. Also let there be no friction between the mass m and the inside of the box. We will disregard the weight of the spring. Suppose the mass m is set into oscillation so that its position, *relative to the box*, oscillates in accordance with the formula

$$\Delta z = z_0 \sin \omega t.$$

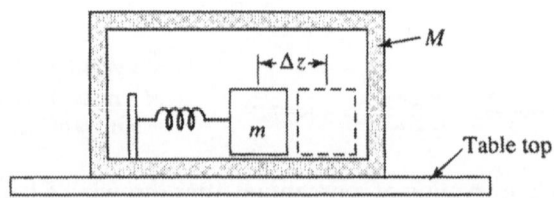

Figure 4.13. The oscillator in a box.

Here ω is the angular frequency of the oscillator. We now ask: What is the motion of the box of mass M as the oscillator executes its motion?

If we were to try to solve this problem using Newton's second law of motion, we would have to compute the force that the spring exerts on the box at the post. However, by using the fact that the system of box, *plus* oscillator, has no net external force acting upon it, we can solve this problem using the fact that the center of mass of the total system moves with constant speed.

Let us measure the position of the center of mass of the oscillator (z_m) and the box (z_b) as shown in Figure 4.14. These positions are measured relative to an inertial coordinate system fixed on the table top. The center of mass of (Z_c), the total system of the box plus oscillator, is at the point

$$Z_c = \frac{M z_b + m z_m}{(M + m)}. \tag{4-37}$$

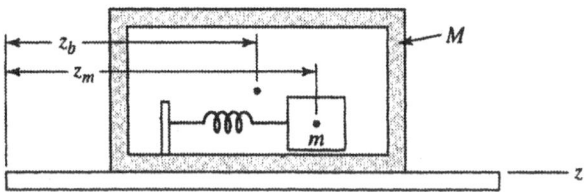

Figure 4.14. Positions of center of mass of the box and oscillator, relative to the table top.

Since the total system of the box plus oscillator have not net force on them Z_c must be in uniform motion, i.e., (dZ_c/dt) is constant. This constant is determined by the initial condition, i.e., how the system motion is started

$$\left(\frac{dZ_c}{dt}\right) = V_0 = \text{constant.} \tag{4-38}$$

Differentiating (4-37) gives

$$\left(\frac{dZ_c}{dt}\right) = V_0 = \left(\frac{M}{M+m}\right)\left(\frac{dz_b}{dt}\right) + \left(\frac{m}{M+m}\right)\left(\frac{dz_m}{dt}\right) \tag{4-39}$$

In this equation we have an expression for the velocity of the box (dz_b/dt) in terms of the velocity of the oscillator relative to the table top. Our information, however, on the position and velocity of the oscillator is given relative to the *box*. We can compute (dz_m/dt) relative to the table top quite simply as follows:

$$z_m = z_{box} + \delta z + z_0 \sin \omega t. \tag{4-40}$$

Here δz is a constant equal to the distance between the center of mass of the box and that of the oscillator when $t = 0$.
    From (4-40) we have that

$$\frac{dz_m}{dt} = \left(\frac{dz_b}{dt}\right) + z_0\omega \cos \omega t. \tag{4-41}$$

Putting this into (4-39), we have

$$V_0 = \left(\frac{M}{M+m}\right)\left(\frac{dz_b}{dt}\right) + \left(\frac{m}{M+m}\right)\left(\frac{dz_b}{dt}\right) + z_0\omega\left(\frac{m}{M+m}\right)\cos\omega t.$$

Solving for dz_b/dt, we have

$$\left(\frac{dz_b}{dt}\right) = V_0 - \frac{z_0\omega m}{(M+m)}\cos\omega t. \tag{4-42}$$

This gives us the motion of the box. We can obtain the motion of the oscillator relative to the table top by using (4-42) in (4-41)

$$\left(\frac{dz_m}{dt}\right) = V_0 + z_0\omega\left(1 - \frac{m}{M+m}\right)\cos\omega t. \tag{4-43}$$

This is the general solution for the velocity of both the box and the oscillator.

In the case that the oscillator mass is small compared to the mass of the box, i.e., $m/(M+m) \ll 1$. Equations (4-42) and (4-43) can be written as

$$\left(\frac{dz_b}{dt}\right) = V_0 - \frac{z_0\omega m}{(M+m)}\cos\omega t, \tag{4-44}$$

$$\left(\frac{dz_m}{dt}\right) \cong V_0 + z_0\omega\cos\omega t. \tag{4-45}$$

Both box and oscillator move with an average speed V_0, the constant initial velocity of the center of mass. In addition, the two bodies oscillate in opposite phase (one moves to the right as the other moves to the left) with different amplitudes. The velocity of the box is smaller at each instant than that of the oscillator by a factor $m/(M+m)$.

In the special case $m = M$, we have

$$\left(\frac{dz_m}{dt}\right) = V_0 + \frac{z_0\omega}{2}\cos\omega t, \tag{4-46}$$

$$\left(\frac{dz_b}{dt}\right) = V_0 - \frac{z_0\omega}{2}\cos\omega t. \tag{4-47}$$

Both box and oscillator move with equal oscillating amplitudes and their motion is completely out of phase.

This problem shows that by knowing the mass of each of the two bodies, a measurement of the oscillating velocity of the encircling box can tell us about the motion of the matter inside it. This, in fact, is the basis of a method of studying the oscillating movement of blood. By measuring the external movement of a subject's body, one can deduce the movement of the center of mass of the blood. This technique is known as ballistocardiography and we shall discuss this technique in the following section.

Our solutions ((4-42) and (4-43)) for the motion of the box and the oscillator were obtained using the law of the conservation of momentum. This was applicable because the system chosen for consideration was the sum of the oscillator and the box. This system has no net external force acting upon it. If we had considered the box alone as the system, we would have had to compute (in effect) the force that the spring exerts on the box in order to obtain the motion.

There is one further wrinkle in the oscillator-in-the-box problem which is interesting to examine. Suppose that the oscillator moves not in a horizontal plane, but moves up and down in the gravitational field—as is indicated in Figure 4.15. Clearly, as the mass m oscillates up and down, the reaction force F will also oscillate. This is precisely the situation which occurs if one stands on a bathroom scale and moves the body up and down. One can, in fact, measure the force F by placing the box on a weight scale.

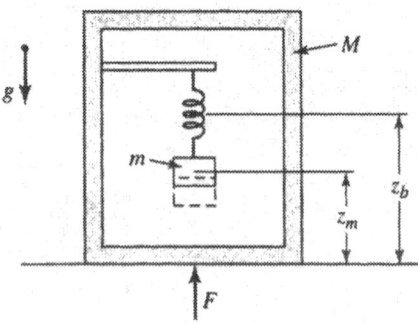

Figure 4.15. Mass oscillates vertically inside a box.

Let us now calculate the reaction force F on the box as the oscillator inside it moves up and down. We shall assume that the box of mass M does not move, i.e., that the table top or the top of the weight scale is essentially stationary. According to (4-31):

$$(F_{\text{tot}})_{\text{ext}} = \frac{d}{dt}\left[(M+m)\frac{dR}{dt}\right]. \tag{4-48}$$

Here M and m are the masses of the box and the oscillating block. Again we neglect the weight of the spring. In (4-48) the *total external* force acting on the box plus oscillator consists of the reaction force from the table (F) and the force of gravity acting on m and on M. Thus we have

$$F - (M+m)g = (M+m)\frac{d^2R}{dt^2}. \tag{4-49}$$

R is the location of the center of mass of the system of the box plus oscillator

$$R = \frac{Mz_b + mz_m}{(M+m)} \tag{4-50}$$

Since the box is assumed not to move (table top is rigid), z_b = constant and only z_m changes with time. Again, as we assumed above, let the oscillator move in accordance with

$$z_m = z_{m_0} + z_0 \sin \omega t. \tag{4-51}$$

z_0 is the amplitude of the oscillation and ω is the angular frequency of the oscillator. z_{m_0} is the average position of the oscillator. Using (4-50) and (4-51), we have

$$\frac{d^2R}{dt^2} = \left(\frac{m}{M+m}\right)\frac{d^2(z_0 \sin \omega t)}{dt^2}$$

or

$$\frac{d^2R}{dt^2} = -\left(\frac{m}{M+m}\right)z_0\omega^2 \sin \omega t \tag{4-52}$$

On putting this into (4-49), we have the following result for the total reaction force.

$$F = (M+m)g - mz_0\omega^2 \sin \omega t. \tag{4-53}$$

The reaction force has two parts: a steady part equal to the weight of the box plus oscillator, and a fluctuating part. The fluctuating part of the reaction force has amplitude $mz_0\omega^2$. This corresponds to the force needed to give the mass m an acceleration $z_0\omega^2$. Alternatively, since $z_0\omega^2 = (z_0\omega)\omega$, and since $z_0\omega$ is the maximum velocity V_m and $\omega = 2\pi/T$, the amplitude of the fluctuating force can be written as $mz_0V_m2\pi/T$. This is the mass of the oscillator times an acceleration equal to 2π times the maximum velocity divided by the period of the motion.

If one were to put such a box with an oscillator inside it on a weight scale, the pointer on the scale would fluctuate in response to the oscillations. The amplitude of the fluctuations of the pointer gives information on the amplitude of the movement of the oscillator.

This example is not an entirely academic one. If one stands on a weight scale with a very accurate dial reading, he will observe that the reading will fluctuate around the average weight reading in response to the heart beat. This will be the case regardless how careful one is to stand perfectly still. The fluctuation of the dial scale is the result of the movement of the center of mass of the blood as the heart goes through its cardiac cycle. In fact, methods have been developed to detect with great accuracy the motion of the center of mass of the blood as a means of measuring the volume of blood pumped by the heart per cycle. These methods fall under the subject of ballistocardiography.

4.5 Ballistocardiography

4.5.A. Introduction

We may begin our discussion of ballistocardiography by estimating the fluctuation in the dial reading as the weight of the body is measured.

During the cardiac cycle, on average, the heart squeezes about 65 g of blood on each contraction into the arterial circulation. The center of mass of this blood moves about 7 cm. During "diastole" the heart refills with blood entering from the vena cava and pulmonary veins. We can expect that while the center of mass of the blood certainly does not oscillate as a perfect sinusoid, that it can so be approximated. The angular frequency ω of the sinusoid is

$$\omega = (2\pi/T),$$

where $T \approx 0.8$ s, the time between heart beats. The amplitude factor mz_0 can be expected to be equal roughly to 65 g $\times$ 7 cm $\sim$ 450 g cm. In fact, detailed experimental and theoretical determinations of the movement of blood in the body show that this estimate for mz_0 is consistent with more accurate determinations [3]. Let us therefore regard the beating of the heart and consequent motion of the blood as equivalent to an oscillator inside the body, for which the product of mass times displacement varies in time in accordance with the equation

$$mz = mz_0 \sin \omega t = 450 \sin\left(\frac{2\pi t}{T}\right) \text{ g cm}, \tag{4-54}$$

$T = 0.8$ s This is indicated schematically in Figure 4.16.

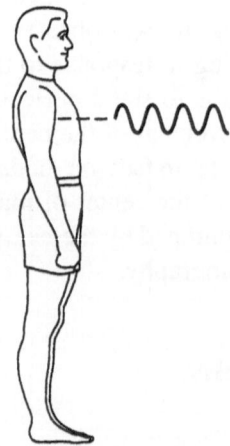

Figure 4.16. Diagrammatic representation of the motion of the center of mass of blood in a standing subject.

The force registered by the weight scale is given according to (4-53) by

$$F = W - mz_0\omega^2 \sin \omega t, \tag{4-55}$$

where W is the total body weight. The amplitude of the fluctuation in the dial reading is

$$(\Delta F) \equiv mz_0\omega^2 = 450 \left(\frac{2\pi}{T}\right)^2 \text{ dyn}. \tag{4-56}$$

Using $T = 0.8$ s, we find that the amplitude of the force fluctuation is

$$\Delta F = (450) \left(\frac{2\pi}{0.8} \right)^2 = 2.76 \times 10^4 \text{ dyn.} \tag{4-57}$$

This is equal to the weight of a mass of 28.5 g. Since the body weight is typically about 75 kg, we see that the fluctuation produced by the movement of blood is a small fraction of the total weight

$$\frac{\Delta F}{F} = \frac{28.5}{75 \times 10^3} \approx 0.38 \times 10^{-3}. \tag{4-58}$$

The fluctuation amplitude then is only about 0.4×10^{-3} of the total weight. In the case of a subject weighing 75 kg = 165 lb, this fluctuation corresponds to 0.06 lb or about 1 oz.

Despite the smallness of this effect it can be observed on a sensitive spring scale. The first evidence of this effect was reported by John W. Gordon in 1877 [4]. A historical investigation [3] of Mr. Gordon reveals that he was an English lawyer with scientific interests who wrote papers on optics and the theory of the gyroscope. He was also King's Counsel in 1914 and Secretary of the Royal Microscopic Society. Gordon's discovery was first interpreted by C. Trotter in the same 1877 volume of the *Journal of Anatomy and Physiology* [5]. Trotter suggested that the observations of Gordon were due to motion of the body's center of gravity.

4.5.B. Elementary Anatomy and Physiology of the Heart

The elementary considerations given above make clear that the motion of the center of mass of the blood contains information on the "stroke volume"; i.e., the amount of blood squeezed into the arterial circulation at each heart cycle. The "cardiac output" is the mean volume flow into the aorta per minute. In the average resting state the cardiac output is 5000 cm^3/min. This implies a stroke volume of 5000 cm^3/75 $\approx$ 65 cm^3/stroke. The actual cardiac output and stroke volume can vary widely depending upon the degree of muscular activity. The stroke volume can be as low as $\sim$ 35 cm^3. The stroke volume is a useful indication of the functioning of the cardiovascular system. As a result, methods have been developed to detect this motion in a more sensitive and direct manner. These techniques permit a measurement of the motion of the center of mass throughout the cardiac cycle.

To obtain the ballistocardiogram, the subject lies down on a light board or cot which is suspended horizontally by means of a nearly friction-free support. We

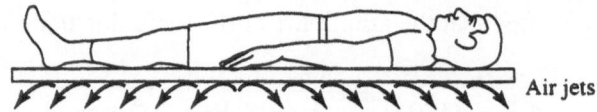

Figure 4.17. Suspension of subject in ballistocardiography

can imagine, for example, that the board or cot is suspended by air jets as in a two-dimensional air track. (See Figure 4.17.) In this horizontal position we have a situation similar to the oscillator moving horizontally in a box. In this case, the body is the enclosing box and the movement of the center of mass of the blood is the oscillating mass. The expulsion of blood from the heart during the cardiac cycle produces a pulsation in the location of the center of mass of the blood.

By referring to the diagram of the heart (Figure 4.18), we may describe the cardiac cycle [6]. We can begin the cardiac cycle as the blood from the venous circulation enters the right atrium through three tubes: the inferior vena cava, the superior vena cava, and the coronary sinus. These vessels deliver, respectively, venous blood from the lower and upper parts of the body and the venous circulation of the heart. After filling, the blood in the right atrium is pumped into the right ventricle through the atrioventricular tricuspid valve. Contraction of the right ventricle closes this valve and forces the blood into the pulmonary artery through the semilunar pulmonary valve. The pulmonary artery branches, sending the venous, partly oxygenated, blood to the left and right lungs. In the lungs the blood is oxygenated and returns to the heart via four blood vessels, two right pulmonary veins and two left pulmonary veins. These four blood vessels deliver oxygenated blood into the left atrium. Contraction of the left atrium forces its contents through the mitral valve into the left ventricle. The left ventricle is the most powerful pumping element in the heart. Its muscular wall is three or four times thicker than that of the right ventricle. The contraction of the left ventricle closes the mitral valve and drives the blood through the aortic valve into the aorta which is the great artery whose branches supply blood to the entire body. The aorta first ascends vertically, then abruptly turns downward forming an inverted U at the aortic arch. Also, a part of the output of the left ventricle goes through the cardiac arteries which provide vitally needed oxygen and metabolites to the heart muscle itself. The development of atheromas, which block the flow of blood through parts of the network of arteries and veins which nourish the myocardium or heart muscle, leads to necrosis of the associated muscle. This is a myocardial infarction.

While the stroke volume can vary widely, let us assume a value of $\sim 65\ \mathrm{cm}^3$. This volume is ejected during the contraction (systole) in a short pulse lasting

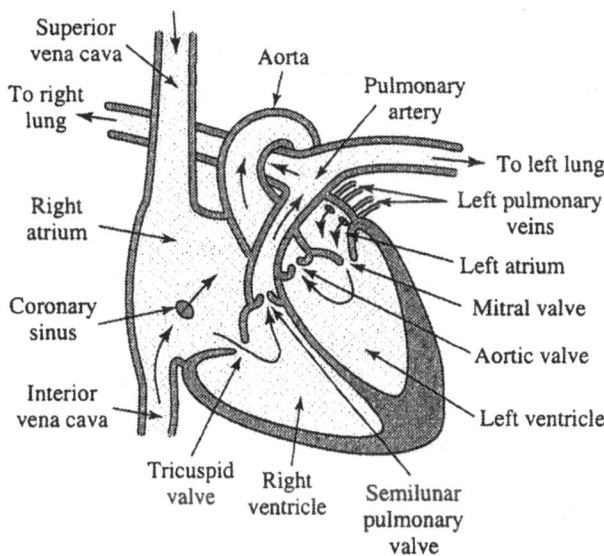

Figure 4.18. Diagram of the heart showing principal structural elements involved in the cardiac cycle.

about 0.13 s. The maximum output rate of flow of blood $\dot{Q}$ from the heart when the stroke volume is 65 cm^3 is roughly 500 cm^3/s. The aorta has a diameter of about 2.5 cm. The speed v with which the blood volume flows through the aorta is related to $\dot{Q}$ by $\dot{Q} = Av$ where A is the cross-sectional area of the aorta. Thus we may estimate that the maximum velocity of blood flow through the aorta is roughly 100 cm/s. These numbers will be useful in later analyses.

4.5.C. The Ballistocardiogram

We now examine how the motion of the center of mass of the blood shows itself in the motion of the frame of the subject's body as he lies on the horizontal, suspended cot. The center of mass of the entire system of subject plus blood is located at the position Z where

$$Z = \frac{M_s Z_s + m_b Z_b}{(M_s + m_b)}. \tag{4-59}$$

Here, all the positions are measured relative to an inertial coordinate system fixed in the laboratory. Z_s is the position of the center of mass of the cot, plus the frame of the subject's body, without considering the blood. M_s is the mass of the subject, plus cot minus the mass of the blood. Z_b is the location of the center of mass of the blood, and m_b is the mass of the blood.

Since the entire system of body plus blood is subject to no net external force, it follows that the center of mass of the entire system must remain fixed. Thus changes in the position of the center of mass of the blood must be reflected in changes in the location of the external frame of the body in accordance with the following equations:

$$\Delta Z = 0. \tag{4-60}$$

Therefore,

$$\Delta(m_b z_b) = -\Delta(M_s Z_s) \tag{4-61}$$

or, ΔZ_s changes in the position of the frame of the body are related to changes in the center of mass of the blood in accordance with the formula

$$\Delta Z_s = -\frac{\Delta(m_b Z_b)}{M_s}. \tag{4-62}$$

As we indicated above in (4-54), an approximate expression for $\Delta(m_b Z_b)$ is given by

$$\Delta(m_b Z_b) \cong 450 \sin(2\pi t/T) \text{ g cm}. \tag{4-63}$$

Thus, we can expect that the motion of the subject plus cot will amount, for a 160 lb man plus cot, to

$$\Delta Z_s = \frac{450 \sin(2\pi t/T)}{(160)(2.2) \times 10^3} \text{ cm}. \tag{4-64}$$

That is,

$$\Delta Z_s \sim 6 \times 10^{-3} \sin\left(\frac{2\pi t}{T}\right) \text{ cm}, \tag{4-65}$$

$$T = 0.8 \text{ s}.$$

The amplitude of the oscillation of the body frame suspended horizontally is thus about 0.06 mm. This is a small distance. Nevertheless, it can be measured quite accurately even in the presence of other disturbing vibrations because the motion is confined to a rather narrow range of frequencies. Thus one can use, in the measurement, filters that do not register frequencies substantially different from those corresponding to the cardiac cycle frequencies. In addition to measuring the position of the body frame, one can use sensitive velocimeters and accelerometers to measure the velocity and acceleration of the body frame. These quantities are proportional to the first and second derivatives of the position of the center of mass of the blood. As a result, the velocity and acceleration are quite sensitive to departures from perfect sinusoidal motion.

In Figure 4.19 we see the changes in position ΔZ_s of the subject frame as a function of time in an actual ballistocardiographic recording [3], [7]. This figure shows that the magnitude of the displacement is about 50 microns (= 0.005 cm). This compares well with the figure of 0.006 cm estimated above. We see also that the displacement versus time is crudely sinusoidal.

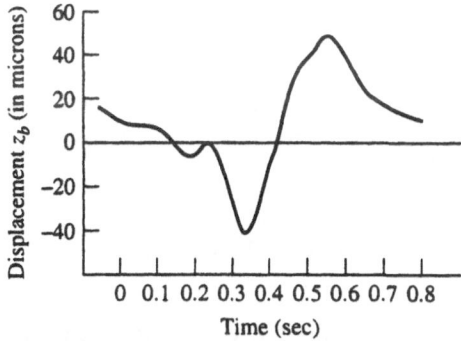

Figure 4.19. Displacement of subject frame versus time (redrawn from Starr and Noordergraaf [3]).

The complexity of the motion of the center of mass of the blood is most clearly seen in the data on the acceleration of the subject frame as a function of time. This is illustrated in Figure 4.20 [3], [7].

We observe from this figure that the magnitude of the acceleration is about 3 "milli-g." Since $g = 980$ cm/s^2, this acceleration is about 3 cm/s^2. It is possible to show that this value is roughly consistent with the stroke volume and the maximum flow rate of blood out of the left ventricle into the aorta and down the descending side of the aortic arch. Let us assume that the heart and aorta are fixed

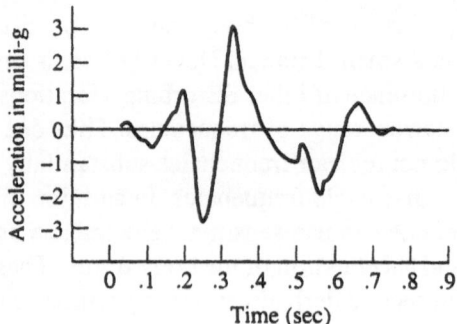

Figure 4.20. Acceleration of subject frame versus time (redrawn from Starr and Noordergraaf [3]).

relative to the subject frame. As the flowing blood rounds the aortic arch each element of mass Δm undergoes a change in velocity equal to $2v$ because the direction of the velocity is reversed. The change in momentum of the element Δm is

$$\Delta P = \Delta m(2v). \qquad (4\text{-}66)$$

Thus each differential element Δm has its momentum changed by this amount. The rate at which the momentum changes is the force that the aorta exerts on the blood to make it turn. This is F_{AB}

$$F_{AB} = 2\left(\frac{dm}{dt}\right)v. \qquad (4\text{-}67)$$

Alternatively, the force that the blood exerts on the aorta is $F_{BA} = -F_{AB}$. Thus, the flowing blood exerts a headward force on the aorta and the body frame as it turns the aortic arch. The resultant acceleration of the subject produced by this motion is

$$\frac{d^2 Z_s}{dt^2} = \frac{1}{M_b}F_{BA} = -\frac{2}{M_b}\left(\frac{dm}{dt}\right)v. \qquad (4\text{-}68)$$

Now the flow rate of the blood, i.e., the volume output of the heart per second, is $\dot{Q}$, as we defined earlier. This is related to (dm/dt) by

$$\left(\frac{dm}{dt}\right) = \rho\dot{Q}, \qquad (4\text{-}69)$$

where ρ is the mass density of the blood, i.e., about 1 g/cm^3. Also, we saw that the velocity of flow in the aorta v is related to $\dot{Q}$ by

$$\dot{Q} = Av, \tag{4-70}$$

where A is the cross-sectional area of the aorta. Thus, we can write the acceleration of the body produced by the flow of blood around the aortic arch in any of the following ways:

$$\frac{d^2 Z_s}{dt^2} = -\frac{2}{M_b} \rho \dot{Q} v \tag{4-71}$$

or

$$\frac{d^2 Z_s}{dt^2} = -\frac{2}{M_b} \frac{\rho \dot{Q}^2}{A}. \tag{4-72}$$

This last formula shows that this contribution to the acceleration of the body frame is proportional to the square of the rate at which blood volume is pumped out of the heart. We can examine the importance of this effect by putting in the numbers for the various quantities

$$M_b \approx 160 \text{ lb} = 72.5 \times 10^3 \text{ g},$$
$$A \approx 4.9 \text{ cm}^2,$$
$$\rho \approx 1 \text{ g/cm}^3,$$
$$\dot{Q} \approx 500 \text{ cm}^3/\text{s}.$$

Thus, we compute

$$\frac{d^2 Z_s}{dt^2} \cong 1.4 \text{ cm/s}^2.$$

Figure 4.19 shows that the maximum acceleration in the ballistocardiograph is in fact about 3 cm/s^2. We, therefore, can conclude that the acceleration of the subject frame produced as the blood rounds the aortic arch is, in fact, an important contribution to the net acceleration of the system frame in the ballistocardiogram. It indicates that the acceleration traces in the ballistocardiogram contain information on the rate of ejection of blood volume into the aorta. It is important to remember, in comparing with the ballistocardiogram experimental traces, that we are assuming in this calculation that the heart itself and aorta remain stationary relative to the subject frame during ejection. This is certainly not true, and this motion of the heart will also play a role in the detailed analysis of the acceleration of the subject frame.

In Figure 4.21 we show on the same time scale the position and acceleration traces of the ballistocardiogram. We also show on this time scale the electrocardiogram and the phonocardiogram records of the heart. The electrocardiogram measures the electrical activity of the heart as measured by the potential differences between two points on the body surface. The phonocardiogram measures the heart sounds as heard with a stethoscope. We discuss those elements of the cardiac cycle associated with each feature of the phonocardiogram.

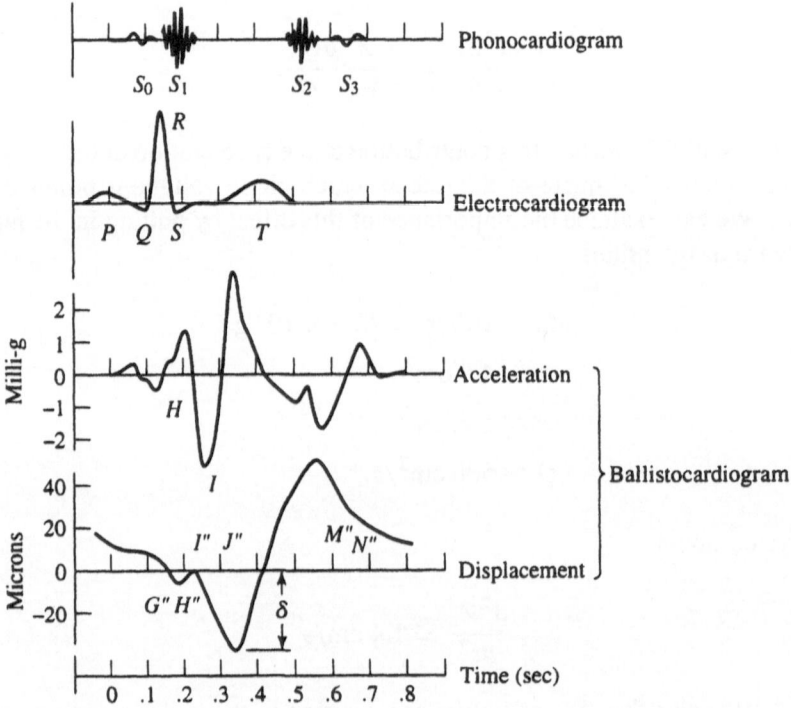

Figure 4.21. Ballistocardiogram, phonocardiogram and electrocardiogram traces (redrawn from Scarborough et al. [7], and Starr and Noordergraaf [3]).

The first heart sound labeled S_1 in the phonocardiogram trace corresponds to the closure of the atrioventricular valves. As the valve starts to close by the equalization of pressure in the ventricle relative to the atrium, the valve starts to vibrate. The mitral valve of the left ventricle and the tricuspid valve of the right ventricle close almost simultaneously giving rise to a single sound which follows immediately after the QRS wave in the electrocardiogram. The second heart sound S_2 in the phonocardiogram is caused by the closure of the aortic and pulmonary valves just after the T wave in the electrocardiogram. After systole (contraction of the left ventricle), when the pressure in the ventricle falls below the arterial pressure, the gradient in pressure closes the outlet valves. The shutting of the valves produces S_2, a vibration which is propagated as waves along the arteries and can be heard easily at the carotid artery. Normally the tissues transmit the vibration of the closing valves and these can be heard at the surface of the thorax.

The electrocardiogram is a measure of the electrical activity associated with the "depolarization waves" which activate the contraction of the muscle cells in the heart. The "P wave" in the electrocardiogram corresponds to the movement of the polarization wave down the right and left atria. The "QRS" segment represents the flow of the electrical excitation into the ventricles. The "T wave" corresponds to the electrical activity associated with electrical repolarization of the depolarized muscle cells. It is, of course, important to realize that the electrocardiogram reflects primarily the electrical excitation of the heart muscle and does not indicate the pumping effectiveness of the heart. This is why the ballistocardiogram can be particularly useful. It indicates the effectiveness of the heart as a pump.

We now examine the significance of some of the features of the ballistocardiogram. It should be kept in mind, in this discussion, that effects such as the recoil of the heart itself, and the filling of blood vessels throughout the body determine the location of the center of mass of the blood. Thus the ballistocardiogram actually represents the sum of several effects and is not so readily interpretable in terms of individual events of the cardiac cycle as are phonocardiograms or electrocardiograms.

At the G'' point in the displacement curve in Figure 4.21, the blood flows from the atria to the ventricles. The segment $H-I$ in the acceleration trace represents the start of the ejection of blood upward and the recoil of the heart downward. Ejection of blood into the aorta continues until about the point M', although the rate at which blood flows out of the heart changes markedly during the period between the points H'' and M''. The point H'' corresponds to the third heart sound S_3. This sound is produced as the blood from the two atria rushes through the atrioventricular valves and fills the ventricles.

It is difficult to make a detailed quantitative analysis of the various features of the displacement and acceleration data in this figure because they measure the position of the center of mass of the blood, and the heart and aorta are not stationary relative to the body, particularly during systole. Nevertheless, careful analysis of these data (see [3, p. 183]) indicates that in the displacement trace the center of mass of blood in fact moves about 7.5 cm as the ballistocardiogram moves to the point J''. It is also known that at this point about half of the stroke volume is ejected [3]. If we, therefore, call δ the distance on the displacement ballistocardiogram between the base line and the minimum at J'', we can write, using the condition that the center of mass of the total system does not move, the following equation:

$$\tfrac{1}{2} \text{ (stroke volume)} \times \text{density} \times 7.5 \text{ cm} = M_s \delta,$$

$$\text{stroke volume} = 2 \left(\frac{M_s}{\rho} \right) \delta / (7.5) \text{ in cm}^3,$$

where δ is measured in cm. In the ballistocardiogram of Figure 4.25, $\delta \cong 40$ microns. Thus, the stroke volume associated with this subject is

$$\text{stroke volume} = 2 \left(\frac{M_s}{\rho} \right) \left(\frac{40 \times 10^{-4}}{7.5} \right).$$

Assuming we deal with a 160 lb man, this corresponds to

$$\text{stroke volume} = 77 \text{ cm}^3,$$

which is in satisfactory agreement with other estimates of the stroke volume.

4.6   Impulse and the Change of Momentum

4.6.A.   Introduction and Definition of Impulse

In the previous section we considered a number of examples in which we calculated the motion of part of a system from a knowledge of the motion of another part of the system. These calculations were possible without knowing the force of interaction between the parts of the system because the total system was chosen so that the net external force on it was zero. In essence then we computed momentum changes while not knowing the forces responsible for these changes.

In the present section we, in essence, follow the reverse procedure. We will compute the forces from observations of the changes in momentum. To be more precise, we will consider systems in which a net force is acting. We will compute this force from a knowledge of the changes in momentum of the system. The means, for this line of analysis, is immediately at hand in Eqn. 4-7, for the rate of change of momentum of the center of mass of a body

$$\vec{F}_{\text{tot}} = \left(\frac{d\vec{P}}{dt}\right). \tag{4-73}$$

A single integration of this equation shows that the change in momentum $\vec{P}$ between the initial momentum $\vec{P}_i$ and final momentum $\vec{P}_f$ is nothing other than the time integral of the force acting viz

$$\int_{t_i}^{t_f} \vec{F}_{\text{tot}}(t)\,dt = (\vec{P}_f - \vec{P}_i), \tag{4-74}$$

where t_f is the time at which the momentum $\vec{P}_f$ is measured and t_i is the time at which the momentum $\vec{P}_i$ is measured. The integral

$$\int_{t_i}^{t_f} F\,dt$$

is called the impulse. Equation (4-74) is of course a perfectly general result. It applies regardless of whether the force is weak and applied over a long time, or strong and applied for a short time. The great utility of (4-74) is found when it is applied to collisions. In essence, a collision is an interaction between two bodies that lasts for a short time compared to the time over which the motion is observed. The time during the collision between a ball and bat is very short compared to the time required for the flight of the ball. In such collisions the forces acting can be vastly larger than any other force like gravity or friction acting on the system. In fact, because of this one can deduce the magnitude of such short-time forces neglecting entirely the other forces which are present. It is often very important to know the magnitude of the forces at work during a collision, for these forces are responsible for damage to the structure of the colliding bodies.

It is clear from (4-74) that even if the change of momentum of the body is known, the deduction of the force during the collision depends on knowing the time of the collision. It can readily be appreciated then that high-speed photography plays a vital role in the analysis of the forces acting during collisions. Such photographs provide the basic information on the momentum changes and the collision time. The use of (4-74) then permits the computation of the force acting during the collision. Because of the widespread use of the automobile, the disastrous effects of collisions on both human and automobile bodies are all too familiar. In view of the fact that about 60,000 Americans die annually in auto crashes, and injuries are occurring at a rate of about 3 million each year [8], we shall discuss the essential factors that determine injury to the human body as a result of collision.

4.6.B.   Forces Acting During Collisions: Illustrative Examples

Let us begin by considering the forces at work when a golf club strikes a golf ball. This sort of collision is familiar in sports such as tennis, baseball, hockey, etc.

(i)   The Driven Golf Ball

In 1933 Harold Edgerton and Kenneth Germeshausen used a high-speed motion picture camera to photograph the collision between the head of a golf club and the ball [9]. The famous golfer, Francis Ouimet, swung the club. The experiment was conducted at MIT. Instead of using a shutter on the camera, Edgerton and Germeshausen illuminated the scene using a stroboscope: a light source which emits light pulses that were about 5×10^{-6} s long and that were spaced by about 1 ms intervals. The photographs showed that a golf ball starting at rest has its velocity increased to about 200 ft/s ($\sim$ 136 mph) during a collision which lasts only about 0.2×10^{-3} s. The ball and club were in contact with one another over a total distance of about 2 in after which the ball travels in its parabolic trajectory in the gravitational field.

If we approximate the time dependence of the force acting on the ball as a square pulse as indicated in Figure 4.22, we can compute this force quite simply.

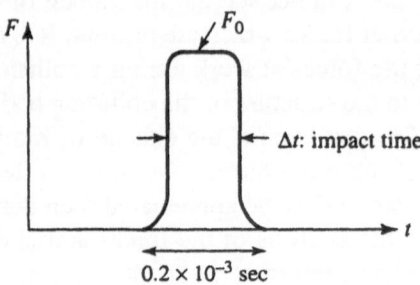

Figure 4.22. Force acting on a ball as a function of time during impact.

The change in momentum of the ball is related to the impulse by

$$\int F \, dt = \Delta P$$

or

$$F_0 t = mV$$

or

$$F_0 = \left(\frac{mV}{\Delta t}\right). \qquad (4\text{-}75)$$

Using the data

$$V = 200 \text{ ft/s} = 6.1 \times 10^3 \text{ cm/s},$$
$$m = \text{mass of golf ball} = 55 \text{ g},$$
$$\Delta t = 0.2 \times 10^{-3} \text{ s},$$

we find that the force applied during the collision is

$$F_0 = 1.6 \times 10^9 \text{ dyn}.$$

Since 1 lb $= 4.45 \times 10^5$ dyn, the force that the club applies to the ball is

$$F_0 = \frac{1.6 \times 10^9}{4.45 \times 10^5} \approx 3600 \text{ lb}. \qquad (4\text{-}76)$$

The great magnitude of this force is directly related to the shortness of the collision time. For a given momentum change of the body, the force acting during impact is inversely proportional to the collision time. A "hard" impact is one for which the impact time is short. Such hard impacts "hurt" because the force produced during the impact is large. Clearly, one prime consideration in reducing deformation or fractures during a collision is to increase the impact time.

It is interesting to remember that in addition to the force produced by the club head, the ball also feels the effect of gravity. During the impact period this force is utterly negligible (the ball weighs about 2 oz) compared to 3600 lb provided by the club. However, if we examine the rest of the motion, after the collision, the weak gravity and air resistance forces acting over the long times associated with the flight of the ball determine precisely the trajectory. One can use $\int F \, dt = P_f - P_i$ throughout the motion. This formula, however, is particularly useful during the collision because it enables us to estimate the forces which can produce deformation and fracture.

(ii)   The Falling Elevator and Bone Fracture

Let us now estimate the force and stress on bones in the human body in a number of possible accident situations. Let us begin with an examination of the forces acting on the bodies of the occupants of an elevator whose supporting cable fails, resulting in the free fall of the elevator down the shaft. Let us suppose that the elevator falls say 60 ft before the air resistance in the shaft brings the elevator to terminal velocity. From the kinematics of a falling body, the velocity V of the elevator on hitting the foot of the shaft will then be

$$V = \sqrt{2gh}, \tag{4-77}$$

where $h = 60$ ft. We now estimate the nature of the collision. In the worst case the elevator bounces up again after hitting the bottom with the same speed as it hit. At best it does not bounce at all. Apart from this factor of 2, the primary element which determines the magnitude of the impact force is the collision time Δt. Let us assume a fairly hard collision, i.e. we assume that the time of collision is ~ 5 ms. In general, the force which the floor of the elevator exerts on the occupants is related to the momentum change by (4-74):

$$\int F \, dt = (mV)_{\text{final}} - (mV)_{\text{initial}}, \tag{4-78}$$

where m is the mass of each person. Using the assumption of full recoil of the elevator from the ground, this gives, for the force,

$$F_0 \Delta t \cong 2mV = 2m\sqrt{2gh} \tag{4-79}$$

or

$$F_0 = \frac{2m\sqrt{2gh}}{\Delta t} = \frac{2mV}{\Delta t}.$$

Using

$$m = 160 \text{ lb} = 72.7 \times 10^3 \text{ g},$$

$$h = 60 \text{ ft} \cong 1830 \text{ cm},$$

$$g = 980 \text{ cm/s}^2,$$

$$V = 1900 \text{ cm/s} = 42.5 \text{ mile/h},$$

$$\Delta t = 5 \times 10^{-3} \text{ s},$$

we find

$$F_0 = 55.2 \times 10^9 \text{ dyn} = 120{,}000 \text{ lb}. \qquad (4\text{-}80)$$

The force associated with the reversal of the momentum of each occupant in the elevator is about 120,000 lb. It is interesting to note that the speed of the elevator after a 60 ft fall is about 42 mph, which is the same order of magnitude as the speed of an automobile driving at a reasonable speed. If a person were driving in such a car and his body collided with the inside of the car at this speed, this force ($\sim$ 120,000 lb) would on average be applied to his body.

Returning to the example of the elevator, we see that the floor of the elevator exerts about 120,000 lb on the passengers feet during the collision. Will fracture occur? Where will the bones break?

The stress on the long bones of the legs during the collision is compressive. In Section 3.4.B we saw that when the compressive stress (force per unit area) exceeded about 1.6×10^3 bar or 23.4×10^3 lb/in^2, the bone breaks in compression. The greatest stress for a given force occurs where the cross-sectional area of bone is smallest. This occurs in the tibia a little above the ankle where the cross-sectional area is about one-half square inch. The stress produced in each of the two tibia during collision on this area is

$$\sigma = \frac{1}{2} \frac{F_0}{A} = \frac{12 \times 10^4 \text{ lb}}{0.5 \times 2} = 120 \times 10^3 \text{ lb/in}^2.$$

Thus, the stress in such a collision is about five times the compressive strength of bone and fracture will occur at this weakest point—the ankles.

What could one do to reduce the chance of bone fracture? Clearly, if the base of the elevator shaft contained a shock absorber which would lengthen the collision time to say 50 ms rather than the hard 5 ms assumed above, the situation would be greatly improved. Another possibility, of course, is to spread out the force over a much wider area on the body than the tibial cross section. This can be accomplished by lying on the floor. However, special care must be taken to reduce the shock to the head.

It is interesting to observe that if one computes the force acting on each horizontal plane of the body during the collision, this force decreases from F_0 at the feet to 0 at the top of the head. This is seen simply as follows. If F_0' is the force on any horizontal mathematical section of the body, then the force during the collision is related to the superincumbent mass m' in accordance with

$$\int F_0' \, dt \approx F_0' \Delta t = m' \Delta V. \tag{4-81}$$

At the feet, the superincumbent mass is that of the entire body. At the neck, m' is the mass of the head, and m' goes to zero at the top of the head.

(iii)   Jumps or Falls to the Ground from a Height

The analysis we have just given for the falling elevator is closely related to that one would make for a subject falling or jumping to the ground from a height. The average force acting on that part of the body which hits the ground is given by

$$F_0 \sim \frac{mV}{\Delta t} \times q, \tag{4-82}$$

where V is the velocity at impact:

> q is a number between 1 and 2 that is determined by the elasticity of the impact,
> $q = 1$ when the body does not bounce,
> $q = 2$ when the body does bounce,
> Δt is the time of collision.

If the jumper falls on his feet he can control the impact time Δt by allowing his knees to bend. We can estimate Δt during the shock-absorbing knee bend as follows. Suppose the jumper decelerates from his landing speed V to zero while the center of mass of his body moves a distance Δh. If we assume constant deceleration during the knee-bending period, we know from kinematics that this mean acceleration ($\bar{a}$) is

$$\bar{a} = \frac{V^2}{2\Delta h}. \qquad (4\text{-}83)$$

The time for deceleration from V to zero is

$$\Delta t = \frac{V}{\bar{a}} = \frac{2\Delta h}{V}. \qquad (4\text{-}84)$$

If we use this in (4-82) with $q = 1$ (no bounce), we see that the reaction force from the ground during the collision is

$$F_0 \cong \frac{mV^2}{2\Delta h}. \qquad (4\text{-}85)$$

However, since the velocity V of the body is related to the height h of the jump by

$$V^2 = 2gh \qquad (4\text{-}86)$$

we see that

$$F_0 = mg\left(\frac{h}{\Delta h}\right). \qquad (4\text{-}87)$$

Equation (4-87) is a simple and quite useful result. It states that the reaction force exerted on the jumper is equal to the jumper's weight times a factor $(h/\Delta h)$, which is the height of the jump h divided by the motion of the center of mass of the body during deceleration by knee bending. (Note Δh is about equal to one-half the change in distance from feet to head during the decelerating crouch.)

In a stiff-legged jump, the deceleration occurs in a small distance, (Δh) is approximately 1 cm. If one lands stiffly with no knee bending the ankle bones can break even with jumps of 6–7 ft. If, for example, $h \sim 2$ m and $\Delta h \sim 1$ cm, $(h/\Delta h) = 200$ and

$$F_0 = 200 \, mg. \qquad (4\text{-}88)$$

For a person weighing 160 lb this corresponds to a force of

$$F_0 = 32{,}000 \text{ lb.}$$

The stress σ on each tibia (area one-half square inch)

$$\sigma = \frac{\frac{1}{2}F_0}{\frac{1}{2}\ \text{in}^2} = 32{,}000\ \text{lb/in}^2.$$

As we mentioned in subsection (ii) above, the ultimate compressive stress for bone is $\sigma \sim 23{,}000\ \text{lb/in}^2$. Thus, such a stiff-legged landing can produce fracture even when the jump height is only ~ 6–7 ft. Of course, if the landing is into loose earth, or sand, or, better yet, water, the situation is completely changed. The distance Δh is tens to hundreds of times larger in such landing situations. The collision time is no longer determined by the decelerating knee bend, but rather by the deceleration produced by the soft sand, earth, or water. However, in a "belly-flop" dive the deceleration distance Δh is again small, F_0 gets large for a given height of dive, and such dives can be painful and even produce internal injury.

By knee bending, Δh can be increased in landing from a vertical jump. It must be borne in mind, however, that the muscle forces needed to provide the deceleration produce great tensile stresses on the ligaments and tendons, and the failure levels for these are much lower than that for bone. Thus, one cannot gain too much in jumping height (h) by trying to increase Δh.

Parachute jumping requires careful understanding of the factors which lead to fractures. In military operations it is essential that the parachute velocity be high so that the vulnerable man quickly reaches the ground. However, the landing speed cannot be too great or bone fracture can result. In fact, the effective landing speed is equivalent to a jump from a second-story window (~ 10 ft). The landing technique for parachutists is designed to increase Δh as much as possible, while also distributing F_0 widely over the body area. Parachutists are trained to make their first contact with the ground on their toes, and use the bending of the ankle joint to begin the process of deceleration. The knees then bend and the body is turned to the side so that the parachutist falls on the leg, then thigh, and side of chest. In this way one maximizes Δh to a value comparable with the size of the body. Also, of course, by falling to the side the force F_0 is distributed as widely as possible over the body, thereby reducing the compressive stress.

(iv)   Tolerance Levels for Whole Body Impacts.
Survival in Falls from a Great Height

In the subsection above we considered the forces at work on the supporting long bones of the leg on landing from a jump. In that case the deceleration distance Δh

can be controlled at least in part by the jumper. If, however, a person falls from a height and lands on his back (supine), his front (prone), or side (lateral), there is no way he can control Δh, and therefore increase the time of collision. In a very useful review entitled "Human Impact Tolerance" [10] by R. G. Snyder there is a tabulation of quantitative information on the falls of children. In the case of lateral impact and prone or supine landings, the primary injuries are to the brain. In Table 4.1 (from [10]) we list the interesting parameters of five recorded falls. These data show that the impact times are of the order of 10 ms. Since the speed of 88 feet per second is that of an automobile traveling at 60 mph, we see that the falls listed above produced an impact velocity roughly corresponding to ~ 20 mile/h. In essence the table informs us that for impact times of about 10 ms full-body collisions at that speed are at the threshold of survival.

Table 4.1: Physical Parameters Associated with Threshold Survival in Full-Body Impacts (from [10]).

Case	Age (Years)	Body orientation at impact	Height of fall (feet)	Impact velocity (ft/s)	Body deformation (Δh) (inches)	Impact time Δh (seconds)
1	14	Supine	45	51	3	0.010
2	5	Supine	23	37	3	0.013
3	13	Lateral	12	25	2	0.007
4	12	Lateral	15	30	3	0.017
5	2	Lateral	10	24	3	0.02

If the body lands either head first or feet first, the survival conditions are somewhat different than those which apply for full-body impact. For example, a feet-first impact on concrete can be survived up to an impact velocity of 40 ft/s. If the landing is on soil one can survive with a landing velocity of 50 ft/s. If the body lands feet first in water, on the other hand, one can survive a landing speed as high as 110 ft/s. A head-first landing on a land surface like concrete or ice leads to very serious injury at speeds exceeding 30 ft/s [10].

It is clear that the damaging effect of impact can be reduced by increasing the collision time, and by distributing the impact force over as wide an area as possible so as to reduce the internal stress (force per unit area). By analyzing data on injuries produced in supine whole-body collisions, A. B. Thompson [10], [11] has been able to provide us with injury criteria which relate the degree of injury to the impact pressure. Thompson [10], [11] finds that if the collision time is less than ~ 70 ms, then the subject will suffer only from shock if the impact pressure (defined as impact force divided by impact area) in whole-body collision is less than ~ 28 lb/in^2. This is about 2 atm. On the other hand, if the impact pressure rises to 50 lb/in^2, there is a 50–50 chance of mortality.

These figures are most helpful in analyzing the survival of persons who have fallen from very great heights onto a soft surface: snow. Let us first examine quantitatively how such survival can occur, then we will compare our analysis with the known, documented accounts of such falls.

As a body falls from a great height in air, its speed does not vary simply as $V^2 = 2gh$. The surrounding air provides a viscous drag that increases with speed. This drag, which is familiar to anyone who has held his hand out of a car window while moving at high speed, has the effect of reducing the acceleration of the falling body and bringing it to a "terminal velocity." The net of the gravity force and the air resistance force become zero and the body ultimately falls with constant speed—the terminal velocity. Near sea level—in the altitude range of 1–10,000 ft—the terminal velocity of a falling person is about 120 mph/hour $\cong$ 176 ft/s. We note at once that this terminal velocity is only three to five times faster than the impact speeds listed in Table 4.1. We are thus led to wonder whether the increase in impact time of a factor of 3 to 5 more, over that in the table, could result in the survival of persons impacting the ground at this terminal velocity.

Let us make these considerations more quantitatively by using Thompson's criteria for impact pressure. According to this criteria the threshold for survival occurs when the impact time Δt, and the impact pressure, satisfy the equation

$$P = \frac{F}{A} = \frac{MV_t}{A(\Delta t)} < 50 \text{ lb/in}^2.$$

Here M is the mass of the falling body,
   V_t is the terminal velocity—the velocity on impact,
   A is the contact area on impact,
   Δt is the impact time.

The impact time Δt, can be related to the deceleration distance Δh just as in subsection (iii) by regarding the deceleration (a) as constant. In this case, we have from kinematics that

$$V_t^2 = 2a(\Delta h),$$
$$0 = V_t - a\Delta t,$$

i.e.,

$$\Delta t = 2\frac{\Delta h}{V_t}.$$

Thus the impact pressure P can be related to the terminal velocity V_t and the depth of the hole (Δh) made in the snow on impact by the formula

$$P \simeq \frac{MV_t^2}{2A\Delta h}.$$

For survival this quantity should be less than P_c the critical threshold value of the impact pressure of 50 psi. We may, therefore, write that Δh must satisfy the criterion

$$\frac{MV_t^2}{2A\Delta h} \gtrsim P_c$$

or

$$\Delta h > \frac{MV_t^2}{2AP_c}.$$

We can now compute numerically the size of Δh to be expected from this criterion. Using

$$Mg = 200 \text{ lb}, \quad g = 32 \text{ ft/s}^2,$$
$$V_t = 176 \text{ ft/s},$$
$$A = \text{area of back} \cong 3.8 \text{ ft}^2,$$
$$P_c = 7.2 \times 10^3 \text{ lb/ft}^2,$$

we have

$$\Delta h \gtrsim \frac{(200)(176)^2}{(32.2)(2)(3.8)(7.2 \times 10^3)} \text{ ft}$$
$$\Delta h \gtrsim 3.5 \text{ ft.}$$

This result is surprising indeed, for it states that if the impact distance, or the hole made by a person into the snow, is about 3.5 ft deep, he/she has a 50–50 chance of survival from a fall in which he/she hits the ground at the terminal velocity of about 120 mph. The gratifying aspect of the result is that it is in agreement with documented reports of the survival of persons who have fallen without the aid of parachutes from airplanes. We quote from one of the accounts contained in a study of terminal velocity impacts into snow made by R. G. Snyder [12]. In describing one such event, he writes of a large airborne parachute jump exercise in Alaska in February, 1955:

> During one of the battalion drops, from 1,200 feet on a clear, relatively warm day, an observer noted what appeared to be an unsupported bundle falling from one of the C-119 airplanes; no chute deployed from the object. The impact looked like a mortar round exploding in the snow. When the aid-men reached the spot, they found a young Negro paratrooper flat on his back at the bottom of a $3\frac{1}{2}$ foot crater in the snow, which consisted of alternating layers of soft snow and frozen crust. He could talk and did not appear injured; nevertheless, he was air evacuated to a hospital. His only injuries were an incomplete fracture of a clavicle, a chip fracture of L-2, and a few bruises. He was released from the hospital in time to return south with his unit.

It is interesting to note the dimensions of the hole found by the aidmen in view of the calculation above.

Sometimes, part of the impact velocity can be reduced by falling through tree branches before the impact with the snow. This can reduce even further the value of Δh needed. Again R. G. Snyder documents a relevant case [12]:

> Perhaps the classic case (of impact survival on snow) on record occurred about midnight on the frigid 23rd of March, 1944, to Flight Sergeant Nicholas Alkemade, an RAF rear gunner. After his Lancaster bomber was set afire by a German night fighter on a raid over Hamburg, he found that he was unable to reach his parachute, stowed forward in the flaming fuselage. Deciding that he didn't care to burn alive, he jumped without a parachute just as the aircraft exploded above him. His altitude was 18,000 feet. Falling at a terminal velocity of about 120 mph during this $3\frac{1}{2}$ mile fall (which lasted 90 seconds), he struck the snowy branches of a pine forest and then landed in less than 18 inches of snow, only twenty yards from the bare open ground. Incredibly, his only reported injuries consisted of superficial scratches and bruises, and burns received prior to the jump. This case is well-documented, was witnessed and thoroughly investigated by the Germans.

The longest "free-fall" yet recorded was from a height of 23,000 ft by a Russian Lieutenant, I. M. Chissov, during World War II [12]. Lieutenant Chissov was forced to bail out at 23,000 ft from his Ilyushin 4 after it was attacked by 12 Messerschmitts and completely disabled. Because of the presence of the fighter planes Chissov delayed opening his parachute, and fell 22,000 ft intending to open his parachute at 1,000 ft. However, he lost consciousness at about that height and hit the ground at the edge of a ravine whose sides were filled with snow 3 ft deep. This, of course, was the ideal geometry for his survival. It permitted a long impact time as his body descended along the slope of the ravine, while being decelerated by the frictional resistance of the deep snow. In fact, he recovered consciousness after about 20 min and was rescued by Russian cavalrymen. He suffered a concussion of the spine and a fractured pelvis, but returned to service $3\frac{1}{2}$ months later as a flight instructor in a military training school.

These examples clearly show that if a body lands in sufficiently deep snow, it is possible to survive even very high-altitude falls. In fact, it is possible to document [12] Russian airborne operations in which troops were dropped into snow without parachutes from low-flying transport aircraft. Snyder reports [12]

> During the Yukov airborne operation of 1942 such cases were observed and reported to German Intelligence by units of the German 4th Army. In one case the airborne troops dropped without parachutes into snow were placed in sacks filled with straw. Nearly a decade earlier the Soviets had apparently experimented with dropping troops without parachutes into snow during the war games of the 1930's and later during the Russo-Finnish War. Sellick states 'Specially built monoplanes carried troops in enclosures in the wings. They were spilled into deep drifted snow from heights of 15 to 50 feet.... Approximately half of the men dropped were still able to fight and the experiments were soon abandoned.'

Snyder [12], in fact, obtained a Sovfoto picture of such an airplane. The enclosures in the wings contain smiling men and women soldiers.

Probably the most detailed quantitative analysis of impact into snow from an airplane concerns the pilot of a Douglas A4E jet fighter [12]. On returning from a proficiency mission on 20 February, 1964, his jet engine failed and the pilot ejected very close to the ground. His parachute did not open, but the pilot survived because he landed on the downward slope of a small hill covered with snow about 21 in deep. His body dug a trough along the surface of the hill for about 31 ft and came to a stop against a pine tree. The impact velocity onto the snow was about 180 ft/s. The deceleration time was relatively long, i.e., about 350 ms. The landing was on his buttocks and back which were protected by the parachute. The pilot suffered a simple fracture of the right femur (probably as a result of impacting the tree) and undisplaced fractures of the ankle. He was returned to duty within 2 months and was on full flight status within 9 months [12].

(v) Deceleration Pulses and the Severity Index: A Criterion for Concussion and Injury to the Brain

In fatal collisions the head injury in 75% of cases is responsible for death. Rapid deceleration of the head, even without fracture, can be fatal. For example, if the head is subject to a deceleration of about 150 g for times longer than a few milliseconds, death can result. If the deceleration is of the order of 50 g, it can be tolerated for as long as 20 or 30 ms. The precise basic damage mechanism at work on the brain is still not understood despite many studies [13]. At present the most likely candidate is the shear strain at the brain stem. Here the nerves of the cervical cord in the vertebral column branch into the brain itself.

Despite the absence of a detailed knowledge of the basic biomechanical mechanism for brain injury in a collision, experimental studies have provided a very valuable criterion of damage in terms of the size and shape of the deceleration pulse of the head. By analyzing experimental data, C. W. Gadd [14] has demonstrated that the severity of a collision can be measured by a weighted integral of the deceleration versus the time pulse. If the cranium is subject to a frontal blow, with an acceleration time course like that shown in Figure 4.23, the "Severity Index" associated with the pulse is defined as [14]:

$$\text{Severity Index} = I = \int a^{2.5}\, dt. \qquad (4\text{-}89)$$

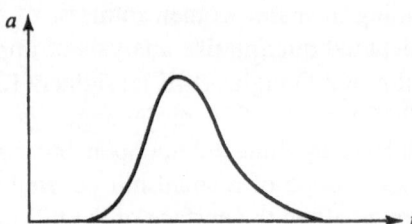

Figure 4.23. Acceleration versus time in an impact.

The trauma associated with the deceleration is dependent not on the maximum acceleration, or the area of the acceleration time pulse. Instead it is measured by the weighted average given above. The severity index weights the high acceleration parts of the pulse much more heavily than the lower parts. The merit of this severity index is that it enables a comparison of the severity of various size and shape acceleration pulses in terms of a single number. If the acceleration a is measured in units of the acceleration of gravity, and if the time is measured in seconds, Gadd has established that when I, the severity index, exceeds 1000 there is a 50–50 chance that the collision is fatal [14].

The disastrous effect of a high-speed collision of the head can be seen quite dramatically using this severity index. If the collision is "elastic," i.e., the head recoils with about the same speed as it hits, we can estimate the *severity index* from the initial velocity V and the impact time as follows:

$$I \sim \left(\frac{2V}{g \, \Delta t} \right)^{2.5} \Delta t. \tag{4-90}$$

Here we have used the definition of the severity index, (4-88) and the assumption that the acceleration pulse is rectangular. This equation demonstrates that the severity of the blow to the head varies very strongly with the initial velocity and inversely as the time of impact raised to the exponent 1.5. Let us put some typical numbers into this expression. Suppose $V = 50$ mph $= 72.5$ ft/s. Impact times range from ~ 2 ms for a very hard collision to $\gtrsim 50$ ms. Let us take as a reasonably hard collision time 10 ms, as was observed in the measurements listed in Table 4.1. With these values of V and Δt we find that the severity index is

$$I \sim \left(\frac{145}{32(10 \times 10^{-3})} \right)^{2.5} (10 \times 10^{-3}) = 43{,}600 \text{ s},$$

which is clearly fatal. Here the acceleration is about 450 g for a duration of $\sim$ 0.01 s. This example shows quite clearly the great danger of high-speed travel. It demonstrates the importance of padding to increase the collision time and seat belts to prevent the impact of the head against the interior of the car.

Values nearer the border line index of $I = 1000$ can be attained in heavyweight boxing. Thanks to the courtesy of Mr. R. L. Le Fevre [15] at the General Motors proving ground in Milford, Michigan, we have data on the acceleration of the head of a dummy which was hit by a heavyweight prizefighter named "Blue Lewis." Figure 4.24 shows approximately the size and shape of this acceleration of the

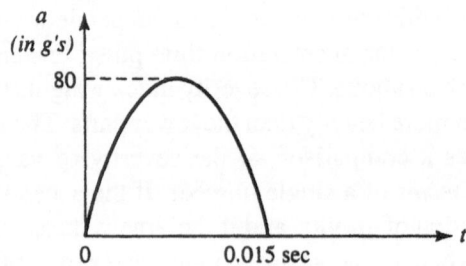

Figure 4.24. Acceleration versus time of the head of a dummy hit by a heavyweight boxer.

head. We may approximate this pulse by a half sine wave of the form

$$a(t) = 80 \sin \left(\frac{\pi t}{0.015} \right) \text{ in units of g.}$$

The severity index for this collision is

$$I = \int a^{2.5} \, dt = \int_{t=0}^{0.015} (80)^{2.5} \left(\sin \frac{\pi t}{0.015} \right)^{2.5} \, dt.$$

Putting the integral in dimensionless units by setting $x = \pi t / 0.015$, we have

$$I = 80^{2.5} (0.015) \left[\frac{1}{\pi} \int_0^{\pi} \sin^{2.5} x \, dx \right].$$

The quantity inside the brackets is equal to 0.458. Thus

$$I = (80)^2 \sqrt{80} \, (0.015)(0.458) \text{ in units of g}^{2.5} \text{ s,}$$
$$I = 395 \text{ in units of g}^{2.5} \text{ s.}$$

This blow by the prizefighter has a severity index of two and a half times smaller than the value of 1000 corresponding to a probably fatal blow. The value of $I \simeq 400$ is indicative of a mild concussion and would cause unconsciousness. These two calculations demonstrate the utility of the severity index in the estimation of the degree of injury in various circumstances. They also demonstrate that, particularly in automobile, bicycle, and motorcycle travel, there is plenty of speed available for serious injury to the brain.

(vi)   Safety Considerations in Automobile Accidents

The quantitative criteria for fracture and brain injury show that at the speeds customary in automobile travel there is great potential for bone fracture, brain damage, laceration, and trauma to internal organs during the "second collision" that can occur between an occupant and the interior of his car. The quantitative criteria also show quite clearly that there are two crucial factors available to designers and to motorists which can markedly affect the degree of injury. The first and paramount factor is the collision time. If impacts can be lengthened from the "hard" region of a millisecond or so to 10–100 times this, we can dramatically reduce injury. Also, we have seen that the distribution of the impact force over a wide area can also be very valuable in reducing the stress that produces internal fractures. Collision-induced injury is not inevitable. The hazards can be markedly reduced by designing and using suitable protective restraint and energy absorbing devices, such as airbags, seat belts, safety glass windshields, and padded interiors. These are very useful in reducing death and injury.

4.7   Flow of Mass and Momentum. Transport in Fluids

4.7.A.   Current Density

We now wish to extend our discussion of momentum to include a description of the flow of momentum and other physical quantities in a continuous medium—a fluid. Suppose that in some small region around a point $\vec{r}$, fluid is flowing with velocity $\vec{v}(r)$ as is indicated in Figure 4.25. The flow of the fluid can be characterized

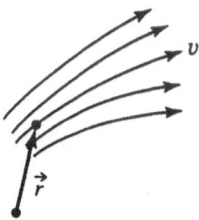

Figure 4.25. Velocity of flow in the vicinity of a point in a fluid.

mathematically by the so-called "mass current density" j_m. If a small area ΔA is imagined oriented so that its normal points directly along the velocity, then the rate of flow of mass across ΔA can be related to the current density j_m by the definition

$$\left(\frac{dm}{dt}\right)_{\Delta A} \equiv j_m \Delta A. \tag{4-91}$$

It is quite simple to relate j_m to the velocity of the fluid at r and the density of the fluid using Figure 4.26.

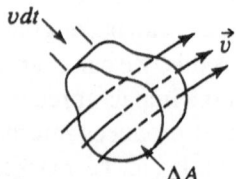

Figure 4.26. Flow across differential area ΔA.

In each differential element of time dt, the volume that passes through ΔA is contained in a cylinder of length $v\,dt$, and base ΔA. Thus the rate of flow of volume through ΔA, (dQ/dt) is

$$\left(\frac{dQ}{dt}\right) = \frac{(v\,dt\,\Delta A)}{dt} = v\Delta A \tag{4-92}$$

and the rate of flow of mass is equal to this times the density (mass per unit volume) ρ, i.e.,

$$\left(\frac{dm}{dt}\right) = \frac{\rho v\,dt\,\Delta A}{dt} = \rho v\Delta A. \tag{4-93}$$

On comparing (4-93) with the definition of the mass current density j_m, we have

$$\vec{j}_m = \rho\vec{v}, \tag{4-94}$$

$$\rho = \text{mass density (mass/volume)}.$$

The mass current density is a vector whose value at each point ($\rho\vec{v}$) tells us the magnitude and direction of the flow of mass.

Having defined the mass current density, it should be mentioned that one can also identify in a perfectly similar way the current density for the flow of momentum, or energy, or a number of particles, or an electrical charge. These quantities are of vital importance in the transport of heat, or the transport of metabolites or the transport of electrical charge carriers. In Volume 2 of this course we shall make frequent use of the concept of current density. Let us widen our list of current density vectors a bit at this point. Suppose the fluid is in fact a gas made up of molecules with a density of n molecules/cm^3. Then if $\vec{v}$ is the velocity of the gas at some point r, the current density for the flow of particles is

$$\vec{j}_n = n\vec{v}, \tag{4-95}$$

n = number density (number/volume).

If the flowing particles are charged electrically with charge e, the electrical current density j_e is

$$\vec{j}_e = ne\vec{v}, \tag{4-96}$$

ne = charge density (charge/volume).

Quite analogously to the definitions above, we can define the current density for the flow of the kth component ($k = x, y, z$) of the momentum density as:

$$\vec{j}(P_k) = (\rho v_k)\vec{v}, \tag{4-97}$$

where ρv_k = kth component of momentum per unit volume. To anticipate a little, we shall see that the "kinetic energy" of a body is $\frac{1}{2}mv^2$. The kinetic energy density of a fluid of particles of mass m is $\rho_E = \frac{1}{2}nmv^2 = \frac{1}{2}\rho v^2$ where ρ is the mass density. Thus the kinetic energy current density is

$$\vec{j}_E = (\tfrac{1}{2}\rho v^2)\vec{v}. \tag{4-98}$$

One of the important uses for the current density vectors is that it enables us to compute the buildup of any physical property of the fluid within any volume inside the fluid. We can see how this is done with the aid of Figure 4.27. This shows a small closed mathematical volume (fixed in the laboratory coordinate system).

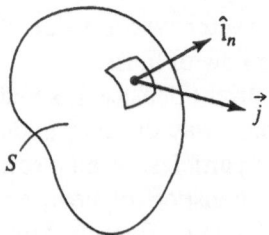

Figure 4.27. Calculation of net flow in and out of a closed surface.

Consider the net flow across some surface area element ΔA of this volume. We see, from the general definition of the current density vector, that the flow of any quantity out of ΔA is equal to the component of $\vec{j}$ along the outward normal of $\Delta A(\hat{i}_n)$ times ΔA. Thus the flow out through the area element shown is $\vec{j} \cdot \Delta \vec{A}$, where $\Delta \vec{A}$ is regarded as a vector pointing in the direction of the outward normal. We can now express the buildup of mass, momentum, particle number, energy, charge, etc., inside the closed surface by integrating $\vec{j} \cdot \Delta \vec{A}$ over the entire surface. Let us think, for concreteness, of the mass current density. The mass that builds up inside S each second is equal to the difference between the amount of mass that flows into and out of the surface S each second. (We assume, of course, that inside S there is nothing that creates or destroys matter.) Thus we see that if M is the total mass inside S, then

$$\left(\frac{dm}{dt}\right) = \int_s \vec{j}_m \cdot d\vec{A} = \int_s \rho \vec{v} \cdot d\vec{A}. \tag{4-99}$$

In a perfectly similar way, if $\vec{P}$ is the momentum inside S, N the total number of particles inside S, E the kinetic energy, and Q the electrical charge inside S, we can express the rate of change of each of these quantities by the formulas

$$\left(\frac{d\vec{P}}{dt}\right) = \sum_k \hat{i}_k \left(\frac{dP_k}{dt}\right) = \sum_k \hat{i}_k \int_s \vec{j}(P_k) \cdot d\vec{A} \tag{4-100a}$$

or

$$\left(\frac{d\vec{P}}{dt}\right) = \int (\rho\vec{v})\vec{v} \cdot d\vec{A} \tag{4-100b}$$

$$\left(\frac{dN}{dt}\right) = \int_s n(\vec{v} \cdot d\vec{A}), \tag{4-101}$$

$$\left(\frac{dE}{dt}\right) = \int_s \left(\tfrac{1}{2}\rho v^2\right)\vec{v} \cdot d\vec{A}, \tag{4-102}$$

$$\left(\frac{dQ}{dt}\right) = \int_s ne\vec{v} \cdot d\vec{A}. \tag{4-103}$$

4.7.B. Pressure Exerted by a Stream of Fluid

The momentum current density vector is useful in calculating the pressure exerted by a fluid stream as it strikes a wall. Let the incoming jet strike the wall obliquely so that an angle θ is made between the incident jet and the normal to the wall. Let the area of impact of the incident jet be A. The rate at which momentum flows onto the surface A is

$$\left(\frac{d\vec{P}_{in}}{dt}\right) = \vec{j}_p \cdot \vec{A} = \rho\vec{v}(\vec{v} \cdot \vec{A}) = \rho\vec{v}vA\cos\theta. \tag{4-104}$$

The rate at which momentum of the reflected fluid comes off the wall is determined by the precise nature of the impact. In a perfectly elastic collision the reflected speed v' is equal to v, and the reflected angle θ' is equal to θ. At this point we shall make no assumption as to v', θ', and we shall even leave the possibility open that the area A' from which the reflected water leaves is different from A. The flow of momentum to the left in Figure 4.28 is

$$\left(\frac{d\vec{P}_{out}}{dt}\right) = -\rho\vec{v}'v'A'\cos\theta'. \tag{4-105}$$

The rate of change of momentum of the fluid is the vector difference between these two equations. The component of the rate of change that interests us is the x-direction. This is

$$\left(\frac{d\vec{P}}{dt}\right)_{total} = (\rho\vec{v}\cos\theta)vA\cos\theta + (\rho\vec{v}'\cos\theta')v'A'\cos\theta'. \tag{4-106}$$

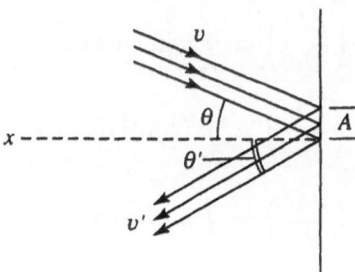

Figure 4.28. Representation of collision between a stream of fluid and a wall.

We can relate v, v', A, A' by the condition that there is no loss of matter. Thus the flow of mass into A must equal the flow of mass backward out of A':

$$\text{i.e.,} \qquad \vec{j}_m \cdot \vec{A} = \vec{j}_m' \cdot \vec{A}', \qquad (4\text{-}107)$$

$$\text{i.e.,} \qquad \rho v A \cos\theta = \rho v' A' \cos\theta'. \qquad (4\text{-}108)$$

Therefore, since $(dP/dt)_{\text{total}}$ is the force exerted by the wall, we have that this force is

$$F = \rho v^2 A \cos^2\theta + \rho v v' A \cos\theta \cos\theta' \qquad (4\text{-}109)$$

or the force per unit area is

$$\left(\frac{F}{A}\right) = \rho v(v \cos^2\theta + v' \cos\theta \cos\theta'). \qquad (4\text{-}110)$$

If the stream of water strikes the wall at right angles to the surface, we have $\theta = \theta'$ and we have

$$\text{pressure on wall} = \rho v(v + v'). \qquad (4\text{-}111)$$

Let us estimate how great this pressure can be. A typical fire hose can send a jet of water about 35 ft high. This corresponds to a nozzle velocity of roughly 15 m/s. Taking then

$$\rho = 10^3 \text{ kg/m}^3$$

$$v = 15 \text{ m/s,}$$

$$v = v' \text{ assuming elastic collision,}$$

$$\text{pressure} = 4.5 \times 10^5 \left(\frac{\text{kg m}}{\text{sec}^2}\right) \frac{1}{\text{m}^2}$$

$$\text{pressure} = 4.5 \times 10^5 (\text{N/m}^2),$$

or

$$\text{pressure} \cong 4.5 \text{ atm.}$$

This is quite a substantial pressure from a fire hose. Yet, much higher velocities can be produced. To produce a speed of 50 ft/s, a fire engine pumper subjects the water in the reservoir to about 16 atm. In the Soviet Union, the Institute of High Pressure Physics has developed pumps which will subject the water to a pressure of 10,000 atm! Such pumps are used to produce high-speed water jets for specialized mining operations.

4.7.C.   Pressure on the Walls of a Curved Hose: The Fire Hose and the Aorta

Suppose that a fluid is moving with velocity v around a curve in the containing hose or tube, or river bed. Since the direction of the flow is changing, forces must act on the fluid. We now calculate the force that the walls of the hose must exert in order to turn the fluid flow. In Figure 4.29 we see the diagram which describes the flow through a small arc $\Delta\theta$. The momentum in the y-direction is changing as the fluid flows around the curve. The change in momentum per unit time $(d/dt)(\Delta P)$ associated with the small angle $\Delta\theta$ is indicated in the vector diagram (Figure 4.30).
    Thus (with $\pi r^2 = A$)

$$\frac{d}{dt}(\Delta P) = \rho v^2 A \Delta\theta \tag{4-112}$$

is the force associated with turning the flow of the fluid inward around the curve. Since $\Delta\theta$ is related to the arc length $\Delta\ell$ and the radius of curvature R by

$$\Delta\theta = \frac{\Delta\ell}{R}, \tag{4-113}$$

and since $(d/dt)(\Delta P)$ is the inward or radial force ΔF exerted by the bounding wall, we have

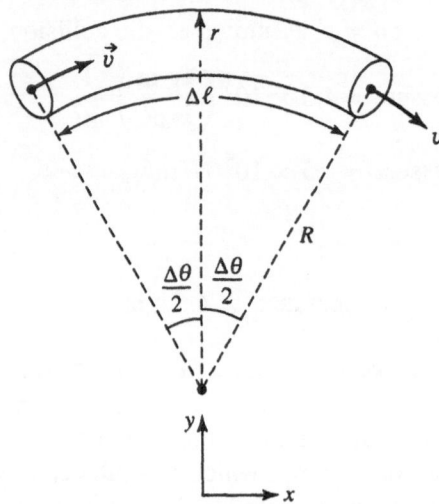

Figure 4.29. Geometry of a flow of fluid around a curve.

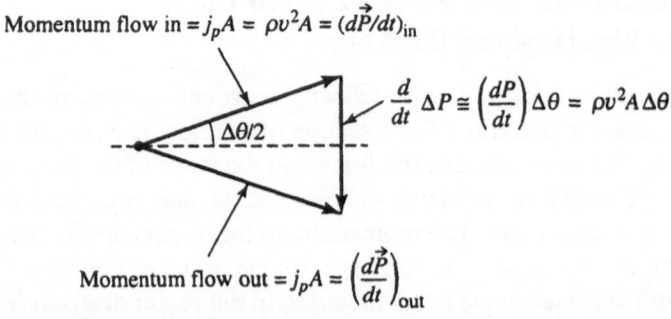

Figure 4.30. Computation of a change in the flow of momentum around curve of angle $\Delta\theta$.

$$\Delta F = \rho v^2 A \frac{\Delta\ell}{R} = \frac{\rho v^2 \pi r^2 \Delta\ell}{R}. \tag{4-114}$$

It is to be noted that $2\pi r \ell$ is the surface area ΔA_s of the hose. Thus we can express (4-114) as

$$\Delta F = \frac{\rho v^2}{2}\left(\frac{r}{R}\right)\Delta A_s. \tag{4-115}$$

Thus the force needed to turn the fluid is proportional to the surface area of the curved section, and, inversely proportional to the radius of curvature.

We can use (4-114) to estimate the additional pressure on the walls of the hose that is needed to cause the fluid to turn the curve. This pressure is in addition to the pressure within the fluid which is associated with its flow along the axis of the hose. The additional pressure associated with the curve can be estimated crudely by assuming that the force on the wall, associated with turning the fluid, is distributed over the outer half of the circumferential area as indicated in Figure 4.31. Under this assumption the average pressure exerted by the outer half of the wall is

$$\text{pressure} = \frac{\Delta F}{(\Delta A_s/2)} = \rho v^2 \left(\frac{r}{R}\right). \qquad (4\text{-}116)$$

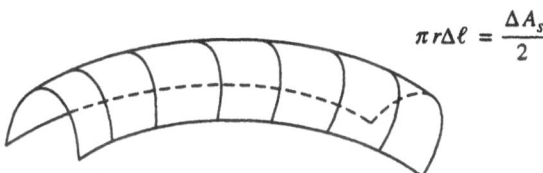

$$\pi r \Delta \ell = \frac{\Delta A_s}{2}$$

Figure 4.31. Area of the upper half of a curving hose.

We can use this result and (4-113) to make some interesting numerical estimates. Consider first the force that firemen must exert on a hose to turn it through some angle $\Delta\theta$. Since $\Delta\theta = (\Delta\ell/R)$, we see that the force required is

$$\Delta F = \rho v^2 \pi r^2 \Delta\theta.$$

Let us use the following values for ρ, v, and r:

$$\rho = 10^3 \text{ kg/m}^3$$

$$v = 15 \text{ m/s (typical nozzle velocity of fire hose)},$$

$$r \cong 2.54 \times 10^{-2} \text{m}$$

$$\Delta F = 4.4 \times 10^2 \Delta\theta \text{ in Newtons}$$

$$1N = .225 \text{ lb}$$

Thus if we have a 90° turn, i.e., $\Delta\theta \sim (\pi/2)$, we find

$$\Delta F \sim 155 \text{ lb}$$

We see, therefore, that a very considerable force is needed to turn the direction of the fire hose. It is, in fact, easy to see that if the diameter of the fire hose gets two or three times bigger than the value we estimated, it will be very difficult to manually direct the hose, provided that the speed of flow is ~ 50 ft/s.

We may next compute the pressure in the hose associated with such a 90° turn. For this we use (4-115).

$$\text{pressure} = \rho v^2 \left(\frac{r}{R}\right).$$

Using now

$$\rho = 10^3 (\text{kg/m}^3)$$
$$v = 15 \text{ m/s,}$$
$$r/R = 1/36$$

we have

$$P = 62.5 \times 10^3 (\text{N/m}^2)$$

Since $1\text{N/m}^2 = 1.45 \times 10^4 \text{ lb/m}^2$ we find:

$$P \cong 0.97 \text{ lb/in}^2.$$

This low value of the pressure in the hose surface signifies that the force computed above is spread out over the hose surface over an area of about 160 in^2, thereby resulting in a low stretching pressure on the hose.

As a final example we can compute the pressure that the aorta experiences in turning the flow of blood around the aortic arch. Once again we use the formula (4-116) for the average pressure

$$\text{pressure} = p v^2 \left(\frac{r}{R}\right).$$

In the case of blood flow we shall use cgs units

$$\rho = 1 \text{ g/cm}^3,$$

$$r \cong 1.25 \text{ cm},$$

$$R \cong 2 \text{ cm},$$

$$v \cong 100 \text{ cm/s}.$$

The latter value of the blood velocity is computed using $500 \text{ cm}^3/\text{s}$ as the maximum rate of expulsion of blood from the heart. (See Section 4.5 on ballistocardiography.) Using these values we find

$$\text{pressure} = 1 \times 10^4 \left(\frac{1.25}{2} \right) \sim 6.25 \times 10^3 \text{ dyn/cm}^2$$

$$\cong 4.7 \text{ mmHg}.$$

Thus the pressure which must be applied by the wall of the aorta to turn the blood around the aortic arch is only about 5 mmHg. This is small compared to the mean of the diastolic and systolic pressure, i.e., ~ 100 mmHg. Thus we cannot expect the direct mechanical effect of the flow around the aortic arch to be of much significance in pathologic conditions such as aneurisms.

To conclude our discussion of the flow of momentum, it should be pointed out that there is a class of phenomena in which the mass of the body under study is changing. This is the case for a rocket which is constantly burning and expelling fuel. It is also true for more mundane examples such as moving railway cars which are dropping coal, say, for example, as they move forward. These problems are quite interesting and challenging. They require a careful analysis and understanding of the "system" under consideration. The book *An Introduction to Mechanics* by Kleppner and Kolenkow [16] has a particularly useful analysis of problems of this type and the interested reader is recommended there for this topic.

4.8   References and Supplementary Reading

[1] *De Motu Animalium* by G. A. Borelli. Lugduni Batavorum (1679).

[2] *Biomechanics of Human Motion* by M. Williams and H. R. Lissner. W. B. Saunders, Philadelphia (1962).

[3] *Ballistocardiography* by I. Starr and A. Noordergraaf. J. B. Lippincott, Philadelphia (1967).

[4] "On Certain Molar Movements of the Body Produced by the Circulation of the Blood," by J. W. Gordon. *Journal of Anatomic Physiology* **11**, 533 (1877).

[5] "On Certain Molar Movements of the Human Body Produced by Circulation of the Blood," by C. Trotter, *Journal of Anatomic Physiology* **11**, 755 (1877).

[6] *The Heart, Its Function in Health and Disease* by A. Selzer. Angus and Robertson, Sydney, London, Melbourne (1966).

[7] "The Nature of Records from Ultralow Frequency Ballistocardiograph Systems and their Relation to Circulatory Events," by W. R. Scarborough, E. E. Folk, P. M. Smith, and J. H. Condon. *American Journal of Cardiology* **2**, 613 (1958).

[8] "Biomechanical Problems Related Vehicle Impacts" by N. Perrone, in *Biomechanics: Its Foundations and Objectives*, edited by Y. C. Fung, N. Perrone, and M. Anliker. Prentice Hall, Englewood Cliffs, NJ (1972).

[9] "Catching the Click with a Stroboscope," by H. E. Edgerton and K. J. Germeshausen. *The American Golfer*, November (1933).

[10] "Human Impact Tolerance" by Richard G. Snyder, in *1970 International Automobile Safety Conference Compendium*, pp. 712–782. Society of Automotive Engineers, New York.

[11] "A Proposed New Concept for Estimating Limit of Human Tolerance to Impact Acceleration," by A. B. Thompson. *33rd Annual Meeting of Aerospace Medical Associates*, Atlantic City, NJ, 9–12 April (1962).

[12] "Terminal Velocity Impacts into Snow," by R. G. Snyder. *Military Medicine* **131**, 1290–1298 (1966).

[13] "Biomechanics of Head Injury" by N. Goldsmith, in *Biomechanics: Its Foundation and Objectives*, edited by Y. C. Fung, N. Perrone, and M. Anliker. Prentice Hall, Englewood Cliffs, NJ (1972).

[14] "Use of a Weighted Impulse Criterion for Estimating Injury Hazard," by J. P. Danford and C. W. Gadd. *Proceedings of the 10th STAPP Car Crash Conference* (1966), pp. 95–100.

[15] R. Le Fevre: General Motors Proving Ground. Private communication (November 6, 1970).

[16] *An Introduction to Mechanics* by D. Kleppner and R. Kolenkow. McGraw-Hill, New York (1973).

4.9   Problems

1. Recoil momentum of a rifle. A rifle bullet has a mass $m = 13$ g, and a muzzle velocity v of 800 m/s.

    (a) Determine the momentum p of the bullet.

    (b) Determine the recoil momentum of the rifle; what would its velocity be, after firing a shot, if it were mounted on a frictionless air track, if its mass is 4 kg?

    (c) Assuming uniform acceleration of the bullet, and a barrel length $\ell = 80$ cm, find:

    (i) the time duration t_0 of the bullet's acceleration; and

    (ii) the force F on the bullet during this time, making use of $F = p/t_0$.

    (d) This is a somewhat harder question: What happens if a person fires a rifle from a standing position? You may consider the upper part of the body as a lumped mass $M = 30$ kg, pivoted about the pelvis; the muscular stabilization of this mass makes it into an oscillator with a natural period T of about 2 s.

    (i) Determine the initial recoil velocity V_0 of the masses $(M + m)$ after the rifle has been fired.

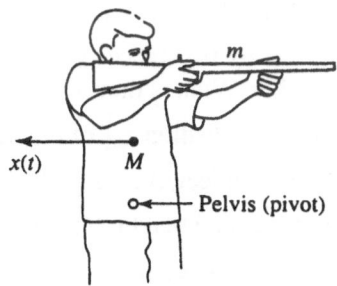

$x(t)$ M

Pelvis (pivot)

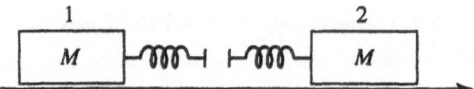

(ii) Assuming that the recoil motion of the body can be described by

$$x(t) = \frac{V_0}{\omega} \sin(\omega t), \qquad \omega = \frac{2\pi}{T},$$

find the recoil distance x_{rec} of the body.

2. A rifle bullet of mass $m = 10$ g and speed $v_0 = 800$ m/s is stopped by a block of wood 1 ft thick. Assume uniform deceleration of the bullet in the block, whose mass is 25 kg, and which is free to slide without friction on a horizontal surface. Find:

(a) The final momentum of the block, which is at rest initially.

(b) Its recoil velocity.

(c) The time required to stop the bullet (impact time).

(d) The force on the block during impact.

3. Elastic collision. Consider two equal masses M, able to slide without friction on a horizontal air track. They are equipped with identical spring bumpers. Mass 2 is initially at rest, and mass 1 moves with initial velocity v_0 toward mass 2.

(a) Describe the motion of the center of mass of this two-body system before, during, and after the collision of the two bodies.

(b) Describe the collision process in the (inertial) frame of reference, moving with the center of mass of the two bodies, and show that after the collision, mass 1 is at rest, and mass 2 moves with velocity v_0.

(c) Replace the two springs by two pieces of putty, so that after impact, the two masses stick together. What is their final velocity?

4. Center of mass. Determine the position z_M of the center of mass of:

(i) a uniform hemispherical object of radius R;

(ii) a cone with circular base, and height H; and

(iii) a hemispherical, thin-walled bowl.

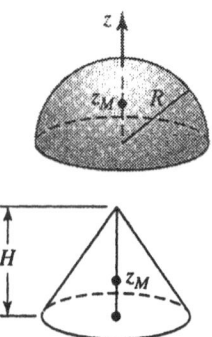

5. Suppose you stand on your bathroom scales, originally in a crouching position, and then, in a quick movement, get yourself into an upright position. The diagram on the left indicates the change of the center-of-mass position z_M with time. Plot, on a similar diagram, the center-of-mass velocity $dz_M/dt = v_M$, and the center of mass acceleration $\frac{dv_M}{dt} = a_M$. Finally, plot qualitatively the time dependence of the "weight"-reading on the scale.

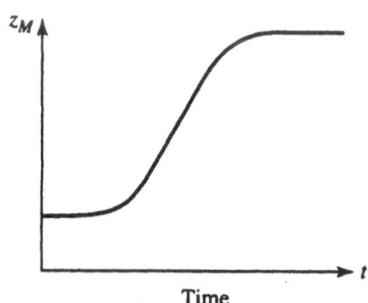

Time

6. Oscillation frequency of mass m, attached to a spring inside a box of mass M. The spring has spring constant k.

   (a) m hangs vertically in box (see figure below). Find the frequency ω_0 of oscillation.

   (b) Both m and M slide without friction. The force exerted by the spring on m is

$$F_x = -k(x_m - x_M)$$

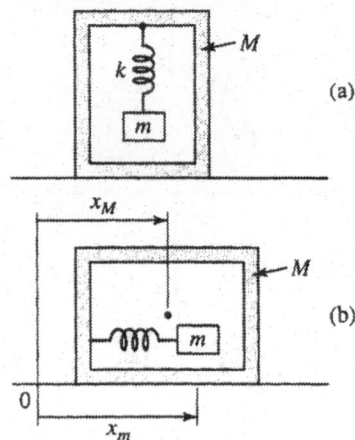

and on M:

$$F'_x = +k(x_m - x_M).$$

Write the equations of motion for both m and M and show that:

(i) $\dfrac{d}{dt}(mv_m + Mv_M) = 0$;

(ii) $\dfrac{d}{dt}(v_m + v_M) = -\left(\dfrac{k}{m} + \dfrac{k}{M}\right)(x_m - x_M)$.

Find the frequency ω of oscillation of the distance $\xi = (x_m - x_M)$, and compare with ω_0, found in case (a). Find an intuitive argument why $\omega > \omega_0$.

7. **Impact time for a tennis ball.** A tennis ball of radius R is inflated to a pressure Δp above atmospheric pressure. On normal impact against a wall, the force exerted by the wall is $F = \Delta p A$, A being the contact area: $A = \pi R^2 \sin^2 \alpha \cong R^2 \alpha^2$ for small α.

   (a) Show that the displacement x of the ball's center, as defined in the figure is given by

   $$x \cong \tfrac{1}{2} R \alpha^2 \quad \text{(for small } \alpha\text{)}.$$

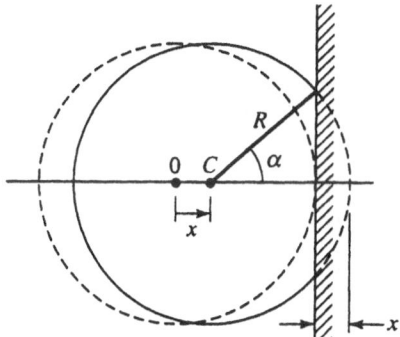

(b) Show that the force F on the ball is given by

$$F_x \cong -kx$$

and determine the constant k in terms of Δp and R.

(c) During impact, the center c of the ball therefore performs a harmonic motion

$$x = a \sin\left(\sqrt{\frac{k}{m}}\, t\right).$$

Show that the time duration of contact, t_0, is equal to one-half the period of this oscillation, and determine the constant a in terms of the ball's initial velocity.

(d) The mass of a regulation tennis ball is $m = 2$ oz. $= 5.67 \times 10^{-2}$ kg; its diameter is $2\frac{1}{2}$ in, and on one ball, we have measured $k = 1.1 \times 10^4$ N/m. Determine the value of its impact time t_0. Does it depend on the speed with which the ball is thrown?

(e) Determine the mean force exerted by the tennis ball on the wall, for an initial velocity v_0, from the total momentum change of the ball, and the impact time t_0. Compare this with the exact time-profile of the force.

(f) Find the numerical value of the mean force, if the ball hits the wall with a velocity $v_0 = 20$ m/s.

8. A walnut may be cracked by throwing it against a brick wall. Let x again be the flattening of the nut due to the contact force. The diagram on the left relates x to the force. Determine the impact velocity needed to crack the nut. Assume $M_{nut} = 5$ g.

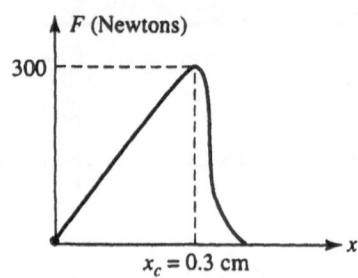

Hint: As in the tennis ball problem, the motion $x(t)$ after impact is given by $x(t) = (v_0/\omega) \sin(\omega t)$; determine ω. The amplitude of this oscillation must be at least equal to x_c for the nut to crack. This gives v_0.

9. Rank the order, according to the severity of impact, of the following three types of car collisions: a car, cruising with speed v_0, collides head on:

   (i) with a stationary car of the same mass;

   (ii) against a solid wall; and

   (iii) against a car moving in the opposite direction, at the same speed v_0.

10. Severity index. A skater falls on ice. Assume that his head hits the ice with a velocity corresponding to free fall from a height of 6 ft.

   (a) Assuming a constant force during impact, show that the severity index for this collision is

   $$I = \left(\frac{t_f}{\Delta t}\right)^{2.5} \cdot \Delta t,$$

   t_f being the time of free fall, and Δt the collision time. Calculate the value of I for collision times $\Delta t = 1, 2, 5, 10$ ms.

   (b) Let $H = 6$ ft be the height of free fall, and Δs the total "compression distance," that is, the amount of "flattening" experienced by skin, and/or protective padding. Show that I can be expressed as

   $$I = \left(\frac{H}{\Delta s}\right)^{1.5} \sqrt{2H/g}.$$

Tabulate I for $\Delta s = 0.1, 0.2, 0.5, 1$ cm. Identify the range of values of Δt (in case (a)) and Δs (in case (b)), for which serious injury must be expected.

11. An engineer wants to build a shock-absorbing bumper, designed to protect a car of mass $M = 1500$ kg up to a speed of $v_{\max} = 5$ mph without damage to the chassis. In his design, the bumper can retract over a maximum distance $D = 15$ cm. The shock absorber is a critically damped spring, so that the position of the *car* after impact (which occurs at $x = 0$, $t = 0$) is given by

$$x(t) = v_0 t e^{-\omega t},$$

v_0 being the impact velocity ($v_0 < v_M$).

(a) Find the expressions for $v_x(t)$ and dv_x/dt.

(b) Find an expression of the impact time t_c (needed to bring the car to a halt).

(c) Determine an expression for the parameter ω (which is $\sqrt{K/M}$, K being the spring constant of the spring in the shock absorber), from the condition that the car comes to rest in a distance D, if $v_0 = v_{\max}$. Determine ω from the data given.

(d) Using these data, plot numerically the force on the car during slow-down, as a function of time.

12. Molecular collisions and pressure. The observed pressure exerted by a gas against the container walls is the net effect of the momentum transfer a gas molecule (or atom) undergoes in a collision with the wall. In the case of helium, the mass m of an atom is $m = 6.8 \times 10^{-24}$ g, and the mean velocity component normal to the wall is $\langle v_1 \rangle_{\mathrm{av}} \cong 600$ m/s.

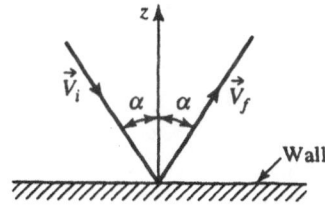

(a) Using these data, find the mean momentum change during a collision of an atom with the wall.

(b) Let n be the number of atoms per unit area, per second, hitting the wall. That is, in a time Δt, and over an area A, there will be

$$N = n\Delta t A \quad \text{collisions.}$$

Give an expression for the total impulse I of all these collisions exerted on the area A in the time interval Δt.

(c) Show that the average force against the area A of the wall is given by

$$\langle F \rangle_{av} = (\text{total impulse in time } \Delta t)/\Delta t.$$

This force, divided by the area A, defines the gas pressure p. Express p in terms of n and the mean momentum change per collision.

(d) For the case that the pressure of helium is equal to 1 atm ($= 10^5$ N/m^2), determine the numerical value of n, the number of collisions/second cm^2.

13. Toy rocket. In this gadget, water is injected into the air-filled rocket body, which, at the time of release, is half-filled with water and half with compressed air. If the nozzle area is a, what is the minimum speed, v, with which the water must be ejected to lift the rocket off the ground? *Hint:* The system, whose total momentum is conserved, consists of two parts: (1) the rocket and (2) the ejected water. Part (2) acquires momentum at the rate

$$\frac{dp_2}{dt} = \frac{d(mv)}{dt},$$

m being the mass of ejected water, v the speed of ejection.

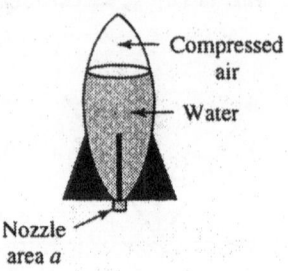

Nozzle
area a

(a) Show that dp_2/dt can be written as $\cong \rho \cdot av^2$, ρ being the density of water.

(b) For liftoff, the force exerted by the ejected water on the rocket ($=$ dp_2/dt), must exceed the weight Mg. From this, find the minimum value of v, assuming $M = 100$ g, $a = 0.1$ cm^2.

CHAPTER 5

Work and Energy

5.1 Introduction

The basic equations of mechanics, Newton's laws of motion, were presented in Chapter 2 in such a way that the *force* is the central quantity that determines the motion. If the force is known as a function of *time*, that formulation permits a quite direct determination of the motion simply by integrating the acceleration twice.

In the previous chapter (Chapter 4) we saw that Newton's laws of motion can be expressed in such a way that the central quantity is not the force but is rather the momentum. The momentum formulation of mechanics is particularly useful if the system under consideration is chosen so that there is no net external force acting on it. Under these conditions it is possible to determine the motion of one part of the system if the motion of a second part is known. This can be done without any knowledge of the forces of interaction between the parts of the system. In the momentum formulation we saw that interesting parts of the motion could be computed without reference to the forces at work.

In the present chapter we wish to continue the process of reexpression and reformulation of Newton's laws. In this chapter the new concept of the "kinetic energy" of a moving body is introduced. It is then shown that a change in kinetic energy is related to the work done on the body. The work–energy formulation of mechanics arises very naturally when the forces acting on the system are known as a function of the position of the body, rather than the time. Since in many physical situations the force is known in terms of the location of the body, it is often much more convenient to solve for the motion using work and energy considerations than to analyze the effects of the force.

On a deeper basis, the work–energy formulation of mechanics has profound implications beyond mechanics itself. This formulation leads to the general idea of the conservation of energy that is at the heart of the science of thermodynamics.

320

Thermodynamics is based on an extended version of the energy conservation principle, in which heat, chemical, and electrical forms of energy are recognized. This enables a systematic discussion of the processes in which energy is transferred from one form to another, while the sum total remains fixed. In Section 5.4.E, we shall apply this idea to the work–energy balance associated with the functioning of the various organs in the human body.

Another point of fundamental importance is the fact that the energy conservation principle remains valid in the domain of molecular, atomic, and nuclear physics, where the classical description of phenomena—as described in this volume—fails and must be replaced by the laws of quantum mechanics. Its range of application thus goes beyond the content in which it originally emerged.

5.2   Work, Kinetic Energy, and Power in One-Dimensional Motion

5.2.A.   A First Integration of Newton's Equation. The Work–Kinetic Energy Formula

By confining ourselves, at first, to one-dimensional motion, we can more readily see the origin, meaning, and utility of the work–energy formulation of the laws of motion. Let us consider a particle of mass m which moves only along the x-axis. The particle is presumed to move under the action of a force ($F(x)$) whose magnitude and direction is known at each position (x). As usual we seek to determine the position and velocity of the particle.

Newton's law of motion for the particle is

$$m\frac{d^2x}{dt^2} = F(x). \tag{5-1}$$

Since the force F is not known as a function of time we cannot proceed to find $x(t)$ directly by two integrations. We can, however, proceed by first obtaining the velocity as a function of position. This is done by making a single integration of (5-1) in the following way. The velocity V is equal to dx/dt. Thus (5-1) can also be written as

$$m\left(\frac{dV}{dt}\right) = F(x). \tag{5-2}$$

Now multiply both sides of this equation by V. This gives

$$mV\frac{dV}{dt} = F(x)V. \tag{5-3}$$

But the left-hand side can be written as

$$V\frac{dV}{dt} = \frac{d}{dt}\left(\tfrac{1}{2}V^2\right). \tag{5-4}$$

Thus we see that (5-3) can be written as

$$\frac{d}{dt}\left(\tfrac{1}{2}mV^2\right) = F(x)V. \tag{5-5}$$

On multiplying by dt and integrating this equation, we have

$$\int d\left(\tfrac{1}{2}mV^2\right) = \int F(x)V\,dt = \int F(x)\,dx. \tag{5-6}$$

Suppose that initially the particle position is x_0 and the velocity is V_0. If we denote the velocity at some position x as $V(x)$, then we can carry out the integration with these quantities in the limits of the integration as

$$\int_{V_0}^{V(x)} d\left(\frac{1}{2}mV^2\right) = \int_{x_0}^{x} F(x)\,dx. \tag{5-7}$$

On carrying out the integration of the perfect differential on the left-hand side, we have

$$\tfrac{1}{2}mV^2(x) - \tfrac{1}{2}mV_0^2 = \int_{x_0}^{x} F(x)\,dx. \tag{5-8}$$

If F is known as a function of x the right-hand side can be integrated explicitly.

The integral of the force over the path of the particle is called the "work" (W) done by the force. The integral on the right-hand side of (5-8) can be written in terms of W as

$$W_{x \leftarrow x_0} = \int_{x_0}^{x} F(x)\,dx. \tag{5-9}$$

This is the "work" done *on* the particle *by* the force in its motion from x_0 to x. This work will also be denoted by the symbol $W_{x x_0}$.

The quantity $(\frac{1}{2}mV^2(x))$ on the left-hand side of (5-8) is called the "kinetic energy" $(T(x))$ of the particle when it is in position x. The quantity $\frac{1}{2}mV^2(x) - \frac{1}{2}mV_0^2$ is the change in kinetic energy as the particle moves from x_0 to x. Thus we can write (5-8) as

$$T(x) - T(x_0) = W_{x,x_0}. \qquad (5\text{-}10)$$

In words, one says that the work done on the body (W_{x,x_0}) is equal to the change in the kinetic energy of the body. Thus, to get the velocity of a body at any position, we need in essence to compute only the "work" needed to move the particle along the path of motion to the position in question.

5.2.B.   Power and the Rate of Change of Kinetic Energy

We may observe that (5-5) also has a direct physical interpretation. The quantity

$$\frac{d}{dt}\left(\frac{1}{2}mV^2\right) \equiv \frac{dT}{dt} \qquad (5\text{-}11)$$

is the rate of change of the kinetic energy. Furthermore, the right-hand side of (5-5), namely, $F(x)V(x) = F\,dx/dt$ is the *rate* at which work is being done on the body. This rate of doing work is called the "power" P, i.e.,

$$P \equiv F(x)V(x). \qquad (5\text{-}12)$$

By putting (5-12) and (5-11) into (5-5), we obtain the relation

$$P = \frac{dT}{dt}. \qquad (5\text{-}13)$$

Physically this states that the power delivered to the body at each point in its trajectory is equal to the rate of change of its kinetic energy.

5.2.C.   Determination of $x(t)$ Using Work–Energy Considerations

The work–kinetic energy formula

$$\frac{1}{2}mV^2(x) - \frac{1}{2}mV_0^2 = W_{x,x_0} \qquad (5\text{-}14)$$

is simply a first integral of the equations of motion. It permits the determination of the velocity at each point x in terms of the work done in taking the particle to that point. It is possible to proceed further and use this work–kinetic energy formula to determine the position x as a function of time. We can see formally, mathematically, how this is done quite directly. Since W_{x,x_0} is presumably known as a function of x, we can rearrange (5-14) into the following form:

$$V_x^2 = \left(\frac{dx}{dt}\right)^2 = \frac{2}{m}W_{x,x_0} + V_0^2 \qquad (5\text{-}15)$$

or

$$\frac{dx}{dt} = \sqrt{\frac{2}{m}W_{x,x_0} + V_0^2}. \qquad (5\text{-}16)$$

On transferring all the variables dependent on x to the left-hand side of this equation, and the time part to the right-hand side, we have

$$\frac{dx}{\sqrt{(2/m)W_{x,x_0} + V_0^2}} = dt. \qquad (5\text{-}17)$$

On integrating this, using the conditions that the position is $x(t)$ when the time is t, and the initial condition that at $t = 0$, $x = x_0$, we have the formal solution of the motion as

$$\int_{x_0}^{x(t)} \frac{dx}{\sqrt{(2/m)W_{x,x_0} + V_0^2}} = \int_0^t dt \qquad (5\text{-}18)$$

or

$$t = \int_{x_0}^{x(t)} \frac{dx}{\sqrt{(2/m)W_{x,x_0} + V_0^2}}. \qquad (5\text{-}19)$$

The actual explicit calculation of the position $x(t)$ then involves carrying out two integrations. The first involves computing the work W_{x,x_0}; the second requires the integration of the right-hand side of (5-19). This equation then gives t as a function of x, and by inversion of this relation one finds x as a function of t.

5.2.D.   Conservative and Nonconservative Forces

To use the work–energy formulation, we see that it is necessary to know the work W_{x,x_0} done as the particle moves from x_0 to x. One would think that this work might depend on precisely how the particle moves from x_0 to x. For example, we might expect that the work could depend upon the speed during the interval from x_0 to x, or on the detailed path. In either of these cases the work–energy formulation of mechanics would be useless. There exists, however, a wide variety of forces for which W_{x,x_0} depends *only* upon the initial and final positions x_0 and x, and *not* on the time sequence of intermediate positions. If the force acting on the particle depends *only* on the position and not on the direction of the motion or on the speed of the motion, then W_{x,x_0} is indeed independent of the path and depends only on x_0 and x. (We will demonstrate this property of W of W_{x,x_0} in the examples to follow in Section 5.2.E, and more generally in Section 5.2.F.) When W_{x,x_0} is independent of the path, it can be evaluated without knowing the motion. The motion can then be determined using W_{x,x_0} as discussed above. Forces which have the property that the work they do is independent of the path are called "conservative" forces. The reason for this name is that if such forces act on a particle, it will move in such a way that its total "mechanical energy" is conserved; that is, the "mechanical energy" of the particle remains *constant* throughout the motion. The existence then of such forces permits the formulation of the law of conservation of mechanical energy.

The nonconservative forces need not be regarded as being "beyond the pale." In fact, by generalizing the work–energy balance to include the "heat" associated with nonconservative forces, it is possible to establish that great generalization: the first law of thermodynamics. This law is nothing other than the law of conservation of energy including "heat" as a form of energy. We shall see later in this chapter precisely what is meant by these terms. At present, however, it is very interesting to recognize that the keys to the further development of the work–energy formulation of mechanics can be found by carefully examining the physical content of a first integral of the motion of a particle in one dimension.

5.2.E.   Examples of One-Dimensional Motion in Simple Force Fields

Using the work–energy formulas, we can analyze the motion of a particle in the following simple conservative force fields:

(i) The gravitational field near the Earth's surface. Here the gravity force is a constant over the motion.

(ii) The elastic restoring force—Hooke's law force. Here the force on the body is linearly proportional and opposite in direction to the displacement x of the body from its stable position.

(iii) The gravitational force at great distances from the Earth's surface. Here the force varies inversely as the square of the distance between the center of the Earth and the particle.

(i)   Constant Force of Gravity

Let us obtain the motion for the following initial conditions: at $t = 0$, $x = 0$, $V = V_0$. The initial velocity is upward and the gravitational force is downward (see Figure 5.1). According to the work–kinetic energy theorem, (5-8), the velocity at each position x is given by

$$\tfrac{1}{2}mV^2(x) - \tfrac{1}{2}mV_0^2 = \int_{x_0=0}^{x} F\, dx = W_{x,0}. \qquad (5\text{-}20)$$

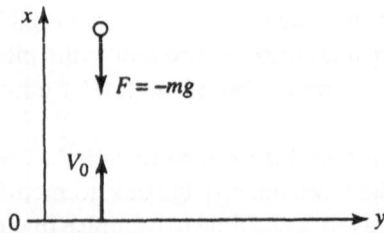

Figure 5.1. Particle moving under a constant gravitational force.

Since $F = -mg$ we see at once that

$$W_{x,0} = \int_{0}^{x} -mg\, dx = -mgx. \qquad (5\text{-}21)$$

Notice that if we had evaluated $W_{x,0}$ over a path like that indicated in Figure 5.2, we would obtain the same result as (5-21).

We can compute $W_{x,0}$ by integrating $(-mg)$, the gravitational force over the alternate path shown in Figure 5.2. This gives

$$W_{x,0} = \int_{0}^{x} (-mg)\, dx = \int_{0}^{x} (-mg)\, dx + \int_{x}^{x'} (-mg)\, dx + \int_{x'}^{x} (-mg)\, dx. \quad (5\text{-}22)$$

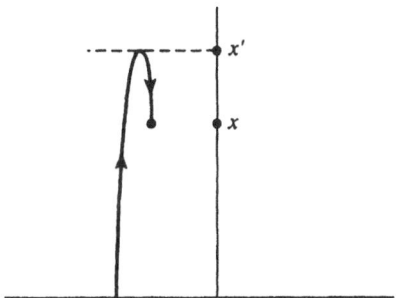

Figure 5.2. Alternative path for calculation of work. Here x' is some arbitrary position greater than the position x for which W_x, x_0 is being calculated.

Since the gravity force always points in the same direction, the integrals from x to x' and x' down to x are equal and opposite. Thus $W(x)$ is the same for this and all other paths between the origin and the position x. Thus, the constant gravitational force is a conservative force. On placing the equation for the work (5-21) into (5-20), we have the following result for the velocity at each height x:

$$V^2(x) = V_0^2 - 2gx. \tag{5-23}$$

This, of course, is precisely the same result which we would expect from the kinematics of a body experiencing a constant negative acceleration g.

To obtain the position of the particle at each moment $x(t)$, we must integrate this equation for the velocity once more. We showed how to do this in general in Section 5.2.C, and obtained the result (5-19) that x is given as

$$t = \int_0^x \frac{dx}{\sqrt{2W_{x,x_0}/m + V_0^2}}. \tag{5-24}$$

Since in the present case $W_{x,x_0} = -mgx$, we have

$$t = \int_0^x \frac{dx}{\sqrt{V_0^2 - 2gx}}. \tag{5-25}$$

This is an elementary integral whose value is

$$t = \left[-\frac{1}{g}\sqrt{V_0^2 - 2gx} \right]_0^x. \tag{5-26}$$

On inserting the upper and lower limit, we find

$$t = \frac{V_0}{g} - \frac{1}{g}\sqrt{V_0^2 - 2gx}. \tag{5-27}$$

This expresses time as a function of x. By rearranging the equation so that x is displayed alone on the left-hand side, and all the terms involving the time are on the right, we obtain the result

$$x = V_0 t - \tfrac{1}{2}gt^2 \tag{5-28}$$

This, of course, is the familiar result for the position of a particle undergoing constant deceleration of magnitude g.

   We see then that the energy approach is a perfectly equivalent alternative for the determination of the motion in this simple case. It is instructive to see how the "energy is conserved" in this particular example of motion under the action of a conservative force. If we return to (5-20) and insert (5-21) therein we find

$$\tfrac{1}{2}mV^2(x) - \tfrac{1}{2}mV_0^2 = -mgx. \tag{5-29}$$

These terms can be rearranged into the form

$$\tfrac{1}{2}mV^2(x) + mgx = \tfrac{1}{2}mV_0^2. \tag{5-30}$$

This equation states that the sum of the kinetic energy $\tfrac{1}{2}mV^2(x)$ and mgx is a constant throughout the motion. Regardless of the position x, as x increases $V(x)$ must decrease so that the sum of these terms is equal to $\tfrac{1}{2}mV_0^2$, the original kinetic energy of the particle. mgx, the negative of the work done by the force, is called the "potential energy" of the particle at position x. When the particle is at $x = 0$ its potential energy is zero. Equation 5-30 represents a conservation of mechanical energy equation in the sense that the sum of the kinetic energy of the particle, plus the potential energy, is always equal to the original sum of the kinetic plus potential energies, i.e., $\tfrac{1}{2}mV_0^2$. The conservation of mechanical energy, (5-30), is at the basis of a graphical analysis of the motion called the energy diagram. In the energy diagram we plot the total energy and the potential energy as a function of the particle position (see Figure 5.3).

   At each point in the diagram the difference between the sloping line and the total initial energy is equal to the kinetic energy. Since the kinetic energy must always be a positive quantity, the particle can never rise higher in the gravitational

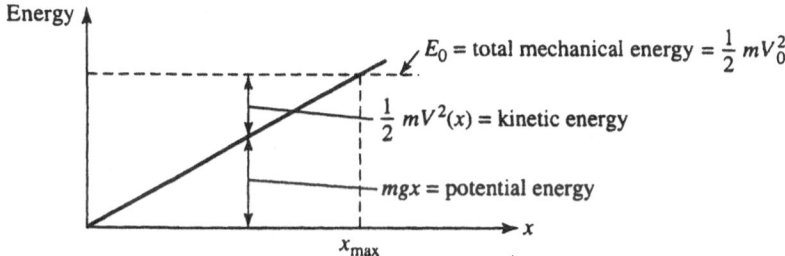

Figure 5.3. Energy diagram for particle in a uniform gravitational field.

field than the value of x_{max} indicated in the diagram. At this height the kinetic energy is zero, and all the mechanical energy is in the potential energy term. Thus

$$mgx_{\text{max}} = \tfrac{1}{2}mV_0^2 \tag{5-31}$$

or

$$x_{\text{max}} = \frac{V_0^2}{2g}. \tag{5-32}$$

(ii)   Linear Restoring Force. The Harmonic Oscillator

We now consider the motion of a particle subject to a force whose magnitude is directly proportional to the displacement x of the particle away from its equilibrium position. The direction of the force is toward the equilibrium position $x = 0$. This type of force in fact occurs quite generally in mechanical and electrical systems when the displacements around the equilibrium points are small. The simplest example is that of a particle attached to a spring (Figure 5.4).

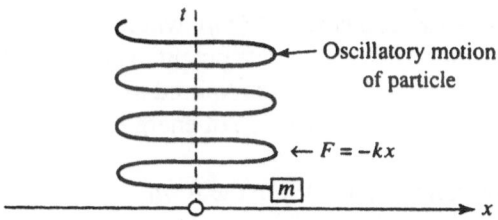

Figure 5.4. A particle of mass m moves under the influence of a restoring force linear in the displacement x.

We have already considered the motion of a particle under the action of a linear restoring force. The motion is oscillatory. The body oscillates sinusoidally and symmetrically about the point $x = 0$. We previously obtained this result in Chapter 2 by solving the second-order differential equation which results from the application of Newton's second law.

Let us now examine the motion from the work–energy point of view. The central question is, of course, the work W_{x,x_0} associated with the particle moving between x_0 and x. This is

$$W_{x,x_0} = \int_{x_0}^{x} F \, dx = \int_{x_0}^{x} (-kx) \, dx, \tag{5-33}$$

$$W_{x,x_0} = -\tfrac{1}{2}kx^2 + \tfrac{1}{2}kx_0^2 = -\tfrac{1}{2}k(x^2 - x_0^2). \tag{5-34}$$

If we take any other path such as the one shown in Figure 5.3, to carry out the integral, the result will be exactly the same; as the reader can readily verify. Thus this linear restoring force is a conservative force. We now use (5-34) in the work–kinetic energy formula and find that the velocity at position x is given by

$$\tfrac{1}{2}mV^2(x) - \tfrac{1}{2}mV_0^2 = -\tfrac{1}{2}k(x^2 - x_0^2). \tag{5-35}$$

Here V_0 is the initial velocity and x_0 is the initial position of the particle. This equation can be reexpressed in two different ways. The first way is appropriate to show the conservation of mechanical energy, viz:

$$\tfrac{1}{2}mV^2(x) + \tfrac{1}{2}kx^2 = \tfrac{1}{2}mV_0^2 + \tfrac{1}{2}kx_0^2 \equiv E_0. \tag{5-36}$$

In this form the equation states that the sum of the kinetic energy $\tfrac{1}{2}mV^2(x)$, plus the "potential energy" $\tfrac{1}{2}kx^2$ is a constant (E_0), where E_0 is equal to the initial sum of the kinetic and "potential energies." The potential energy of the particle is, again in this example, the negative of the work required to move the particle to the point x. We shall presently use this conservation of energy expression to further analyze the motion.

Alternatively, one can write (5-35) in the form

$$V^2(x) = V_0^2 + \frac{k}{m}x_0^2 - \frac{k}{m}x^2. \tag{5-37}$$

This equation tells us that the maximum velocity occurs at $x = 0$ and at that point it is equal to

$$V_m = \sqrt{V_0^2 + \frac{k}{m}x_0^2}. \tag{5-38}$$

As the particle moves either to the right or to the left, the velocity decreases with displacement until it is equal to zero at the end points corresponding to the maximum displacement at

$$x_m = \sqrt{x_0^2 + \frac{mV_0^2}{k}}. \tag{5-39}$$

This is the value of x at which $V(x) = 0$.

We observe that V_m and x_m are related by the equation

$$V_m = \sqrt{\frac{k}{m}}\, x_m. \tag{5-40}$$

Since, as we shall soon see, $\sqrt{k/m}$ is the angular frequency of the oscillatory motion, we shall denote this quantity as

$$\omega = \sqrt{\frac{k}{m}}. \tag{5-41}$$

We continue our analysis of the motion of the particle by integrating the velocity formula (5-37) to find the position as a function of time. As we saw in general in Section 5.2.C, we can compute $x(t)$ using (5-19). This is the same as substituting $(dx/dt) = V(x)$ in (5-37) and integrating. In either case we have that $x(t)$ is found from

$$t = \int_{x_0}^{x(t)} \frac{dx}{\sqrt{(k/m)(x_0^2 - x^2) + V_0^2}} \tag{5-42}$$

or, since $\omega^2 = k/m$, we have

$$\omega t = \int_{x_0}^{x(t)} \frac{dx}{\sqrt{(V_0^2/\omega^2 + x_0^2) - x^2}}. \tag{5-43}$$

Since, as we have already seen, the maximum displacement x_m is given by $x_m^2 = x_0^2 + V_0^2/\omega^2$ (see (5-39)). We can reexpress (5-43) as

$$\omega t = \int_{x_0}^{x(t)} \frac{dx}{\sqrt{x_m^2 - x^2}}. \tag{5-44}$$

This can be integrated at once since $\int dx/\sqrt{a^2 - x^2} = \arcsin(x/a)$. Thus, we find

$$\omega t = \arcsin\left(\frac{x}{x_m}\right)\Bigg|_{x=x_0}^{x(t)} = \arcsin\left(\frac{x(t)}{x_m}\right) - \arcsin\left(\frac{x_0}{x_m}\right). \tag{5-45}$$

The quantity $\arcsin(x_0/x_m)$ is an angle ϕ which is shown in the triangle in Figure 5.5. With this definition for ϕ, which is conventionally called the "phase angle," we have from (5-45) the solution of the motion

$$\omega t + \phi = \arcsin\left(\frac{x(t)}{x_m}\right) \tag{5-46}$$

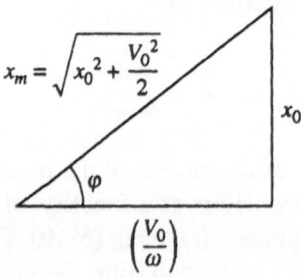

Figure 5.5. Definition of the angle ϕ. $\phi = \arcsin(x_0/x_m)$.

or, finally,

$$x(t) = x_m \sin(\omega t + \phi). \tag{5-47}$$

The motion of the particle is sinusoidal. The particle oscillates symmetrically about $x = 0$. The angular frequency of the oscillation is $\omega = \sqrt{k/m}$. The amplitude of the oscillation, i.e., the maximum displacement, is

$$x_\mathrm{m} = \sqrt{x_0^2 + \frac{V_0^2}{\omega^2}}. \tag{5-48}$$

The velocity also oscillates. The velocity as a function of time is

$$V(t) = x_\mathrm{m}\omega\cos(\omega t + \phi). \tag{5-49}$$

The maximum velocity is $x_\mathrm{m}\omega$.

Having obtained this detailed solution of the motion of the particle, let us examine the energy diagram and the conservation of mechanical energy formula to see how we could obtain all the essential features of the motion simply by inspection. In Figure 5.6 we show a plot of the various contributions to the energy of the oscillating particle.

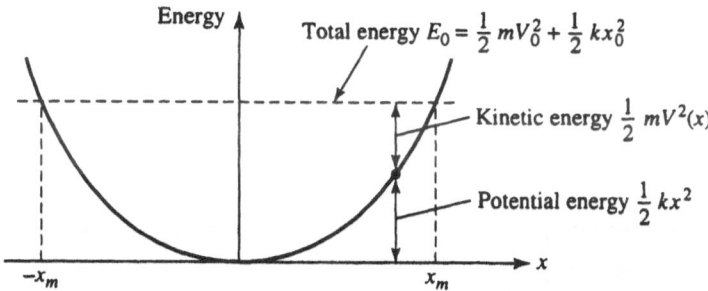

Figure 5.6. Energy diagram for the harmonic oscillator.

The conservation of energy formula (5-36) shows us that at each point in the motion the sum of the kinetic and potential energies is a constant. We see from this at once that the velocity is maximum at $x = 0$, and its value there is

$$V_\mathrm{m} = \sqrt{\frac{2E_0}{m}} = \sqrt{V_0^2 + \left(\frac{k}{m}\right)x_0^2}. \tag{5-50}$$

We also see at once that the maximum displacement is at the point at which the potential energy curve intersects the total energy line. The body cannot move further from the origin than this, or else its velocity would be imaginary. The intersection between the parabola ($\frac{1}{2}kx^2$) and the line E_0 occurs at the maximum displacement

$$x_m = \sqrt{x_0^2 + \left(\frac{m}{k}\right) V_0^2} = \sqrt{\frac{k}{m}} V_m. \qquad (5\text{-}51)$$

We can visualize the motion of the particle in Figure 5.6. The position of the particle oscillates back and forth between $-x_m \leq x \leq x_m$. During this oscillating motion the total energy with which the system is started is constantly shifting between the kinetic energy of the particle and the potential energy stored in the system which provides the restoring force.

It is interesting to observe that simply by examining the expression for the total energy of the system we can obtain all the principal features of the motion. Namely, from

$$E = \tfrac{1}{2} m V^2 + \tfrac{1}{2} k x^2 \qquad (5\text{-}52)$$

we find $x_m = \sqrt{2E/k}$ (since $V = 0$ at $x = x_m$), and also we find $V_m = \sqrt{2E/m}$ (since $x = 0$ at $V = v_m$). Finally, we find that the frequency ω and the period τ of the motion are given by

$$\omega = 2\pi/\tau = \left(\frac{V_m}{x_m}\right) = \sqrt{\frac{k}{m}}. \qquad (5\text{-}53)$$

The reason that we have spent so much time discussing this linear restoring force system is that a very wide class of physical systems has a total energy which has the same mathematical form as does the mechanical system discussed above. In all these systems the total energy can be written in the general form

$$E_0 = A \left(\frac{dq}{dt}\right)^2 + B q^2. \qquad (5\text{-}54)$$

The physical meaning of the variable denoted by the symbol q is different in each physical system. Nevertheless, in each system, q varies in time exactly as does the displacement x in the mechanical example which we have discussed so carefully. In each case the angular frequency is $\omega = \sqrt{B/A}$. The maximum value of q is

$$q_m = \sqrt{\frac{E_0}{B}} \qquad (5\text{-}55)$$

and the maximum value of (dq/dt) is

Table 5.1: A Few Examples of Physical Systems That Behave Like Harmonic Oscillators.

System	General displacement variable (q)	Significance of displacement variable
Pendulum or tuning fork	θ	Angular position of pendulum bob Angular displacement of tines
Oscillating electrical "L–C circuit"	Q	Electrical charge on capacitor
Piezoelectric crystal oscillator	ϵ	Compressional or shear strain of crystal
Vibration of atoms bound in molecules	q	Displacement of vibrating atom
Electromagnetic field in resonant cavity	E	Electric field vector
Musical wind instruments, organ pipes, brasses	p	Pressure of air in the instrument

$$\left(\frac{dq}{dt}\right)_{\text{max}} = \sqrt{\frac{E_0}{A}} = \omega q_{\text{m}}. \tag{5-56}$$

In Table 5.1 we give a list of some of the physical systems that behave like the harmonic oscillator, i.e., systems whose total energy has the form of (5-54). We also show in this table the physical interpretation of the variable q in each case.

(iii) Motion in a Nonuniform Gravitational Field

Suppose that we consider the motion of a particle that is shot with great speed from the Earth's surface. Let the speed be so great that the object can travel distances comparable to the radius of the Earth or even larger before being pulled back by gravity to the Earth's surface. We can analyze the motion of this problem very conveniently using the work–kinetic energy formulation of mechanics. In fact, we can quite simply determine the maximum height reached by the body as a function of its initial velocity V_0, and can also easily find the velocity needed to "escape." (See Figure 5.7.)

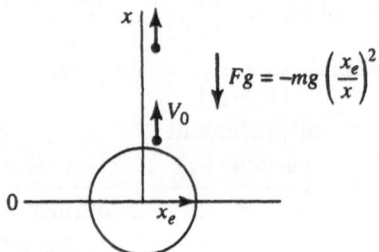

Figure 5.7. Diagram of the motion of a particle thrown upward from the Earth's surface with initial velocity V_0.

We proceed in precisely the same manner for this example as we did in the previous two examples. The work–kinetic energy formula is

$$\tfrac{1}{2}mV^2(x) - \tfrac{1}{2}mV_0^2 = \int_{x_0}^{x} F\,dx. \tag{5-57}$$

In the present problem, the initial position x_0 is at the surface of the Earth, $x = x_e$. The force acting is the force of gravity which varies inversely as the square of the distance from the center of the Earth

$$F = -mg\left(\frac{x_e}{x}\right)^2, \tag{5-58}$$

where m is the mass of the body and g is the acceleration of gravity at the Earth's surface.

The work W_{x,x_e} is equal to

$$W_{x,x_e} = \int_{x_e}^{x} F\,dx = -mgx_e^2 \int_{x_e}^{x}\left(\frac{dx}{x^2}\right). \tag{5-59}$$

Carrying out the elementary integral gives

$$W_{x,x_e} = mgx_e^2 \left(\frac{1}{x}\right)\Big|_{x_e}^{x}, \tag{5-60}$$

i.e.,

$$W_{x,x_e} = mgx_e^2 \left(\frac{1}{x} - \frac{1}{x_e}\right) \equiv W(x) - W(x_e). \tag{5-61}$$

Notice that if this integral was evaluated along any path (like that in Figure 5.7), the result would be identical to that given above. (This is because the integral beyond x to x' will cancel against the integral from x' back down to x.) W_{x,x_e} depends only on x and x_e and is independent of the path followed between them. Thus this gravitational force is a "conservative" force. Because of this, the work–kinetic energy formula can be written in the form of the conservation of total mechanical energy. This is seen as follows. Using (5-61) in (5-57), we find

$$\tfrac{1}{2}mV^2(x) - \tfrac{1}{2}mV_0^2 = mgx_e^2\left(\frac{1}{x} - \frac{1}{x_e}\right). \tag{5-62}$$

This can be rearranged into the form representing the conservation of mechanical energy, viz:

$$\tfrac{1}{2}mV^2(x) - \frac{mgx_e^2}{x} = \left(\tfrac{1}{2}mV_0^2 - mgx_e\right) = E_0. \tag{5-63}$$

As in our other examples, the work associated with moving the particle between the two points x_e and x can be written as the difference between the two terms $W(x)$ and $W(x_e)$ shown in (5-61). If we call $-W(x)$ the "potential energy" associated with x, and $-W(x_e)$ the potential energy associated with the point x_e, then (5-63) can be interpreted as follows. The sum of the kinetic energy $\tfrac{1}{2}mV^2(x)$ and the potential energy $-W(x) = -mgx_e^2/x$ at any point in the motion is equal to a constant (E_0). This constant is equal to the total kinetic and potential energies of the particle at the beginning of the motion.

The conservation of energy formula can be represented graphically by an energy diagram (Figure 5.8) in which one plots the potential energy and the total energy of the body as a function of the distance x. We can now determine the velocity of the particle at each position of the particle. From (5-63) the velocity $V(x)$ is related to x, and the original energy by the formula

$$\tfrac{1}{2}mV^2(x) = E_0 + \frac{mgx_e^2}{x} \tag{5-64}$$

or

$$V^2(x) = \frac{2E_0}{m} + \frac{2gx_e^2}{x}. \tag{5-65}$$

The turning points of the motion can be found very simply by referring to the energy diagram (Figure 5.8). Suppose first that the total initial mechanical energy

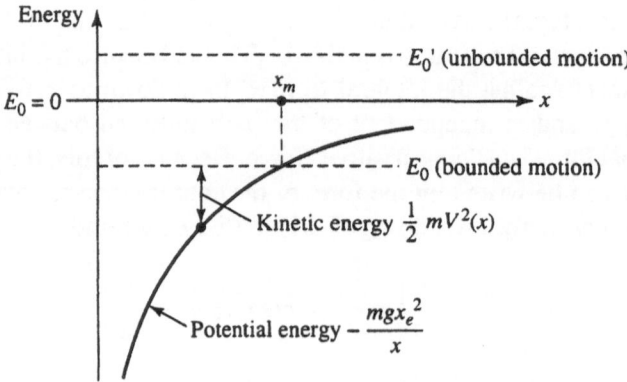

Figure 5.8. Energy diagram for a body in the Earth's gravitational field.

E_0 is a negative number, i.e., that the potential energy $-mgx_e$ at the Earth's surface is larger than the initial kinetic energy $\frac{1}{2}mV_0^2$ given the particle. In that case, the total energy E_0 is represented as a horizontal line beneath the $E_0 = 0$ axis as shown above. The kinetic energy of the body is the difference between E_0 and the potential energy curve. For negative values of E_0 the particle cannot move higher than the position $x = x_m$. Above this point, in the region shaded in Figure 5.8, its kinetic energy would be negative, and its velocity imaginary. The position $x = x_m$ is the turning point of the motion. The body rises to this point and then falls back down to Earth.

The actual value of x_m is the value of x in (5-64) for which $V(x) = 0$, i.e.,

$$\frac{mgx_e^2}{x_m} = -E_0. \tag{5-66}$$

We see that when E_0 is negative we can solve for x_m. One says in this case that the motion is "bounded"—the particle only goes to some maximum height and then returns to Earth. The maximum height is

$$x_m = \frac{mgx_e^2}{-E_0}. \tag{5-67}$$

On using the expression (5-63) for E_0, we can write this formula for x_m as

$$x_m = \frac{mgx_e^2}{(mgx_e - \frac{1}{2}mV_0^2)}. \tag{5-68}$$

From this formula we see at once that the particle will just escape from the Earth when its initial kinetic energy is equal to the gravitational potential energy mgx_e at the Earth's surface, i.e.,

$$mgx_e = \tfrac{1}{2}m\left(V_0^2\right)_{escape}. \tag{5-69}$$

Thus the initial velocity needed for escape is

$$(V_0)_{escape} = \sqrt{2gx_e}. \tag{5-70}$$

Since the radius of the Earth is

$$x_e \cong 3.95 \times 10^3 \text{ mile}$$

and the acceleration of gravity is

$$g = 32.2 \text{ ft/s}^2 = 6.10 \times 10^{-3} \text{ mile/s}^2.$$

The numerical value of the escape velocity is

$$(V_0)_{escape} = \sqrt{2(6.10 \times 10^{-3})(3.95 \times 10^3) \text{ mile}^2/\text{s}^2}.$$
$$(V_0)_{escape} = 6.94 \text{ mile/s},$$

or

$$(V_0)_{escape} \cong 25,000 \text{ mile/h}. \tag{5-71}$$

Suppose the initial velocity in V_0 is some fraction f times the escape velocity, i.e.,

$$V_0 = f(V_0)_{escape}. \tag{5-72}$$

Here f is a number between 0 and unity. We can easily express that the height of the particle will rise in terms of this number f, since the initial kinetic energy can be written as

$$\tfrac{1}{2}mV_0^2 = \tfrac{1}{2}mf^2(V_0)^2_{escape}$$
$$= f^2 mgx_e.$$

On inserting this into (5-68), we find that x_m can be expressed as

$$x_m = \frac{mgx_e^2}{mgx_e - f^2mgx_e} \qquad (5\text{-}73)$$

or, finally,

$$x_m = \frac{x_e}{(1 - f^2)}. \qquad (5\text{-}74)$$

Thus if the initial velocity is, say, one-half the escape velocity, the particle will rise to a distance $x_m = \frac{4}{3}x_e$ above the center of the Earth, or a distance equal to $\frac{1}{3}x_e$ above the Earth's surface. When the velocity of the body V_0 is small compared to the escape velocity, i.e., $f \lesssim 0.2$, we can obtain a very simple expression for the height that the particle rises over the surface of the Earth. This height is $h = (x_m - x_e)$. Since for small f we have approximately that

$$\frac{1}{1 - f^2} \cong 1 + f^2, \qquad (5\text{-}75)$$

we can write (5-74) as

$$x_m \cong x_e(1 + f^2).$$

or

$$h = (x_m - x_e) \cong x_e f^2 \qquad (5\text{-}76)$$

Thus, for small values of f, the particle rises above the Earth a distance equal to the Earth's radius times the square of the fraction f.

The energy diagram of Figure 5.8 also shows that if the initial energy E_0 is positive, i.e., the initial kinetic energy exceeds the potential energy at the Earth's surface, the horizontal line E_0' never intersects the potential energy hyperbola. The difference between these two curves never goes to zero, and the particle always has a finite kinetic energy. This means that the projectile never stops going. It slows down, but it never stops. It escapes the Earth. The motion in this case is said to be unbounded. The crossover between unbound and bounded motion occurs when E_0 is equal to zero.

5.2.F.   Conservative Forces and the Conservation of Mechanical Energy

In each of the examples discussed above we saw that the work done by the forces was independent of the path between two points and only depended on the coordinates of the points. Because of this it was possible to reexpress the work–kinetic energy formula in (5-14) in such a way that a quantity called the total mechanical energy is constant throughout the motion. Let us formulate this conservation of energy law generally so that it applies for any form of conservative force. We continue to confine ourselves to one-dimensional motion for convenience

$$\tfrac{1}{2}mV^2(x) - \tfrac{1}{2}mV_0^2 = \int_{x_0}^{x} F(x)\,dx = W_{x,x_0}. \tag{5-77}$$

Now, if the force $F(x)$ is conservative, then as we saw in each example, W_{x,x_0} is independent of the path between x_0, and x and is dependent only on the coordinates x_0 and x. This means that W_{x,x_0} can be written as the difference between the value of some function $(-U)$ evaluated at x and x_0. (The reason for the negative sign in the definition of the function (U) will presently be quite clear.) That is,

$$W_{x,x_0} = \int_{x_0}^{x} F(x)\,dx \equiv -\big(U(x) - U(x_0)\big). \tag{5-78}$$

The function $U(x)$ is called the potential energy, and $U(x) - U(x_0)$ is the negative of the work required to bring the body, regardless of the path, from x_0 to x. Equation (5-78) is the definition of the potential energy for a conservative force field. If we use the definition of U above in (5-82), we see at once that the work–kinetic energy relation can be arranged into the form

$$\tfrac{1}{2}mV^2(x) + U(x) = \tfrac{1}{2}mV_0^2 + U(x_0). \tag{5-79}$$

The right-hand side of this equation is a constant equal to the initial sum of the kinetic plus potential energies. This sum is denoted by E_0, the total energy,

$$E_0 = \tfrac{1}{2}mV_0^2 + U(x_0). \tag{5-80}$$

Equation (5-79) is the law of conservation of mechanical energy. It states that if a particle moves under the influence of a conservative force, the sum of the kinetic energy T and potential energy U is the same throughout the motion

$$T(x) + U(x) = E_0. \tag{5-81}$$

In the examples discussed above we have seen in each case how the law of conservation of mechanical energy, combined with the energy diagram, permits one to obtain quite simply the principle features of the motion.

It is useful to examine a little more fully the potential energy function U and its relation to the force. First, however, note that the negative sign introduced in the definition of this force permits U to enter with positive signs on both sides of (5-79). In this way one can write the total energy as the *sum* of the kinetic and potential energies. Now, let us show how one can quite readily determine whether or not a force is conservative, from (5-78). Here we have

$$-U(x) + U(x_0) = \int_{x_0}^{x} F(x)\,dx. \tag{5-82}$$

Let us now compute $-dU(x)/dx$, i.e.,

$$-\frac{dU(x)}{dx} = \frac{d}{dx}\int_{x_0}^{x} F(x)\,dx. \tag{5-83}$$

But from the definition of a differential, we have that the right-hand side can be written as

$$-\frac{dU(x)}{dx} = \lim_{\Delta x \to 0} \frac{\int_{x_0}^{x+\Delta x} F(x)\,dx - \int_{x_0}^{x} F(x)\,dx}{\Delta x}. \tag{5-84}$$

Now since both the integrals have the same value regardless of the path between x_0 and x, we see that the two integrals nearly cancel, leaving only the integral between x and $x + \Delta x$, i.e.,

$$-\frac{dU(x)}{dx} = \lim_{\Delta x \to 0}\left\{ \frac{\left(\int_{x}^{x+\Delta x} F(x)\,dx\right)}{\Delta x}\right\} \tag{5-85}$$

in the limit as $\Delta x \to 0$, $F(x)$ is constant in the integration, and $\int_{x}^{x+\Delta x} F(x)\,dx \cong F(x)\Delta x$. Thus

$$-\frac{dU(x)}{dx} = \lim_{\Delta x \to 0}\left(\frac{F(x)\Delta x}{\Delta x}\right) = F(x). \tag{5-86}$$

Thus we see that when $F(x)$ is conservative, its value at each point in space can be computed as the first derivative of the potential energy function

$$F(x) = -\left(\frac{dU(x)}{dx}\right). \tag{5-87}$$

Any force that can be expressed as the derivative of the "potential" U is thus a conservative force. In the three examples discussed above in Section 5.2.E the potential energy U had the following forms:

$$
\begin{aligned}
U(x) &= mgx && \text{(constant gravitational field),} \\
U(x) &= \tfrac{1}{2}kx^2 && \text{(linear restoring force)} \\
U(x) &= -\frac{gx_e^2}{x} && \text{(gravitational field far from the} \\
&&& \text{center of the Earth).}
\end{aligned}
\tag{5-88}
$$

Let us now step back a bit, and review the idea content of the law of conservation of mechanical energy [(5-79)]. This law contains no new laws of physics beyond that already contained in Newton's second law of motion. We found, by integrating this law of motion, that the kinetic energy of a moving particle changes during motion by an amount equal to the work done by the forces acting. If, in addition, the forces are of a particular kind (i.e., "conservative"), then the work they do does not depend upon the path between the initial and final particle positions (x_0 and x). The work done by such forces is equal to the change $-(U(x) - U(x_0))$ in a mathematical function called the potential energy. Thus, as the particle moves to different points (x) in the space where the force acts, we can imagine that the potential energy associated with the system of forces assumes the values $U(x)$. If we keep, in separate mental compartments, the particle that feels the force, and the system that provides the force, then we must say that the change in kinetic energy of the particle is equal to the negative of the change of the potential energy of the forcing system. The separation of the total dynamical system into two parts, the responsive particle part, and the motive force part is, in fact, awkward and unnecessary. Instead, in the conservation of energy formulation we, in fact, consider, as *one* system, both the moving particle and the system that provides the force. If we choose this as our system, then we see that, taken as a whole, a gain in kinetic energy is always accompanied by a loss in potential energy, in such a way that the total mechanical energy remains equal to the value it had when the motion began.

The proper choice of "the system" was, of course, also at the basis of the law of conservation of momentum, which we also found often permitted a convenient solution to problems of mechanics. We saw there that in many cases one could choose the system under consideration in such a way that the total external force on the system is equal to zero. Under this condition, the center of mass of the system moves with constant momentum regardless of the complexity of the motion of its individual parts.

The conservation of energy law may be used as a bookkeeping device. It tells us that during the motion certain quantities must add up in such a way that they are constant throughout the motion. When this is combined with the knowledge of the potential energy as a function of position, one can often quite readily use the conservation principle to determine the velocity of the particle versus position. It should be stressed that the energy formulation of mechanics is not always the most useful means for analysis. In particular, when the acting forces are not known as a function of position, or when the acting forces are nonconservative, then the energy approach may not be a useful one.

As a final remark in this discussion, we must generalize the implicit thought that we are considering the motion of just a single particle. In fact, if the system contains many particles or parts, and if each part moves under the influence of only external forces, then

$$\frac{d}{dt}(m\vec{v}_i) = (\vec{F}_i)_{\text{ext}},$$

where $(\vec{F}_i)_{\text{ext}}$ is the total external force acting on the ith part of the system, we can carry out the analysis presented for each part of the system. Such an analysis will show again that the kinetic plus potential energies of the total system is a constant. Namely, that

$$\sum \tfrac{1}{2}m_i v_i^2(x_i) + \sum U_i(x_i) = E_0. \qquad (5\text{-}89)$$

This formula applies of course if each of the $\vec{F}_i$ are conservative forces. $U_i(x)$ is the potential energy of the ith particle when it is in position x_i. E_0 is the total energy of the system and is equal to the sum of the kinetic plus potential energies associated with all the particles and force fields at the beginning of the motion.

Practically, of course, in many particle systems the interaction *between* parts of the system is crucial. In this case, there will also be forces $\vec{F}_{ij}$ between the parts of the system. If these individual interaction forces are conservative, one can demonstrate the existence of the energy conservation law for the total system. A demonstration of this is included in the problems section at the end of this chapter.

5.2.G. Nonconservative Forces; Power, and the Law of Conservation of Energy and Heat

Suppose that the forces acting on a particle consist of both conservative and non-conservative types. We can then write

$$m\frac{dV}{dt} = F^c + F^{nc}. \qquad (5\text{-}90)$$

If we integrate as in Section 5.2.F, we find that we can express the total mechanical energy at each position x as follows:

$$\left(\tfrac{1}{2}mV^2(x) + U(x)\right) - \left(\tfrac{1}{2}mV_0^2 + U(x_0)\right) = \int_{x_0}^{x} F^{nc}\,dx. \qquad (5\text{-}91)$$

Here the potential energy is associated with the conservative parts of the forces acting on the particle, i.e.,

$$-[U(x) - U(x_0)] = \int_{x_0}^{x} F^c\,dx. \qquad (5\text{-}92)$$

Previously, there were no nonconservative forces and the right-hand side of (5-91) was zero. The mechanical energy $E(x) = \tfrac{1}{2}mV^2(x) + U(x)$ was then equal to the initial mechanical energy $E(0)$ with which the system was started. In the present case, however, we have that $E(x)$ is not equal to $E(0)$. The relation between the two is given in accordance with (5-90) by

$$E(x) - E(0) = \int_{x_0}^{x} F^{nc}(x)\,dx. \qquad (5\text{-}93)$$

This states that as the system moves from the initial to the final state the total mechanical energy differs from the initial energy by an amount equal to the work done by the nonconservative forces.

Equivalently, we may examine at each point in the motion the *rate* at which the total mechanical energy is changing. We can obtain this rate by starting again with Newton's second law, (5-90). On multiplying by the velocity, and expressing the left-hand side of the resulting equation as a derivative of the kinetic energy, we have

$$\frac{d}{dt}\left(\tfrac{1}{2}mV^2(x)\right) = F^c(x)V + F^{nc}V. \tag{5-94}$$

Now the conservative force $F^c(x)$ is related to the potential energy $U(x)$ by (5-91), i.e.,

$$F^c(x) = -\frac{dU}{dx}. \tag{5-95}$$

Thus, we can express $F(x)V(x)$ as

$$F^c(x)V = -\left(\frac{dU}{dx}\right)\left(\frac{dx}{dt}\right) = -\frac{dU}{dt}. \tag{5-96}$$

On inserting this into (5-94), we find that the total mechanical energy $E(x)$ is changing at the rate

$$\frac{d}{dt}\left(\tfrac{1}{2}mV^2(x) + U(x)\right) = F^{nc}V \tag{5-97}$$

or

$$\frac{d}{dt}E(x) = F^{nc}(x)V. \tag{5-98}$$

When the nonconservative force is friction, as is generally the case, $F^{nc}(x)V$ is a negative number and is equal to the rate at which work is done against the nonconservative force. Equation (5-98) states that the total mechanical energy decreases at a rate $(dE(x)/dt)$ exactly equal to the rate at which work is done against the nonconservative force. This type of nonconservative force "dissipates" or removes mechanical kinetic and potential energies from the system.

At first sight we might regretfully conclude that the nonconservative forces restrict the generality of the principle of conservation of energy. If, however, the "lost" mechanical energy actually reappears in some other guise or form, it might be possible, by measuring the quantity of this new form, to recover the conservation law. It is a common observation that friction produces heat, and it is natural to inquire if the amount of heat is quantitatively related to the lost mechanical energy.

The work of Joseph Black (1728–1799) demonstrated that there is a conservation law for heat. In any heat exchange between two bodies the amount of heat (ΔQ_1) lost by one is equal to the amount gained (ΔQ_2) by the other, i.e.,

$-\Delta Q_1 = \Delta Q_2$ or $\Delta Q_1 + \Delta Q_2 = 0$. This conservation law makes possible the definition of the concept of specific heat C, as the amount of heat gained (or lost) ΔQ by a body per unit temperature change (ΔT) and per unit mass (m) of the body, i.e.,

$$\Delta Q = Cm\Delta T.$$

We have then, on the one hand, a conservation law of mechanical energy (in the absence of frictional forces). On the other hand, we have a conservation law of heat in purely calorimetric processes (where no mechanical work is involved). Is it possible that the work done by the frictional forces, that is equal to the loss in mechanical energy, is also in some way equal to the heat generated? Could it be that *heat* is the alternate guise or form which appears in place of the "lost" mechanical energy? The answer to this suggestive question lies in the demonstration of a precise, quantitative relationship between the amount of mechanical work done by the dissipative forces (as measured in joules) and the amount of heat produced (as measured calorimetrically).

As we will describe more fully in Section 5.4.C, the great pioneers of thermodynamics, particularly James Joule, were able to demonstrate experimentally that there is a definite "mechanical equivalent of heat." Regardless of the nature of the nonconservative, dissipative force, if it does a certain amount of mechanical work, that work always produces the same amount of heat.

This enables us to extend the narrow concept of the conservation of mechanical energy to a broader one; namely, the conservation of energy and heat. Let us denote the heat generated by nonconservative forces as ΔQ, i.e.,

$$\Delta Q = -\int_{x_0}^{x} F^{\mathrm{nc}}(x)\,dx. \tag{5-99}$$

The minus sign is due to the fact that F^{nc} is opposite to the displacement dx, yet we want the heat generated ΔQ to be a positive number. Thus ΔQ represents an increase in the heat content of the system. Let us also denote the change in the total internal energy of our system as

$$\Delta E = E(x) - E(0). \tag{5-100}$$

Then we can write (5-93) as:

$$\Delta E + \Delta Q = 0. \tag{5-101}$$

This is a particular form of the law of conservation of mechanical energy and heat that applies when a system under consideration does no work on another system. It says that when a mechanical system having nonconservative forces changes from one state (x_0) to another (x), the decrease in mechanical energy is balanced exactly by an increase in the heat content of the system.

(i)   Motion of a Body Falling Through a Viscous Liquid

The role of nonconservative forces can be seen clearly by considering the case of a body falling through a fluid under the action of gravity (minus the force of buoyancy) and under the action of a viscous drag force. Let the body have mass m, volume Ω, and density ρ. Let the liquid have density ρ_ℓ. The free-body diagram of the forces acting is shown in Figure 5.9.

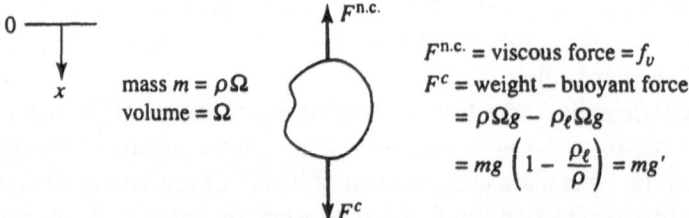

Figure 5.9. Free-body diagram for a body falling through a viscous medium. The effective weight of the body including buoyancy is mg' where $g' = g(1 - \rho_\ell/\rho)$.

The viscous resistance force has the property that its magnitude is proportional (f is a constant of proportionality) to the velocity V of the body, and its direction is opposite to the velocity. Clearly, this force is not known as a function of the position (x) of the body. Also, clearly, it is not a conservative force because it cannot be written as the spatial derivative $(-dU/dx)$ of some potential. To obtain the motion, we cannot conveniently use the work–kinetic energy formulation. We must return to Newton's second law of motion. The body moves in accordance with the equation

$$m\frac{dV}{dt} = mg' - fV. \tag{5-102}$$

This equation can be solved directly to provide the velocity of the particle as a function of time. We assume that at $t = 0$, $x = 0$ and that the initial velocity V_0 is equal to zero. Equation (5-102) can be rearranged into the form

$$\frac{dV}{(g' - (f/m)V)} = dt. \tag{5-103}$$

We integrate this equation between $t = 0$ and t and $V = 0$ and $V = V(t)$ to obtain

$$\int_{V=0}^{V(t)} \frac{dV}{(g' - fV/m)} = \int_0^t dt = t, \tag{5-104}$$

$$-\frac{m}{f} \ln\left(g' - \frac{fV}{m}\right)\Big|_0^{V(t)} = t, \tag{5-105}$$

or

$$\ln\left(1 - \frac{fV}{mg'}\right) = -\frac{ft}{m}. \tag{5-106}$$

Let us now define

$$V_\infty \equiv \frac{mg'}{f}. \tag{5-107}$$

This is the "terminal" velocity that the body would assume when the viscous drag is equal to the weight, and no further acceleration takes place. (Set $(dV/dt) = 0$ in (5-102).) Also define the "time constant" τ as

$$\tau = \frac{m}{f}. \tag{5-108}$$

In terms of these constants we have

$$\ln\left(1 - \frac{V(t)}{V_\infty}\right) = -\frac{t}{\tau} \tag{5-109}$$

or

$$\left(1 - \frac{V(t)}{V_\infty}\right) = e^{-t/\tau}$$

or, finally,

$$V(t) = V_\infty \left(1 - e^{-t/\tau}\right). \tag{5-110}$$

We plot in Figure 5.10 the velocity of the body as a function of time. The velocity of the body approaches V_∞ asymptotically for times $t \gg \tau$. At $t = \tau$ the velocity is equal to $(1 - 1/e) \approx 0.63$ times its final value of V_∞.

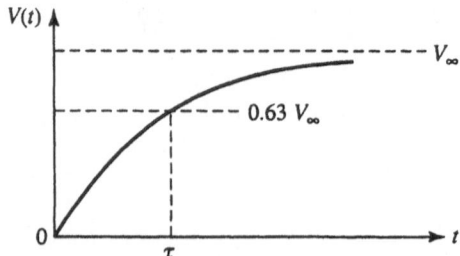

Figure 5.10. Velocity of a body falling through a viscous liquid.

For small times, the velocity increases linearly with time as

$$V(t) \rightarrow V_\infty \left(\frac{t}{\tau}\right), \qquad \frac{t}{\tau} \ll 1. \tag{5-111}$$

Let us now study the energy balances in this motion. For this purpose it is most convenient to consider the rates at which each of the energy terms are changing. According to (5-97) we have that the sum of the kinetic energy T and the potential energy change at a rate equal to the power dissipated by the nonconservative force. This is expressed mathematically as

$$\frac{dT}{dt} + \frac{dU}{dt} = \frac{d}{dt}\left(\tfrac{1}{2}mV^2\right) + \frac{d}{dt}U = F^{nc}V. \tag{5-112}$$

We will compute each of these terms separately and interpret their significance in the energy balance equation.

Let us begin with the potential energy term. First we observe that the potential energy $U(x)$ associated with the gravity force is found from the relation between force and potential energy, viz:

$$-\frac{dU(x)}{dx} = mg',$$

$$U(x) = -mg'x. \tag{5-113}$$

The rate of change of potential energy is

$$\frac{dU}{dt} = -mg'\left(\frac{dx}{dt}\right) = -mg'V(t). \tag{5-114}$$

Using (5-110) for $V(t)$ we have

$$\left(\frac{dU}{dt}\right) = -mg'V_\infty\left(1 - e^{-t/\tau}\right). \tag{5-115}$$

As $(t/\tau) \gg 1$ the falling body approaches its terminal velocity V_∞, and the constant rate of fall in the gravitational field implies a steady loss in potential energy at the rate $dU/dt \rightarrow -mg'V_\infty$, as is given in (5-115) for $t \rightarrow \infty$. For short times, $V(t) \rightarrow 0$ and the rate of change of the potential energy goes to zero. In Figure 5.11 we show, as one of the curves, the rate of change of potential energy.

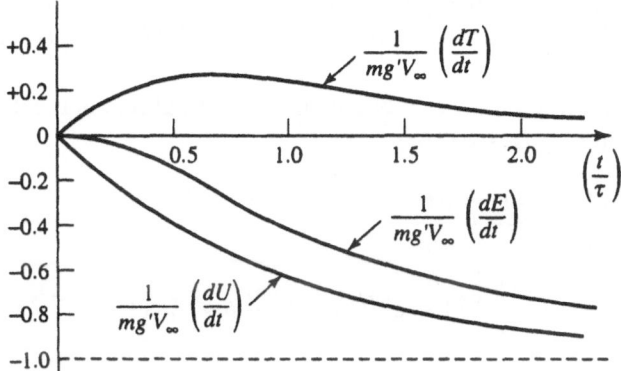

Figure 5.11. Rate of change of kinetic energy and potential energy; and the rate at which power is dissipated in the motion of an object falling through a viscous medium.

It is useful to observe that the final rate of change of potential energy $(mg'V_\infty)$ can be alternatively expressed in the following ways:

$$mg'V_\infty = \frac{mV_\infty^2}{\tau} = fV_\infty^2. \tag{5-116}$$

These can be verified quite easily using the definitions

$$V_\infty = \frac{mg'}{f} \quad \text{and} \quad \tau = \frac{m}{f} \quad \text{(Equations (5-108), (5-107))}.$$

We can next compute the kinetic energy term directly

$$\frac{dT}{dt} = \frac{d}{dt}\left(\tfrac{1}{2}mV^2\right) = mV\left(\frac{dV}{dt}\right).$$

Again, upon using (5-110), we find

$$\left(\frac{dT}{dt}\right) = \frac{mV_\infty^2}{\tau}(1 - e^{-t/\tau})e^{-t/\tau}. \tag{5-117}$$

Notice that for $t/\tau \gg 1$, when the particle has reached close to terminal velocity, the kinetic energy is no longer changing and $dT/dt \to 0$. Similarly, at $(t/\tau) \ll 1$, the velocity is very small, going to zero as $t \to 0$ so that here too $dT/dt \to 0$. The rate of change of kinetic energy is maximum at $t/\tau \cong 0.695$ and its value there is $dT/dt = 0.247mV_\infty^2/\tau$ or $0.247mg'V_\infty$. In Figure 5.11 we also show a graph of dT/dt.

We can now compute the rate of change of the total mechanical energy, as it is equal to the sum $dT/dt + dU/dt$:

$$\frac{dE}{dt} = \frac{dT}{dt} + \frac{dU}{dt}$$

$$= \frac{mV_\infty^2}{\tau}(1 - e^{-t/\tau})e^{-t/\tau} - mg'V_\infty(1 - e^{-t/\tau}). \tag{5-118}$$

Since $mV_\infty^2/\tau = mg'V_\infty$ from (5-116) we can write (5-118) as

$$\left(\frac{dE}{dt}\right) = -mg'V_\infty(1 - e^{-t/\tau})^2. \tag{5-119}$$

This is also shown in Figure 5.10. dE/dt is always negative. As $(t/\tau) \to 0$, $(dE/dt) \to 0$, and as $(t/\tau) \gg 1$, $(dE/dt) \to -mg'V_\infty$.

We can now compare the rate of decrease of the total energy with $F^{nc}V$, the rate at which the nonconservative forces do work. Since

$$F^{nc} = -fV,$$

we see at once that

$$F^{nc}V = -fV^2 = -fV_\infty^2(1 - e^{-t/\tau})^2. \tag{5-120}$$

But from (5-116) $fV_\infty^2 = mg'V_\infty$ so that we see that

$$F^{nc}V = -mg'V_\infty(1 - e^{-t/\tau})^2, \tag{5-121}$$

which is exactly equal to (dE/dt).

Thus, we see explicitly in this example that the rate of decrease of mechanical energy is exactly equal to the power dissipated by the nonconservative force acting on the system. Mechanics alone, however, does not tell us how the dissipated power manifests itself. In fact, the work done by the nonconservative force appears as an increase in the *heat content* of the viscous fluid. The temperature of the fluid increases by an amount such that the increase in heat content is equal to the nonconservative work done. If we could examine the microscopic state of the fluid we would see that the heat rise is nothing more than an increase in the average kinetic and potential energies of the interacting molecules of the fluid. Thus, what we recognize as heat when we look at the system macroscopically, is in fact once again mechanical energy when the system is viewed microscopically. The disorderly character of the heat–energy limits its convertibility into mechanical work. The second law of thermodynamics deals, in fact, with the limits that the microscopic disorder places on the recoverability of the heat energy.

Quite apart from these considerations, the example considered above shows us in principle how one could determine the mechanical equivalent of heat. We can simply measure the rate of loss of mechanical energy, after terminal velocity is reached, and also measure the rate dQ/dt at which heat is produced—using a calorimeter. By equating these two by

$$\left(\frac{dE}{dt}\right)_{\text{mechanical}} = K\left(\frac{dQ}{dt}\right)_{\text{heat}} \tag{5-122}$$

we can obtain the relation between the mechanical energy unit and the thermal heat unit (the calorie). This in fact was first done, by a method quite similar to that proposed above, by James Joule.

5.2.H.   Units and Conversion Factors

Work and energy have units of force times distance. Power has units of energy per unit time. In the cgs and mks systems the energy units have the names erg and joule, respectively:

In cgs units:

$$1 \text{ dyn} - \text{cm} = 1 \text{ erg}. \tag{5-123}$$

In mks units:

$$1 \text{ N-m} = 1 \text{ J}. \tag{5-124}$$

In British engineering units:

$$1 \text{ foot-pound} = 1 \text{ ft-lb}. \tag{5-125}$$

From the definitions of these units, the conversions between them are

$$1 \text{ J} = 10^7 \text{ erg} = 0.7376 \text{ ft-lb}. \tag{5-126}$$

In the various systems of units the power units are:

In cgs units:

$$1 \text{ dyn} - \text{cm} = 1 \text{ erg/s}. \tag{5-127}$$

In mks units:

$$1 \text{ J/s} = 1 \text{ W}. \tag{5-128}$$

In British engineering units the power unit is 1 ft lb s. Since this is a rather small unit practically, James Watt suggested that a convenient unit of power would be the maximum power that a horse could deliver. This was chosen to be 550 ft-lb/s. Thus,

$$1 \text{ horsepower} \equiv 1 \text{ hp} = 550 \text{ ft-lb/s} \tag{5-129}$$

or

$$1 \text{ ft-lb/s} = 1.82 \times 10^{-3} \text{ hp}. \tag{5-130}$$

From these definitions we can obtain the following relations between the various units of power:

$$1 \text{ W} = 1.341 \times 10^{-3} \text{ hp} = 0.7376 \text{ ft-lb/s}. \tag{5-131}$$

Heat energy is generally measured in units of the calorie. One calorie is the amount of heat needed to raise the temperature of one gram of water one degree centigrade from 14.5 to 15.5°C. The kilocalorie is equal to one thousand calories, and is the unit generally used to describe the calorie content of foods. The British thermal unit, or Btu, is the unit generally used in engineering practice. One Btu is the heat needed to raise the temperature of one pound of water from 63 to 64°F. The Btu and the calorie are related, according to their definition, by

$$1 \text{ Btu} = 232 \text{ calories}. \tag{5-132}$$

According to the law of conservation of heat and energy, we can express energy in the units in which heat is measured. We need only know the mechanical equivalent of heat, i.e., the connection between mechanical energy and heat. This was found by James Joule as a result of increasingly refined measurements carried out over many years. He found that the dissipation of 4.186 joules generated 1 calorie of heat, i.e.,

$$4.186 \text{ J} = 1 \text{ cal}. \tag{5-133}$$

With this equivalence we can relate each of the mechanical units of work to the heat units of energy. In this way we find the following conversions:

$$1 \text{ J} = 10^7 \text{ erg} = 0.7376 \text{ ft-lb} = 0.239 \text{ cal} = 9.48 \times 10^{-4} \text{ Btu}. \tag{5-134}$$

In a similar way we can relate the mechanical and heat units of power as follows:

$$1 \text{ W} = 3.41 \text{ Btu/h} = 1.34 \times 10^{-3} \text{ hp} = 0.239 \text{ cal/s}. \tag{5-135}$$

5.3   Work, Energy, and Power in Three Dimensions

In the previous section (5.2) we obtained the principal features of the work–energy formulation of mechanics simply by considering one-dimensional motion. To use work–energy methods in two and three dimensions, we must use a more general expression for the work than applies to the one-dimensional case. In three dimensions we also obtain a deeper understanding and interpretation of the potential energy, and its connection with the force.

5.3.A. The Work–Kinetic Energy Formula

We begin, just as we did in the one-dimensional case, by carrying out an integration of Newton's second law of motion. In the case of a particle of mass m, or a rigid body of mass m, whose center of mass moves with velocity $\vec{V}$, we have

$$m\frac{d\vec{V}}{dt} = \vec{F}(\vec{r}).$$ (5-136)

Here $\vec{V}$ is the vector velocity and $\vec{F}(\vec{r})$ is the net external force acting on the body, when its center of mass is at the position $\vec{r}$. Again we assume that the forces acting on the body are known as a function of position $\vec{r}$.

To obtain a first integral of this equation, we take the scalar product of both sides of this equation with the velocity vector $\vec{V}$. (The scalar product of two vectors $\vec{A}$ and $\vec{B}$ is denoted as $\vec{A} \cdot \vec{B}$. $\vec{A} \cdot \vec{B}$ is a scalar whose magnitude is $AB\cos\theta$ where θ is the angle between $\vec{A}$ and $\vec{B}$.)

$$m\vec{V} \cdot \left(\frac{d\vec{V}}{dt}\right) = \vec{F}(\vec{r}) \cdot \vec{V}.$$ (5-137)

In three dimensions $\vec{V}$ has components V_x, V_y, and V_z along the x-, y-, and z-directions. If $\hat{i}$, $\hat{j}$, and $\hat{k}$ are unit vectors along the x-, y-, and z-directions, then $\vec{V}$ can be written as

$$\vec{V} = \hat{i}V_x + \hat{j}V_y + \hat{k}V_z \qquad \text{and} \qquad \frac{d\vec{V}}{dt} = \hat{i}\frac{dV_x}{dt} + \hat{j}\frac{dV_y}{dt} + \hat{k}\frac{dV_z}{dt}.$$

Since $\hat{i} \cdot \hat{j} = \hat{k} \cdot \hat{j} = \hat{k} \cdot \hat{i} = 0$ and $\hat{i} \cdot \hat{i} = \hat{j} \cdot \hat{j} = \hat{k} \cdot \hat{k} = 1$, we see that

$$\vec{V} \cdot \left(\frac{d\vec{V}}{dt}\right) = V_x\frac{dV_x}{dt} + V_y\frac{dV_y}{dt} + V_z\frac{dV_z}{dt}$$ (5-138)

$$= \frac{1}{2}\frac{d}{dt}(V_x^2 + V_y^2 + V_z^2)$$ (5-139)

or

$$\vec{V} \cdot \left(\frac{d\vec{V}}{dt} \right) = \frac{1}{2} \frac{d}{dt} (V^2).$$

(5-140)

Using this in (5-137), we find

$$\frac{d}{dt} \left(\tfrac{1}{2} m V^2 \right) = \vec{F} \cdot \vec{V}.$$

(5-141)

The left-hand side of this equation is equal to the rate of change of the kinetic energy, where the kinetic energy (T) is

$$T = \tfrac{1}{2} m (V_x^2 + V_y^2 + V_z^2)$$

(5-142)

or

$$T = \tfrac{1}{2} m V^2.$$

(5-143)

The right-hand side of the equation is the power (P). Thus in three dimensions the power generated by the applied force is

$$P \equiv \vec{F} \cdot \vec{V}.$$

(5-144)

The power is a scalar, whose magnitude is equal to the product $F V \cos\theta$ where θ is the angle between the velocity vector and the force vector. This is shown in Figure 5.12. At each instant of time, the rate of change of the kinetic energy of the particle (or center of mass of a rigid body) is equal to $V F \cos\theta$. This is the product of the velocity of the body and the component of the force along the velocity. Equation (5-144) tells us then that the increase of the kinetic energy of the body is determined by the component of the force along the direction of motion. That part of the force which is perpendicular to the motion does not change the kinetic energy. Here we see one of the important generalizations associated with three-dimensional motion. Only the component of the force along the motion does "work." We can see this quite clearly if we recognize that for one, two, or three dimensions, power is the rate at which work is done. If W symbolizes the work, then

$$P \equiv \frac{dW}{dt}$$

(5-145)

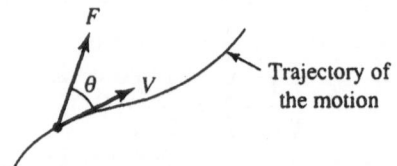

Figure 5.12. Force acting on the particle, its trajectory and velocity.

by definition. From (5-145) we have that

$$P = \vec{F} \cdot \vec{V} = \frac{dW}{dt}.$$
(5-146)

But

$$\vec{V} = \frac{d\vec{r}}{dt}.$$
(5-147)

Thus

$$\vec{F} \cdot \frac{d\vec{r}}{dt} = \frac{dW}{dt}.$$
(5-148)

Thus, we conclude that the work done (dW) as the particle moves a distance $d\vec{r}$ along its trajectory is

$$dW = \vec{F} \cdot d\vec{r}.$$
(5-149)

The differential work (dW) done is $F\, dr \cos\theta$ where θ is the angle between the direction of the force and the path of the motion (Figure 5.13). The total work done

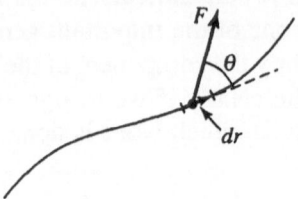

Figure 5.13. Work done by the applied force as the particle moves a distance $d\vec{r}$.

as the particle moves from some initial position to a final position $\vec{r}$ is found by integrating (5-141) using (5-146) through (5-148), i.e.,

$$\frac{d}{dt}\left(\tfrac{1}{2}mV^2\right) = \vec{F} \cdot \frac{d\vec{r}}{dt}$$

or

$$\int_{V_0}^{V(r)} d\left(\tfrac{1}{2}mV^2\right) = \int_{\vec{r}_0}^{\vec{r}} \vec{F} \cdot d\vec{r} = \int_{\vec{r}_0}^{\vec{r}} dW \equiv W_{r,r_0}. \qquad (5\text{-}150)$$

We take the velocity as V_0 at $r = r_0$ and $V(r)$ when $r = r$. The left-hand side of (5-150) is therefore the total change in kinetic energy, and the right-hand side is the work done by the force as the particle moves along the trajectory from r_0 to its final location r. The work–kinetic energy formula in three dimensions is then

$$\tfrac{1}{2}mV^2(r) - \tfrac{1}{2}mV_0^2 = \int \vec{F} \cdot dr \equiv W_{r,r_0}. \qquad (5\text{-}151)$$

The work done in moving the particle from $\vec{r}_0$ to $\vec{r}$ is the integral, or the sum of $F \Delta r \cos\theta$ over small elements Δr of the path. If a force acts so that it is always perpendicular to the path of the particle, it does no work. It cannot change the speed of the particle. It can only change the direction of the velocity vector.

5.3.B.   Conservative and Constraint Forces and the Conservation of Mechanical Energy

In three-dimensional motion we see that the work done on the particle is produced entirely by the component of the applied force along the path of the particle. The integral expression for the work

$$W_{r,r_0} = \int_{r_0}^{r} \vec{F} \cdot d\vec{r}$$

is called a path integral. Its value, in general, will depend upon the path along which it is evaluated.

Just as in one-dimensional motion, the work–energy formulation of mechanics is not particularly useful if the work done by the applied forces depends upon the path. If the nature of the force is such that regardless of the three-dimensional path between the initial and final positions (r_0 and r) the work is the same, then such a force is called conservative and the work–energy formulation is useful in finding the motion. In three dimensions, even if the applied force is nonconservative, the work–energy formulation is useful provided that the force always acts perpendicular to the path of the motion. Such forces are sometimes referred to as "constraint forces" because they usually are the forces that constrain the particle to move along a given path. The normal reaction force on a particle moving along a fixed track like a roller-coaster track is a constraint force. The tension in a string holding a particle which moves perpendicularly to the string is a constraint force.

If the forces acting on the particle are conservative or constraint forces, the work will depend only on the initial and final positions of the particle. If, as in the one-dimensional case, we define the potential energy change ($U(r) - U(r_0)$) as the negative of the work done in moving the particle between r and r_0, we can express the work–kinetic energy formula in terms of a law of energy conservation. Defining then, the potential energy change as

$$U(r) - U(r_0) = -\int_{r_0}^{r} \vec{F} \cdot d\vec{r} = -W_{r,r_0}. \tag{5-152}$$

This equation applies if the force F is a conservative or constraint-type force. From (5-152) we can express the work–kinetic energy formula (5-151) as

$$\tfrac{1}{2}mV^2(r) - \tfrac{1}{2}mV_0^2 = -\bigl(U(r) - U(r_0)\bigr). \tag{5-153}$$

By transferring all the terms dependent on r to the left and those dependent on r_0 to the right, we can write that at each point in the motion the sum of potential plus kinetic energy of the system is a constant and equal to the initial value of the kinetic plus potential energy, i.e.,

$$\tfrac{1}{2}mV^2(r) + U(r) = \tfrac{1}{2}mV_0^2 + U(r_0), \tag{5-154}$$

which is the law of conservation of mechanical energy. It has the same form in three dimensions as in the one-dimensional case. Here, of course, $V^2(r) = V_x^2 + V_y^2 + V_z^2$.

It is, of course, important in principle to be able to determine at the outset whether the forces acting on the system are conservative or not. In one-dimensional motion we saw that if the force at each position could be expressed as $F = -dU/dx$, then the force was conservative, and U was the potential energy. In three dimensions the precise identification as to whether or not F is conservative is mathematically more intricate. In fact, the subject of vector calculus provides a direct means to test any given force law for its "conservativeness." An exceptionally lucid and detailed account of the mathematical properties of conservative forces can be found in *An Introduction to Mechanics* by D. Kleppner and R. Kolenkow [1]. In Section 5.3.D we will briefly shown how one can relate the force and potential energy in three dimensions. We shall generally, however, confine our attention to a limited number of forces. In each of these cases it will be simple to verify directly from the definition (5-152) that the force is conservative, and to obtain the potential energy.

Our discussion above has been framed in terms of the motion of a single particle. If the system under consideration consists of many particles, we can express the work–energy theorem and the law of conservation in terms of the total work done on each of the particles as follows. If T_i is the kinetic energy of the ith part of the system, and if W_i is the work done on this part by all the forces acting on it, we have, by adding together the work–energy formula for each part, that

$$\sum_i \left(T_i(r) - T_i(0)\right) = \sum_i W_i(r) - W_i(0). \tag{5-155}$$

When the forces are conservative

$$W_i(r) - W_i(0) = -\left(U_i(r) - U_i(0)\right) \tag{5-156}$$

and the conservation of energy for the entire system can be written as

$$\sum_i \left(T_i(r) + U_i(r)\right) = \sum_i \left(T_i(0) + U_i(0)\right) = E_0. \tag{5-157}$$

Thus the kinetic and potential energies of each part of the system may change during the motion, but the total mechanical energy of the entire system remains constant provided that the forces acting are either conservative or constant forces.

Let us apply this work–energy formula to the analysis of motion in a few examples.

5.3.C. Mechanical Examples. The Recoiling Block and Springboard Diving

(i)   The Recoiling Block

Consider the situation shown in Figure 5.14. A mass m is released, and slides without friction down an arbitrarily curved path under the force of gravity. It leaves the large block (mass M) horizontally. The large block slides without friction on the hatched surface. Compute the velocity of the block and the slider (m) when they separate. If we were to solve this problem from the force point of view, we would have to obtain the equations of motion of both the block and slider in terms of both the unknown reaction force between them and the gravity force. The interaction force would then be determined from the two equations and then integrated to determine the velocity along the path. As you can see, this is a rather difficult process. Let us use instead the work–energy formulation of mechanics.

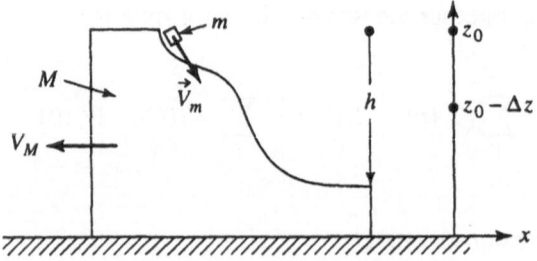

Figure 5.14. The recoiling block.

We choose as our system the sum of the block and the slider together. We compute the work done during the motion and equate it to the change in kinetic energy of the two parts of the system. If we regard the system as block plus slider, then the work on this system is done solely by gravity. The work done as the block falls a distance Δz from its initial position downward is given by

$$W = mg\,\Delta z. \tag{5-158}$$

Since the force and displacement are in the same direction, the work done during the fall is positive. Also because the gravity force is a constant which always points in the same direction, the work done is dependent only on the vertical height fallen and is independent of the path.

The slider at each moment has two velocity components V_{m_z}, and V_{m_x} along the vertical or z-axis and the horizontal axis. These velocities, of course, are measured relative to an inertial system fixed on the table top. The block has a velocity (V_{M_x}) and, of course, no vertical component. At $t = 0$, $V_{m_x} = V_{m_z} = V_{M_x} = 0$.

The work–kinetic energy formula for the total system tells us that after the slider has dropped from rest a distance Δz, the kinetic energy of the system is equal to the work done, i.e.,

$$\tfrac{1}{2}MV_{M_x}^2 + \tfrac{1}{2}mV_{m_x}^2 + \tfrac{1}{2}mV_{m_z}^2 = mg\Delta z \tag{5-159}$$

Thus at any arbitrary position in the motion there are three unknowns V_{m_x}, V_{m_z}, and V_{M_x}. At the point at which the slider leaves the block, however, the track is horizontal so that $V_{m_z} = 0$ when $\Delta z = h$. The assumption that the track is horizontal at the point where the velocity is needed, therefore, considerably simplifies the problem. After solving this simple case we will see how to obtain V_{m_x}, V_{m_z}, and V_{M_x} at any point during the motion.

Since $V_{m_z} = 0$ when $\Delta z = h$, we see that when the slider leaves the block that

$$\tfrac{1}{2}MV_{M_x}^2 + \tfrac{1}{2}mV_{m_x}^2 = mgh. \tag{5-160}$$

To obtain a second relation for the two unknowns here, we make use of the fact that there is no net external force acting on the system in the x-direction. As a result the momentum in this direction must be conserved, and we have

$$M\vec{V}_{M_x} + m\vec{V}_{m_x} = 0, \tag{5-161}$$

since the initial momentum of the system in the x-direction is zero. Using this relationship between V_{m_x} and V_{M_x}, we find

$$V_{M_x} = -\frac{m}{M}V_{m_x}. \tag{5-162}$$

The block and slider move in opposite directions. Using this in (5-160), we can obtain expressions for V_{m_x} or V_{M_x} directly. These are

$$V_{m_x}^2 = \frac{2gh}{(1 + m/M)} = \text{velocity of slider at end of track} \tag{5-163}$$

and

$$V_{M_x}^2 = \frac{2gh}{(M/m)(1 + M/m)} = \text{velocity of block as slider leaves.} \tag{5-164}$$

If we wish to consider the more general problem of the values of V_{m_x}, V_{m_z}, and V_{M_x} at any point in the motion, then we have at our disposal the two equations (5-159) and (5-161) which represent the conservation of energy and the conservation of momentum in this case. To obtain a third equation, we must use the fact that during the motion the slider is always on the track in the block. We characterize the track in terms of its slope $\tan\alpha$ at any position Δz below the starting point (Figure 5.15).

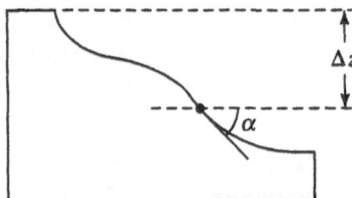

Figure 5.15. Definition of the slope of a track.

If the block did not move, the constraint that the slider remains on the block would give the following equation relating V_{m_z} and V_{m_x}:

$$\frac{V_{m_z}}{V_{m_x}} = \tan\alpha.$$

When the block does move, an observer fixed on the block would say that the apparent velocity of the slider in the x-direction is $\vec{V}_{m_x} - \vec{V}_{M_x}$. Thus the condition that the slider must remain on the block is

$$\frac{V_{m_z}}{V_{m_x} - V_{M_x}} = \tan\alpha. \tag{5-165}$$

(Of course, V_{m_x} and V_{M_x} have opposite signs according to (5-161).) This kinematical constraint ((5-165) above) permits a full solution to the problem. That is, we can compute the velocity of the block and the slider at any point in the motion. The solution is found as follows. First use (5-165) to express V_{m_z} in terms of V_{m_x} and V_{M_x} in (5-158) which gives the kinetic energy of the system as a function of the distance Δz that the slider falls. This gives

$$\tfrac{1}{2}MV_{M_x}^2 + \tfrac{1}{2}MV_{m_x}^2 + \tfrac{1}{2}m(V_{m_x} - V_{M_x})^2\tan^2\alpha = mg\,\Delta z. \tag{5-166}$$

Now, we can use (5-161) to obtain either V_{m_x} or V_{M_x} as a function of Δz and $\tan \alpha$. If one carries out this computation, one finds the following results for V_{m_x} and V_{M_x}:

$$V^2_{m_x} = \frac{2g\,\Delta z}{(1 + m/M)(1 + (1 + m/M)\tan^2 \alpha)} \qquad (5\text{-}167)$$

and

$$V^2_{M_x} = \frac{2g\,\Delta z}{\frac{M}{m}(1 + m/M)(1 + (1 + m/M)\tan^2 \alpha)}. \qquad (5\text{-}168)$$

In applying this solution to a particular shape for the track, we remember that the specification of the shape of the track tells us what the slope α is at each position Δz. Thus V_{M_x} and V_{m_x} are functions of Δz both explicitly in the numerator, and implicitly in the denominator through the factor $(\tan^2 \alpha(\Delta z))$.

It is testimony to the power of the work–energy approach to mechanics that this problem, even in the general case that the track has arbitrary shape, can be solved so directly. If we had to solve this problem by computing first the forces at work on the block and slider, the solution would have been very much more difficult to reach.

(ii)   Springboard Diving

a) Maximum Height and the Conservation of Energy.   Figure 5.16 shows a photograph of a champion diver, Professor Charles Batterman of MIT, during various points in the execution of a "swan dive." The photograph is a multiple exposure taken by Professor Harold Edgerton using a high-intensity stroboscopic light source.

For the spectator the most crucial parts of the dive are the graceful and precise movements of the body as it twists through the air and cleanly enters the water. Equally important, however, is the approach and takeoff from the end of the board. For the diver to be successful he must properly use his body and the board to spring to the maximum attainable height. The height which he attains entirely determines the time available for the execution of the required maneuvers during the free flight portion of the dive. Clearly the diver must maximize the time in flight in order to complete his aerial maneuvers.

In the present subsection we focus our attention on the approach and take-off portion of the dive. We shall see that it is possible to understand quantitatively

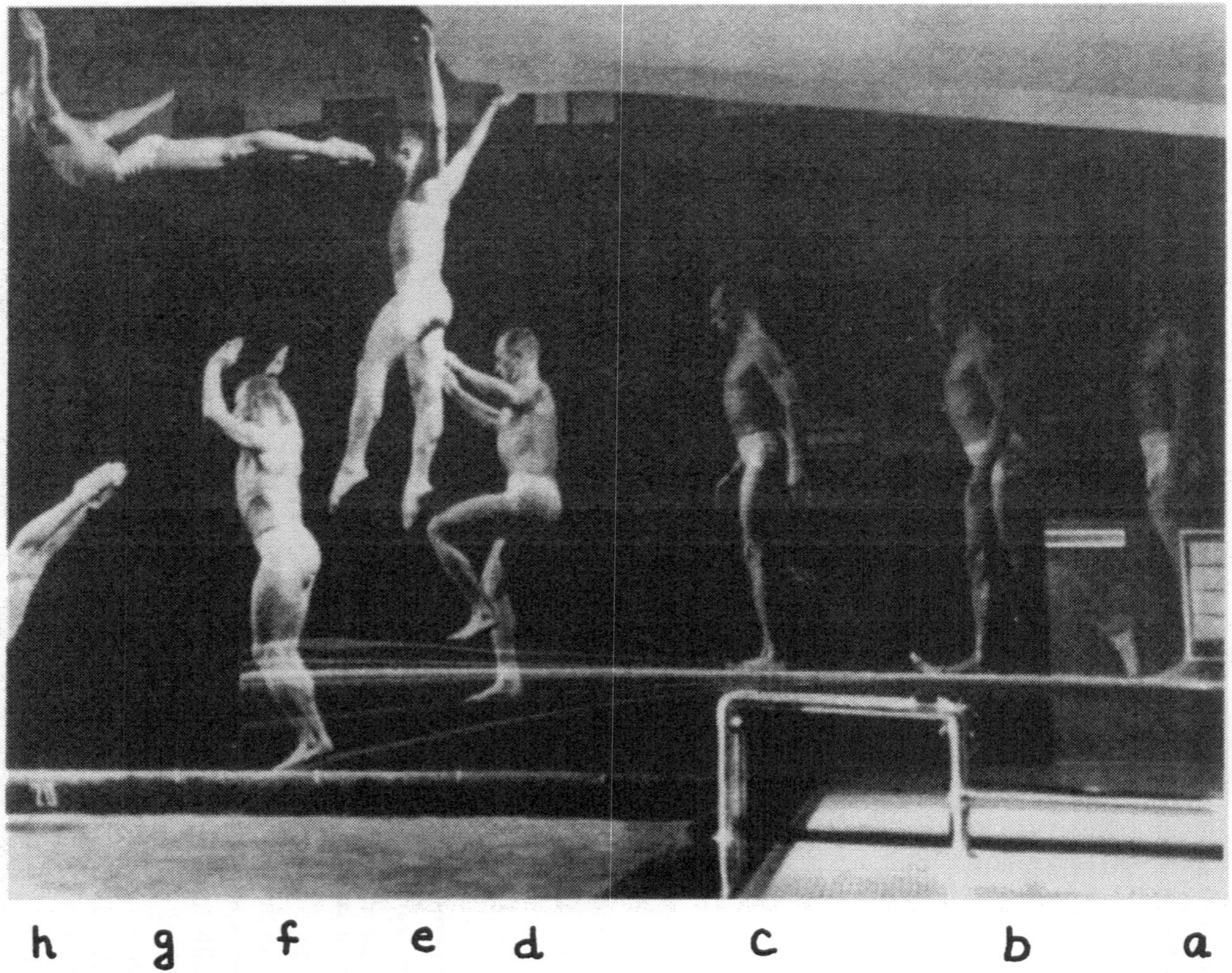

Figure 5.16. A springboard diver.

366

a number of essential features of the dive using the work–energy formulation of mechanics.

A very clear and complete account of the various stages of the forward approach of the diver to the end of the board is to be found in the excellent book, *The Techniques of Springboard Diving* by C. Batterman [2]. Let us analyze the photograph [2] in Figure 5.16 as a physicist would.

The diver begins in the "address" standing at attention. He then takes two steps forward $a \rightarrow b$ and $b \rightarrow c$. His third step $c \rightarrow d$ concludes at the position d as the diver begins the process of making a great upward jump called the "hurdle." During the hurdle the diver leaps upward with an initial kinetic energy W_1 and his center of mass rises a distance H_1. At the top of the hurdle (at position e) the energy is entirely in the form of potential energy and

$$W_1 = mg H_1. \tag{5-169}$$

Here m is the mass of the diver's body. In the dive shown in the photograph $H_1 \cong$ 74 cm.

The diver, after reaching maximum height, returns to the board, and on touching it begins the process of jumping upward once again. The board is depressed downward. At its maximum depression the board and diver are stationary. All the previous kinetic energy of the diver is now stored as potential energy in the spring-board. The best that the diver can do mechanically is to recover from the board the initial kinetic energy with which he landed, and to add to it an amount W_2 equal to the work he does during his leg extension on the second leap—the "takeoff." He can do this provided that he leaves the board in its horizontal position and in a stationary state. Under these conditions the total kinetic and potential energies of the board are the same as they were when he landed on it at the end of the hurdle. In this way, the board takes away none of his energy, and the total mechanical energy of the diver will be $W_1 + W_2$.

In essence then, in a properly executed approach and takeoff, the springboard acts on the one hand to store, and then return to the diver's body the energy W_1 his muscles provided during the hurdle. To this he adds an energy W_2 produced by his body during the second leap. Of course, if he is clumsy during the second leap, he may allow the board to retain some of his energy, and the board will bounce and rattle. In general, the diver will retain only a fraction f of his first jump energy, and will begin free flight with a total energy W_T given by

$$W_T = W_2 + f W_1. \tag{5-170}$$

In a perfect dive $f = 1$, and he gets maximum energy out of each jump $W_2 \cong W_1 \cong mgH_1$. Now when the diver leaves the end of the board, depending on the type of dive, some of his kinetic energy is in the form of the rotation of his body about the center of mass. That is, if T_R is the rotational kinetic energy of the body about the center of mass, and if T_c is the kinetic energy of the center of mass, then if the diver leaves the board near its unbent position, we have

$$W_T = T_R + T_c = W_2 + fW_1. \tag{5-171}$$

On leaving the board the second time, the center of mass of the diver rises a height ΔH determined by the equation

$$mg\Delta H = T_c. \tag{5-172}$$

At this point, the turning point in his vertical motion, the kinetic energy of the center of mass is fully converted into potential energy. On combining (5-171) and (5-172), we obtain the following general equation for the increase in height of the center of mass:

$$mg\Delta H = W_2 + fW_1 - T_R. \tag{5-173}$$

The maximum height is achieved when $T_R = 0$ (as in the "jacknife"), and when $f = 1$ and $W_1 = W_2 = mgH_1$. Under optimal conditions then

$$mg\Delta H = 2mgH_1 \tag{5-174}$$

or

$$\Delta H = 2H_1. \tag{5-175}$$

Therefore, in the optimal case, the diver's center of mass rises a distance equal to twice the height the diver can achieve during a single leap. This result is quite independent of the stiffness of the springboard and depends entirely on the height H_1 that the athlete can achieve in a single leap.

Let us compare this prediction with the information on ΔH and H_1 contained in the photograph. In the "address" position, the center of gravity of the body is located at about the center of the bathing suit. (More precisely it is located at the height of the anterior superior spine of the right ilium (see Figure 5.16).) The diver shown is 170 cm tall (5 ft, 7 in). This figure provides a scale of height in the photograph. His center of mass is measured to be ~ 93 cm above the level of the board before the jumps begin. After takeoff, the maximum height occurs at position h in the photograph. Because of the arching of the back the location of the center of mass is a little difficult to locate precisely. Examination of the figure indicates it to be at about 230 cm above the level board. Thus the increase in position of the center of mass at the top of the dive is

$$\Delta H = 230 \,\text{cm} - 93 \,\text{cm} \approx 137 \,\text{cm}.$$

According to the prediction of (5-175) this should be equal to $2H_1$. The photograph indicates $H_1 \sim 74$ cm. Thus $2H_1 \sim 148$ cm which, considering the errors involved in precisely locating the center of mass is in good agreement with (5-175). The performance of the diver in this dive is very nearly optimal.

To obtain optimum height in a springboard dive, it is necessary, as mentioned before, that the springboard is left nearly horizontal and having nearly zero speed. In this way the springboard retains zero potential and kinetic energies, and the diver succeeds in extracting from the board all the energy he gave it during the hurdle and takeoff. In the photograph of Figure 5.16 it is difficult to determine the degree to which this was achieved because the stroboscopic light was not properly synchronized to the characteristic oscillation frequency of the board. Examination of the many still photographs in [2] shows that good divers do indeed leave the board nearly stationary after takeoffs. A board left rattling (final velocity not zero) after takeoff is a well-known symptom of a poor dive.

How does a diver leave the board behind having "zero" kinetic and potential energies? In a way it seems paradoxical that a diver can at one and the same time leave the board, with which he is in contact, stationary, while his center of mass is moving upward with maximum velocity. To do this the diver pivots his feet and toes downward as he comes near to leaving the board. Part of his body (feet and toes) move down while his center of mass moves upward. By this "pointing the toes," the diver extracts the last bits of energy from the moving board and leaves it nearly without potential or kinetic energy.

b) Mechanics of the Vertical Jump.   To understand more fully further features of the springboard dive and other athletic events, such as the high jump, we must examine more fully the mechanics of the vertical jump. In the vertical jump from a stationary floor, the athlete crouches down, then jumps upward by straightening both legs and throwing his arms upward. A useful review article on the physics of the vertical jump has been written by E. J. Offenbacher in the *American Journal of Physics* [3]. The jump can be divided into two parts. The first is the extension phase. The function of this phase is to impart maximum velocity, and therefore kinetic energy, to the center of mass of the jumper. The straightened body leaves the ground at the end of this phase. The second phase is the free flight of the center of mass of the jumper no longer in contact with the ground. These phases are shown diagrammatically in the stick figures in Figure 5.17.

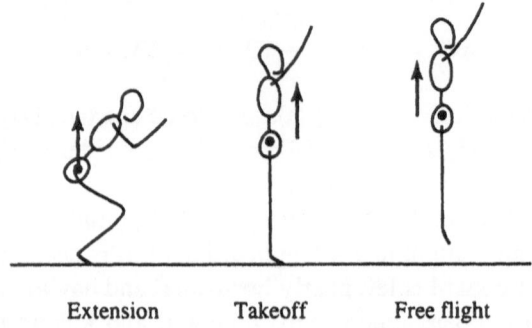

Extension   Takeoff   Free flight

Figure 5.17. Phases of the vertical jump.

The maximum increase in height (ΔH) of the center of mass attainable in a vertical jump can be related quite simply to the velocity V_0 of the center of mass at the moment of takeoff. At maximum height the potential energy of the body is equal to the kinetic energy at takeoff, i.e.,

$$\tfrac{1}{2}mV_0^2 = mg\,\Delta H.$$

This follows directly from the conservation of energy for a body moving in a uniform gravitational field. Thus, regardless of the mass (m) of the jumper,

$$\Delta H = \frac{V_0^2}{2g}. \tag{5-176}$$

How big is ΔH for a human being? What determines the relative performance in the jump of different individuals? We can answer these questions in a simple way. Let us assume that during the extension, from the bottom of the crouch to takeoff, the acceleration of the center of mass is roughly constant. Under those conditions the take-off velocity V_0 is related to the depth of the crouch (h) and the time for the full extension (τ) by the equation $V_0 = 2V_{\text{ave}} = 2(h/\tau)$. Numerically, the depth of the crouch is about 1.2 ft. The more variable feature in the jump, however, is the time (τ) for the full extension. A typical value for (τ) is $\cong 0.25$ sec. Using these values, we find the velocity V_0 to be

$$V_0 \sim 10 \text{ ft/s} = 305 \text{ cm/s}.$$

An experimental study [3], [4] of the vertical jump shows this to be about the observed value of the take-off velocity. Clearly, since by and large the heights of various individuals range rather narrowly from 5 ft 5 in to 6 ft 0 in, there is not much variation in h (the crouch depth) from person to person. An individual, however, who can decrease the extension time τ appreciably below 0.25 s can increase his speed V_0 and increase his height by the square of V_0. How high can one expect to go? Let us put numbers in (5-176):

$$\Delta H = \frac{V_0^2}{2g} = \frac{(10 \text{ ft/s})^2}{2(32.2 \text{ ft/s}^2)} = 1.67 \text{ ft}$$

$$\Delta H = 1.67 \text{ ft} \cong 51 \text{ cm}.$$

In fact, after making experimental studies of the height ΔH that could be achieved by 270 male Columbia University students, Gerrish [4] found that in all the jumps ΔH fell in the narrow range between 1 and 2 ft! This indicates that the speed with which muscle can shorten does not vary substantially from one man to another, and consequently τ does not vary much from person to person. This finding is in agreement with much more wide-ranging studies by A. V. Hill [5] who has shown that the speed of shortening of muscles of the same type is the same from species to species.

It is interesting to calculate the mechanical work and the mechanical power produced by the muscles to produce this vertical jump. The work done in each jump is

$$W = \tfrac{1}{2}mV_0^2.$$

The average power developed is

$$\overline{P} = \frac{W}{\tau}.$$

A person with a weight of 155 lb has mass

$$m \cong 70 \text{ kg}.$$

The work W is

$$W = \tfrac{1}{2}(70) \text{ kg } (3.05 \text{ m/s})^2,$$

$$W = 325 \text{ J}.$$

The body muscles are not very efficient producers of mechanical energy. If ΔE is the energy they expend doing an amount of mechanical work W, then $W/\Delta E$ is called the efficiency of the muscle. If we estimate muscle efficiency at about 25%, we see that to produce 325 J, the muscle must use up 1300 J of chemical energy to make one jump. Since 4200 J is equal to the food energy available from one "calorie" of food (one food calorie is equal to 1 kcal of the physicist), we see that one vertical jump uses only about one-quarter of a food calorie.

The average power generated is

$$\overline{P} = \frac{W}{\tau} = \frac{325 \text{ J}}{0.25 \text{ s}} = 1300 \text{ W}.$$

Since 1 hp is 746 W, we see that during the vertical jump the average mechanical power developed is about 2 hp. Of course, this is done for only $\tfrac{1}{4}$ s.

c) Timing of the Takeoff: "Tight" Versus "Loose" Springboards   On taking off from the end of the springboard, the diver is executing a vertical jump. To obtain maximum height, he must synchronize the jump movements correctly in relation to the movements of the board. Let us examine how this is done. At the end of the hurdle the diver lands on the end of the board with the same kinetic energy as he left it. The kinetic energy is translated into the potential energy of the board as it is depressed. The upward leap of the diver must start when the board is nearly stationary at its maximum depression. If the jumper should begin the extension while the board is moving downward, the velocity of his center of mass V_0 in space is less than the extension velocity of his leg muscles V_e by an amount equal to the downward speed of the board V_b; i.e., $V_0 = V_e - V_b$. Since the diver must optimize V_0, he must start his upward jump when V_b is zero and complete it during the period in which the board is rising upward. As the board rises upward, the speed V_b begins from zero, takes on a maximum negative value, and then becomes zero again as a result of the toe-pointing maneuver. Thus on landing after the hurdle, the diver is in a crouch position. At the bottom of the depression of the board he begins to throw his arms upward and extends his legs.

If τ is the time required for the extension phase of the vertical jump, and if the oscillation period of the board loaded with the diver is T, then clearly the diver and board must be synchronized in such a way that

$$\tau = \tfrac{1}{4}T. \tag{5-177}$$

The period T of the loaded board depends on the weight of the diver. Also different divers have slightly different acceleration times τ. Competition diving boards are constructed in such a way that T can be varied by turning a wheel which locates the fulcrum of the board. By moving the fulcrum back or forward, one can effectively change the length of the board and hence its period. We can see from (5-177), and the fact that $\tau \sim 0.25$ s, that the characteristic period of a *loaded* diving board must be about 1 s.

There is controversy among divers as to whether a "tight" board or a "loose" board is most advantageous; i.e., whether the fulcrum should be adjusted for values of T that are smaller or larger than the average. A tight board forces the diver to execute the vertical jump more rapidly as he strives to satisfy the synchronization condition $\tau = \tfrac{1}{4}T$. A small value of τ is advantageous because it requires a rapid leg extension. On the other hand, the shorter value of τ gives the diver less time to execute such important maneuvers as toe pointing near the end of the takeoff. Also it should be kept in mind that unless the diver has a fast muscle extension capacity he may not be able to synchronize his leap to the quarter period of the hard board. In fact, in practice, better divers prefer a softer, "loose" board.

d) Connection with the High Jump   The mechanics of the vertical jump, combined with the limited extension velocity of muscle shows that the height achievable in such a jump is, on average, ~ 1.6 ft or 51 cm, and ranges generally between 1 and 2 ft (2 ft $\simeq 62$ cm). We saw by measuring the photograph in Figure 5.16 that a champion diver can achieve an elevation H_1 of ~ 75 cm by using the one-legged jump of the hurdle. These numbers raise two interesting questions which we shall discuss in this subsection. First, how does the diver achieve such a good performance (75 cm) in the one-legged hurdle jump compared to the two-legged vertical jump (62 cm)? Second, how does a champion high jumper clear a bar that is located about 105 cm above the position of his center of mass, considering the fact that on jumping upward he can only raise his center of mass about $H_1 \sim 70$ cm! In fact, one is forced to conclude that during the high jump the center of mass remains about 35 cm below the position of the high bar.

The excellent performance of the diver in the hurdle jump is probably due to two factors. First, the diver actually takes a bit of a hop during his second step. The work done in this hop is stored and returned to the diver by the board. Also the single-knee rise of the hurdle acts to elevate the center of gravity more than is possible in the two-legged vertical jump.

Let us now examine the performance of the high jumper. Consider a jumper 6 ft 0 in tall (183 cm). His center of gravity when standing erect is about 3 ft 7 in (109 cm) above the ground. Even if he is a very fine athlete, he can raise his center of gravity only about 2 ft 4 in (70 cm) higher than this. Thus, his center of gravity during the jump does not exceed 5 ft 11 in. Nevertheless, if he is a very good athlete he can clear a bar 7 ft above the ground. In executing such a jump, the center of mass must pass 13 ft (33 cm) *underneath* the bar. The successful high jumper has mastered the art of draping and curling his body in such a way that his center of mass is low while each individual part of this body in turn passes over the bar. The most successful way of executing this maneuver is the "Fosbury Flop." In this jump the athlete leaps upward, turning his back to the bar. The extended legs keep the center of gravity low until the back gets over the bar. The head and shoulders then fall well under the level of the bar and the legs finally curl over following the body's downward flight. The old style scissors jump requires that the trunk of the body be vertical as the jumper passes over the bar. This is very disadvantageous. The "eastern" or "western" rolls arrange that the body is parallel to the bar, and the legs and arms straddle the bar as the jumper passes over. Clearly, the Fosbury Flop configuration keeps the center of gravity low and the passing part of the body high as the jumper goes over the bar.

5.3.D.   Relation Between Force and Potential Energy—A Mathematical Note

In the one-dimensional case we saw that if the force was conservative, one could derive the force from the potential energy $U(x)$ using the relation

$$F(x) = -\frac{dU}{dx}. \tag{5-178}$$

In three dimensions the force has three components F_x, F_y, and F_z. We shall show that in this section that if a force is conservative then one can relate the force components in three dimensions to the potential energy $U(r)$ by the equations

$$F_x = -\left(\frac{\partial U(r)}{\partial x}\right),$$

$$F_y = -\left(\frac{\partial U(r)}{\partial y}\right),$$

$$F_z = -\left(\frac{\partial U(r)}{\partial z}\right). \tag{5-179}$$

The operation $\partial/\partial x$ is the "partial derivative." That is, if the function U on which it operates has several variables x, y, z, etc., the partial derivative $(\partial/\partial x)$ is the derivative of the function with the variables other than x regarded as being constants.

We may derive (5-179) as follows. If $\vec{F}$ is a conservative force, then in general the integral

$$\int_{r_a}^{r_b} \vec{F}(f) \cdot d\vec{r}$$

is independent of the path and depends only on the end points $\vec{r}_b$ and $\vec{r}_a$. The potential energy of a conservative force is defined by

$$U(\vec{r}_b) - U(\vec{r}_a) = -\int_{r_a}^{r_b} \vec{F}(\vec{r}) \cdot d\vec{r}. \tag{5-180}$$

U is, of course, a scalar function. It has a magnitude which varies from one point to another, but there is no direction associated with the potential energy. Suppose the position $\vec{r}_b$ differs by a small amount $\Delta \vec{r}$ from the position $\vec{r}_a$. If Δr is small enough $F(r)$ is essentially constant in the integrand of (5-180), and we can write

$$(U(\vec{r}_a + \Delta \vec{r}_a) - U(\vec{r}_a)) = -\int_{\vec{r}_a}^{\vec{r}_a + \Delta \vec{r}} \vec{F}(\vec{r}) \cdot d\vec{r} \cong -\vec{F} \cdot \Delta \vec{r}. \tag{5-181}$$

We may now expand the left-hand side in a three-dimensional Taylor's series about $\vec{r}_a$. On retaining only the lowest-order terms in the expansion, we have

$$U(\vec{r}_a + \Delta \vec{r}) - U(\vec{r}_a) \cong U(\vec{r}_a) + \left(\frac{\partial U}{\partial x}\right)\Delta x + \left(\frac{\partial U}{\partial y}\right)\Delta y + \left(\frac{\partial U}{\partial z}\right)\Delta z - U(\vec{r}_a). \tag{5-182}$$

Combining this with (5-181), we have

$$\left(\frac{\partial U}{\partial x}\right)_{r_a}\Delta x + \left(\frac{\partial U}{\partial y}\right)\Delta y + \left(\frac{\partial U}{\partial z}\right)\Delta z = -\vec{F}(\vec{r}_a)\cdot\Delta\vec{r}. \qquad (5\text{-}183)$$

Since $\Delta\vec{r} = \hat{i}\Delta x + \hat{j}\Delta y + \hat{k}\Delta z$, this equation can be expressed as

$$\left(\frac{\partial U}{\partial x}\right)_{r_a}\Delta x + \left(\frac{\partial U}{\partial y}\right)_{r_a}\Delta y + \left(\frac{\partial U}{\partial z}\right)_{r_a}\Delta z$$

$$= -F_x(\vec{r}_a)\Delta x - F_y(\vec{r}_a)\Delta y - F_z(\vec{r}_a)\Delta z. \qquad (5\text{-}184)$$

Since Δx, Δy, and Δz can be varied independently of one another, and the equation above must still be satisfied, we can conclude that each component of the force is related to the potential energy U by the equations

$$F_x(\vec{r}_a) = -\left(\frac{\partial U}{\partial x}\right)_{r_a},$$

$$F_y(\vec{r}_a) = -\left(\frac{\partial U}{\partial x}\right)_{r_a},$$

$$F_z(\vec{r}_a) = -\left(\frac{\partial U}{\partial z}\right)_{r_a}. \qquad (5\text{-}185)$$

Other mathematical properties that U and $\vec{F}$ possess when $\vec{F}$ is a conservative force are discussed rather fully in the book by D. Kleppner and R. Kolenkow [1].

The mathematical result we have obtained above has an important graphical interpretation. This interpretation can be seen most simply if we consider the case that the potential energy U is a function of only two variables x and y. In this case we can construct in three dimensions a representation of $U(xy)$. At each value of x and y in the plane we associate a point in the vertical direction that is equal to U. The function $U(x, y)$ is then represented as a surface above the x–y plane as shown in Figure 5.18. We can establish "contour lines" on the potential energy surface. Along any contour line the potential energy is some constant (λ). In Figure 5.18 we show various contour lines corresponding to a number of different constant values of the potential energy (λ_1, λ_2, etc.).

If we now imagine a horizontal projection of these potential energy contours, we can determine the direction of the force at any point (x, y). It is easy to show (see the problems section at the end of this chapter) that the direction of the force

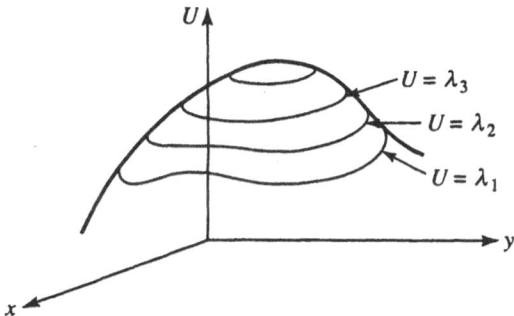

Figure 5.18. Three-dimensional representation of $U(x, y)$.

$\vec{F} = \hat{i} F_x + \hat{j} F_y = -\hat{i}(\partial U/\partial x) - \hat{j}(\partial U/\partial y)$ is perpendicular to the contour line of constant U at each value of xy. This is illustrated in the graph in Figure 5.19.

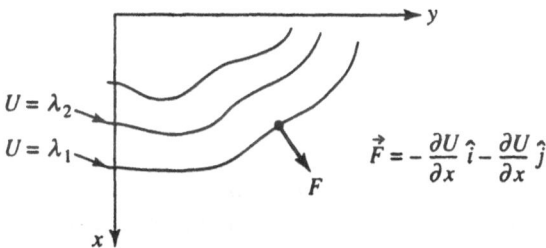

Figure 5.19. Contour map of constant potential energy lines.

This direction of force is also the direction of the steepest gradient down the $U(xy)$ surface in Figure 5.18. Thus the quantity

$$\hat{i}\left(\frac{\partial U}{\partial x}\right)_{r_a} + \hat{j}\left(\frac{\partial U}{\partial y}\right)_{r_a} + \hat{k}\left(\frac{\partial U}{\partial z}\right)_{r_a}$$

in three dimensions is called the "gradient" of the function U. The direction of this vector is that for which the potential energy U changes most rapidly in the vicinity of the point V_b. In two dimensions the force at (x, y) is perpendicular to the contour line of U = constant. When the potential energy is a function of the three variables (xyz), then the U = constant *lines* in the two-dimensional contour map are replaced by constant energy *surfaces*.

5.3.E.   Energy Diagrams and Mechanical Stability

The concept of "stability" is an important one in subjects as diverse as physics, physiology, economics, politics, and psychology. In each field, stability refers to the response of a property (or variable) of the system to a small change (a perturbation). Generally, if a small perturbation produces a large change in the variable, that variable, or the system as a whole, is said to be "unstable." On the other hand, if a small perturbation produces only a small change in the variable followed by a return to the unperturbed position, the system is said to be stable.

Examples of stability or instability in the fields mentioned above are legion— here are a few: In 1914 the political relations between the nations of Europe were so unstable that the assassination of the Archduke Ferdinand and his wife triggered the tragic conflagration of World War I. A simple and familiar example of instability in psychology is the behavior of a tired child or adult, who on hearing a mildly humorous remark breaks into a fit of uncontrolled laughter. In economics, the price structure of stocks on the New York Stock Exchange in 1929 was so artificially inflated and unstable (partly due to inadequate margin requirements) that the perturbation of one day's heavy selling on Black Thursday (October 24, 1929) led to the great stock market crash. In two weeks this produced a paper loss of $26 billion and ushered in the Great Depression.

In physiology the need for the maintenance of stable levels of such quantities as blood volume, ionic strength, pH, and glucose concentration in blood serum, hormone levels, etc., is so important that life cannot exist without this stability. The precise means by which the organ systems maintain the stable levels of physiologic variables against continual internal and external perturbations is at the very heart of the subject of physiology.

In physics we can easily find and examine in minute detail the stability of mechanical and electrical systems. By clearly understanding the criteria and conditions for stability or instability in physical systems, we are better prepared to analyze the more complex questions of stability in biological, economic, and social systems.

A very early and important example of mechanical stability was studied by Archimedes. This is the criterion for the stability of a ship as it rolls around its "beam" axis under the action of waves. The profile of the hull of a ship must be designed in such a way that as the ship rolls the buoyant forces of the water act to correct that roll and turn the deck back toward the horizontal position. Archimedes showed that if the cross section of the hull was in the shape of a parabola, it would be stable under roll.

We can easily describe the conditions for stability in mechanics in terms of the departures from "equilibrium." A mechanical system is said to be in equilibrium if there is no net force acting on the system. It is in "stable equilibrium" if a perturbation from the equilibrium position results in forces that push the system back toward the equilibrium position. It is in unstable equilibrium if a perturbation from the equilibrium state produces a force that pushes the system *further away* from the equilibrium position.

A very simple example of a mechanical system that has two equilibrium positions, one stable and the other unstable, is the fork and cork teeter–totter shown in Figure 5.20. The teeter–totter is really a pendulum. If the pendulum is suspended at a height ℓ *above* its center of mass, it will oscillate around its equilibrium position with period $T = 2\pi\sqrt{\ell/g}$. In this case, the pendulum bob is in stable equilibrium. If, on the other hand, the pendulum suspension is *below* its center of mass, it is called an inverted pendulum. The equilibrium position is that for which the center of mass is exactly above the suspension. If the axis of the inverted pendulum is perturbed from this position, the gravity forces act so as to keep it falling downward. The inverted pendulum is unstable about its vertical equilibrium position. That is what happens to the fork and cork teeter–totter of Figure 5.20.

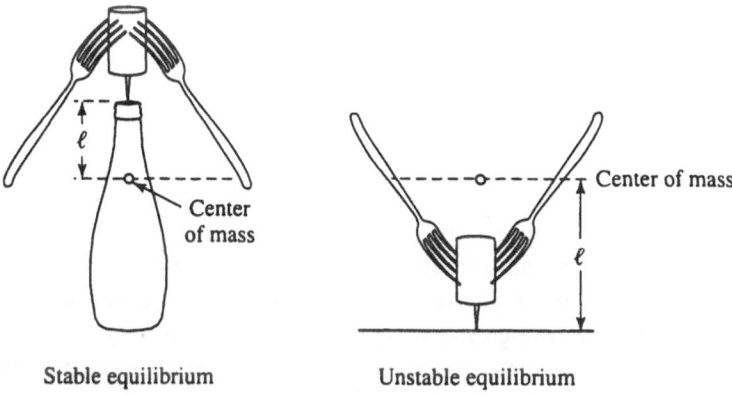

Stable equilibrium   Unstable equilibrium

Figure 5.20. A teeter–totter made of two forks and a cork in positions of stable and unstable equilibrium.

It is interesting to recognize that the human body in the erect position is mechanically an inverted pendulum. The vertical position is mechanically unstable. We would in fact be constantly falling over without the aid of an elaborate sensing and control system that exerts torques on the legs constantly in directions so as to maintain the vertical position stably. If a person has a brain tumor in the cerebellum, the control system can fail, and he or she no longer maintains the ability to overcome the intrinsic mechanical instability of the standing position. The afflicted person falls easily, and then fractures and further complications can readily follow. The maintenance of stable postural equilibrium is in fact a vital requirement for a normal life.

The work–energy formulation of mechanics provides a particularly useful and general means of characterizing the stability of mechanical systems. According to the conservation of energy theorem

$$T(\vec{r}) + U(\vec{r}) = E_0, \tag{5-186}$$

where $T(\vec{r})$ is the kinetic energy of the system at position $\vec{r}$ and $U(\vec{r})$ is the potential energy at position $\vec{r}$. E_0 is the constant value of the total energy of the system.

The equilibrium states of the system are located at points for which the spatial derivatives of the potential energy are equal to zero. At these points the forces are zero, i.e.,

$$-\left(\frac{\partial U}{\partial x}\right) = F_x = 0,$$

$$-\left(\frac{\partial U}{\partial y}\right) = F_y = 0,$$

$$-\left(\frac{\partial U}{\partial z}\right) = F_z = 0.$$

We can locate such positions simply on an energy diagram. Such a diagram is shown in Figure 5.21. For simplicity we plot U as a function of just a single variable (x). In this diagram the position x_1 is one of stable equilibrium. We can see this quite easily by examining the energy diagram in the vicinity of x_1 as indicated in Figure 5.22. If the system is perturbed so that the energy is increased to $E_{0_1} + \Delta$, the system can move anywhere in the small region from $x_1 - \Delta x < x_1 < x_1 + \Delta x$, but not beyond these limits. These are the turning points of the motion. For values outside this range the kinetic energy $T(x)$ would have to be negative. The position x_1 is one of stable equilibrium basically because $\partial^2 U/\partial x^2$ is positive at x_1. That is, $U(x)$ has a local minimum around x_1 and the potential energy rises on either side of this minimum. Another way to see this is to note that if the particle moves to the *right*, it comes to a position where $-(\partial U/\partial x)$ is negative. That means the particle feels a force acting toward the *left*, pushing the particle back toward its unperturbed position. A similar situation occurs if the particle moves to the left.

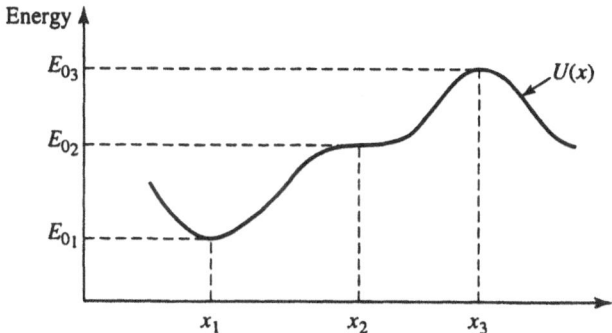

Figure 5.21. General energy diagram showing stable, metastable, and unstable equilibrium positions.

The second derivative $(\partial^2 U/\partial x^2)_{x_1}$ is the crucial mathematical characterization of the nature of the stability around each equilibrium position. In the present case its positive sign indicates stable equilibrium. Its magnitude determines the frequency with which the system oscillates back and forth between turning points. This is seen as follows. For small changes around the position x_1 we can expand $U(x)$ in a Taylor's series as follows:

$$U(x) = U(x_1) + \left(\frac{\partial U}{\partial x_1}\right)(x - x_1) + \tfrac{1}{2}\left(\frac{\partial^2 U}{\partial x^2}\right)_{x_1}(x - x_1)^2 + \cdots. \qquad (5\text{-}187)$$

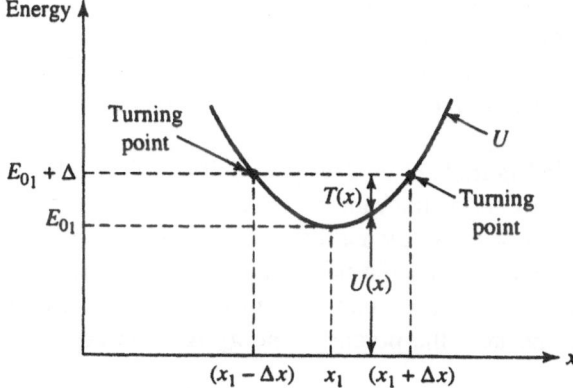

Figure 5.22. Energy diagram of a system near stable equilibrium.

Since $(\partial U/\partial x)_{x_1} = 0$, we see that at any point in the motion the total energy in one dimension is

$$\tfrac{1}{2}m\left(\frac{dx}{dt}\right)^2 + U(x_1) + \tfrac{1}{2}\left(\frac{\partial^2 U}{\partial x^2}\right)_{x_1}(x - x_1)^2 = E_0. \tag{5-188}$$

This is the expression for the energy of a harmonic oscillator. In Section 5.2.E(ii) of this chapter we showed that the angular frequency corresponding to such an oscillator is

$$\omega = \sqrt{\frac{1}{m}\left(\frac{\partial^2 U}{\partial x^2}\right)_x}. \tag{5-189}$$

The second derivative of the potential energy is in essence "the spring constant" that determines the frequency of the oscillator. A large value of the second derivative means a high oscillation frequency for the perturbed system and a small amplitude of motion. The result shown in (5-189) can also be obtained by differentiating (5-188) once with respect to the time. On simplifying the resulting equation, one finds the following expression:

$$\frac{d^2(x - x_1)}{dt^2} + \frac{1}{m}\left(\frac{\partial^2 U}{\partial x^2}\right)_{x_1}(x - x_1) = 0. \tag{5-190}$$

This is the equation of motion of a harmonic oscillator whose frequency is related to the coefficient of the $(x - x_1)$ term as $\omega^2 = (\partial^2 U/\partial x^2)_{x_1}/m$.

Of course in three dimensions the requirement for stability is that each of the three second derivatives $(\partial^2 U/\partial x^2)$ and $(\partial^2 U/\partial y^2)$, and $(\partial^2 U/\partial z^2)$ must be positive.

The point x_3 in Figure 5.23 is clearly a position of unstable equilibrium. This is seen clearly in the energy diagram in the vicinity of x_3. Suppose a particle originally at the position of x_3 is given a little kinetic energy raising the energy of the system to $E_{0_3}+\Delta$. Then clearly the subsequent motion of the system is unbounded. The particle never comes to rest again. Regardless of its position its kinetic energy is always positive because the potential energy is never equal to the total energy $E_{0_3} + \Delta$.

Suppose, on the other hand, the total energy is reduced below the value of E_{0_3}. This can be done by placing the particle with zero kinetic energy at either the point

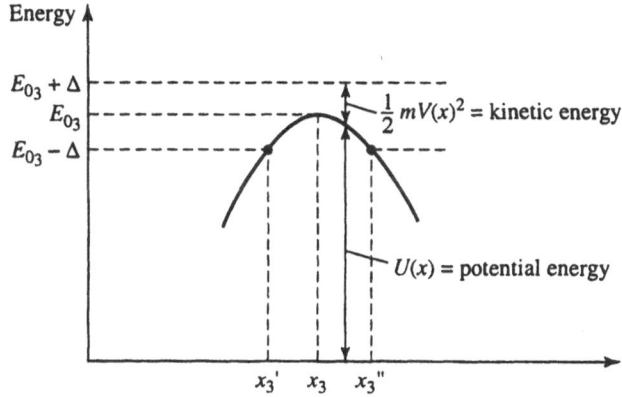

Figure 5.23. Energy diagram near a position of unstable equilibrium.

x_3' or x_3''. At both these positions the particle will feel a force $-(\partial U / \partial x)$ which pushes the particle further away from x_3. For example, at x_3'' the particle will feel a force to the right. The system must move so as to reduce its potential energy. The force $(\partial U / \partial x)$ on the particle drives it this way. Thus a perturbation either to the left or right of x_3 leads to unbounded motion away from x_3. Even if the particle comes back toward x_3'' from some rise in potential energy at large x, the particle will just momentarily stop at x_3'' and then turn around and accelerate to the right once again. Thus x_3 is a position of unstable equilibrium. It is so because here $(\partial^2 U / \partial x^2)$ is a negative number. In three dimensions, for a position to be one of unstable equilibrium in all directions, we must have $\partial^2 U / \partial x^2$, $\partial^2 U / \partial y^2$, and $\partial^2 U / \partial z^2$ all negative. Of course, at a point where $(\partial U / \partial x) = (\partial U / \partial y) = (\partial U / \partial z) = 0$ the second derivatives may not all have the same sign. For example, $\partial^2 U / \partial x^2$ and $(\partial^2 U / \partial y^2)$ could be negative while $\partial^2 U / \partial z^2$ might be positive. This means simply that the position is stable for motion in the z-direction, but unstable for motion in the x- or y-directions.

The position x_2 in Figure 5.21 is called a position of metastable equilibrium. x_2 is an inflection point of $U(x)$ and $\partial^2 U / \partial x^2 = 0$. Thus, if the system is displaced a very small amount from x_2, there will be neither a restoring nor a repelling force acting on the particle, and the particle if placed at rest will remain at rest. However, if the particle is given any kinetic energy its motion will be unbounded to the left and bounded on the right. Thus, here the system is stable under small perturbations in position but unstable against small perturbations in kinetic energy. The system is metastable.

5.4   The First Law of Thermodynamics

5.4.A.   The Conservation of Mechanical Energy, External Work and Heat

In Section 5.2.G we considered (in one dimension) the balance between the change in mechanical energy (ΔE^{mech}) and the work done by nonconservative, dissipative forces. We considered the particular example of a body falling through a viscous fluid and showed that the change in kinetic energy ΔT, plus the charge in potential energy ΔU, was at each instant of time equal to the work done by the frictional forces

$$\Delta E^{\text{mech}} = \Delta T + \Delta U = \Delta W^{\text{nc}}. \tag{5-191}$$

We then discussed briefly the fact that the nonconservative work manifests itself in an increase of the temperature of the system (falling body plus fluid). This temperature increase in turn corresponds to the generation of a well-defined quantity of heat ΔQ. How much heat is generated? In Section 5.4.C we will give a brief historic account of the demonstration that the heat ΔQ generated is strictly proportional to the work done by the nonconservative forces (ΔW^{nc}). The factor of proportionality between ΔQ and ΔW^{nc} is a universal constant 4.186 cal/J, i.e.,

$$\Delta Q \text{ (in calories)} = 4.186 \text{ cal/J} \times \Delta W^{\text{nc}} \text{ (in joules)}. \tag{5-192}$$

This universal conversion factor between work and heat signifies that heat produced (ΔQ) in a process actually represents energy, and may be entered in the "energy balance sheet" for the process as follows:

$$\Delta T + \Delta U = \Delta W^{\text{nc}} \equiv \Delta Q. \tag{5-193}$$

Here we recognize that ΔW^{nc} is a negative number, and if the heat produced ΔQ is to be positive, we must include a minus sign in the relation between them. Rearranging (5-193), we find

$$(\Delta T + \Delta U) + \Delta Q = \Delta E^{\text{mech}} + \Delta Q = (\Delta E)^{\text{total}} = 0. \tag{5-194}$$

This equation states that the total energy change—potential energy change, plus kinetic energy change, plus thermal energy produced in the system—is equal to zero. The total energy is *conserved* or unchanged. More generally, this is true for any "closed system." A closed system is one in which all the forces, of whatever kind they may be, act or operate *within* the system. In such a closed system the total energy (kinetic, plus potential, plus thermal) is always conserved or constant, regardless of what happens within the system.

In the present subsection we wish to go beyond the discussion of "closed systems." We wish now to formulate a conservation law for energy which applies to "open systems." As we have seen again and again, our description of the physics of a problem depends crucially on our choice of "the system." Nowhere is this more true than in the field of thermodynamics.

An "open system" is a part or a subsystem of a closed system. Through the boundaries of an open system there pass *links* between it and other subsystems. The links permit the transfer of energy between the open subsystems. As an example of the concept of an open system, consider the mechanically activated electric water heater illustrated in Figure 5.24 on page 387. Taken as a whole, it is a closed system whose total energy is conserved. We may divide it into two parts, each part being an open system. The first open system consists of the heat bath, plus the electric generator. The second open system is the weight. The cable from the weight to the generator penetrates the boundaries of each open system and provides the *link* between them. By means of this link, energy can be transferred from system 1 to system 2. Thus, if the weight Mg descends with constant speed by an amount h, the cable tension F will be equal to the weight and the cable will do an amount of work

$$\Delta W = Mgh, \tag{5-195}$$

on the open system (1). The total energy within the open system (1) then increases by an amount $\Delta E^{(1)}$:

$$\Delta E^{(1)} = \Delta W = Mgh. \tag{5-196}$$

In considering such open systems, we must recognize a second means of energy transfer that is familiar from calorimetry, namely, the process of heat exchange. We already pointed out in Section 5.2.G that in a thermally isolated system, if heat is transferred between two systems (1) and (2), then the heat lost by system 1 is equal to that gained by system 2, or vice versa, i.e.,

$$\Delta Q^{(1)} = -\Delta Q^{(2)}$$

or

$$\Delta Q^{(1)} + \Delta Q^{(2)} = 0. \tag{5-197}$$

Recognizing heat as a form of energy, this heat exchange is another form of energy exchange. The "thermal contact" between two subsystems is another *link* (in addition to the cable link mentioned in the example above) by means of which energy can be transferred through the boundaries of an open system. In recognizing these two transfer modes: one, a linkage through which work can pass; and two, a linkage through which heat can pass; we may conclude quite generally for any open system that the increase in its total energy ΔE is equal to the amount of heat ΔQ which flows *into* the system, plus the work ΔW done *on* the system by external forces. That is,

$$\Delta E = \Delta Q + \Delta W. \tag{5-198}$$

This is a statement of the first law of thermodynamics.

In our discussion we have, until now, defined ΔW as the work done on the system. The convention usually employed in thermodynamic textbooks, however, is to define ΔW as the work done *by* the system on other open systems to which it is linked. Clearly, this is the negative of ΔW as previously defined. Thus, if we wish to express the energy balance in terms of the work done *by* the system, the sign of the ΔW term in (5-198) must be made negative, i.e.,

$$\Delta E = \Delta Q - \Delta W \tag{5-199}$$

It is, of course, entirely immaterial whether we identify (5-199) or (5-198) as *the* first law of thermodynamics. If we keep in mind the meaning of ΔW in each case the equations are identical.

We shall use (5-199) in subsequent work to remain consistent with the usual convention. Further, since we will be considering energy exchanges between the human body and external systems, one frequently measures directly the work done *by* the person. In this case it is, in fact, convenient to use the expression of (5-199) as the first law of thermodynamics.

5.4.B.   Heat, Work, and Energy Changes in an Electrical Water Heater

Let us imagine that we consider the energy exchanges involved in the operation of an electric water heater. In this "dropping-weight–water-heater" the fall of the weight causes the generator shaft to turn (Figure 5.24). This in turn drives an electric current through the heater wire. The heat produced in the wire is then transferred to the water because of the temperature difference between them. This

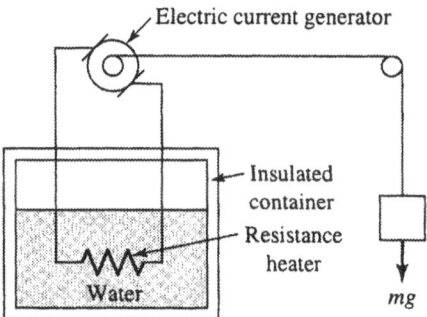

Figure 5.24. An electric water heater. The electric current is produced by the electric generator turned by the falling weight. The resistor produces heat.

simple-minded device demonstrates that our identification as to what is work, heat, and change in energy depends in detail upon what we regard as "the system."

Case I. The system is the water.

   Let us examine how the first law of thermodynamics applies to the water as the weight drops a height h with constant speed. What are the values of ΔE, ΔQ, and ΔW in the equation

$$\Delta E = \Delta Q - \Delta W \qquad (5\text{-}200)$$

Clearly the water does no work so that $\Delta W = 0$. The heat content of the water increases. The way in which this occurs is as follows. The flow of current through the electrical resistance raises the temperature of the wire above that of the water. Heat then flows from the wire and continues to flow until both wire and water come to the same temperature. At this new temperature the water's heat content has increased by an amount ΔQ. If the wire is very light so that its "heat capacity" is small, essentially all the heat developed as the current passed through the resistor will appear in the water. Thus, essentially all the energy mgh associated with the change in position of the weight goes into ΔQ:

$$\Delta Q \cong mgh.$$

Since $\Delta W = 0$, we have that

$$\Delta E = \Delta Q \cong mgh, \qquad (5\text{-}201)$$

according to the first law of thermodynamics. What is the meaning of this equation? In what sense has the internal energy of the water increased? In a molecular picture of heat, we would say that the energy of the assembly of water molecules has, in fact, increased by this amount mgh. The molecules are moving a little faster, and the potential energy of interaction between molecules has also fallen a little.

Thus, if the water is the "system" no work is done on the system as the weight falls—only the internal energy of the water increases an amount which at maximum is equal to mgh.

Case II.　The system is the water plus the resistor.

With this choice of system, since there is no flow of heat into the system through the insulated walls, we have that

$$\Delta Q = 0.$$

On the other hand, the total change in internal energy of the system (ΔE) must be equal to mgh. This change in internal energy is produced by the work $-\Delta W$ done by the moving electrons as they move through the wire against the resistance of the atoms in the wire

$$\Delta W = mgh.$$

Thus

$$\Delta E = -\Delta W = mgh.$$

Note that this is true regardless of the "heat capacity" of the resistor. The increase in internal energy of the heater plus water is precisely equal to mgh.

Thus, with this choice of system no heat is transferred to the system. The internal energy of the system changes simply because the moving electrons do work on the resistor, and the resistor indirectly increases the internal energy of the water in which it is placed.

Case III.　The system is the resistor.

If the resistor is regarded as comprising the system, none of the terms ΔE, ΔQ, and ΔW in the first law of thermodynamics is zero after the weight falls a distance h. In this process the resistor does work on the moving electrons. This work is

$$-\Delta W = mgh.$$

Since the force of resistance between the resistor and the passing electrons is non-conservative, the resistor temperature rises above that of the surrounding water. As a result an amount of heat (ΔQ) will flow from the resistor until the water temperature and the resistor temperature become equal. (In the special case that the mass of water is so large that the water temperature changes but little, the heat flow ΔQ is

$$\Delta Q \cong -mgh.$$

After the flow of heat between the resistor and its surrounding is complete, the temperature of the resistor will be higher than at the outset. Thus the internal energy of the resistor will have risen (call the rise ΔE). Regardless of the final temperature of the system we have

$$\Delta E - \Delta Q = mgh.$$

In the special case that the final temperature is close to the initial temperature because of the large mass of water, we have

$$\Delta E \simeq 0,$$
$$-\Delta Q \cong mgh.$$

If the final temperature of the resistor differs from the initial temperature the heat flow from the resistor is reduced an amount δQ from the maximum of mgh:

$$-\Delta Q = (mgh - \delta Q).$$

This reduction of outflow remains in the form of internal energy in the resistor. That is,

$$\Delta E = \Delta Q + mgh$$

or

$$\Delta E = \delta Q.$$

In this choice of the system then, the meaning of the terms heat flow, work done, and internal energy change are again quite different from that which applies with other choices of "the system."

Having extended our discussion of the first law of thermodynamics with this simple example, let us express it in the following general form which makes it applicable to any system whatsoever.

Let a system be brought from some initial state (i) to some final state (f):

If, in the process, there is no flow of heat into or out of the system, then the change in energy of the system ($E_f - E_i = \Delta E$) is independent of the path and is equal to the work done *on* the system ($-\Delta W$). (Remember that ΔW was chosen earlier as the work done *by* the system.) Thus, $\Delta E = -\Delta W$.

If the system is brought from (i) to (f) by a process in which heat is exchanged with its surroundings, then the change ΔE in the energy of the system is the sum of the heat flow *into* the system, and the work done *on* the system ($-\Delta W$), viz:

$$E_f - E_i = \Delta Q - \Delta W. \tag{5-202}$$

5.4.C. Elements in the Historical Development of the Law of Conservation of Energy

The principal conceptual elements of the law of conservation of energy are:

(1) an understanding of the microscopic nature of heat;

(2) the recognition that energy in physical and living systems can be converted from one form to another and into heat, but that the total amount of energy, including heat, is always conserved; and

(3) the quantitative conversion factor between mechanical (or electrical or chemical, etc.) energy and heat.

Our present understanding of these concepts is the result of work done principally during the eighteenth and nineteenth centuries. We know of no better short account of the development of these ideas than that contained in the fascinating book *Introduction to Concepts and Theories in Physical Science* by G. Holton [6]. In this fine book, and the references to be found therein, the reader will come to realize the gaps in knowledge, the conceptual uncertainties, the initially unappreciated great forward leaps, and the eventual great clarification connected with this important advance in human thought. The fascination of these developments is that they typify the pattern of scientific growth. The clarity and inescapable "rightness" of today's knowledge emerged from a matrix of confusion, uncertainty, and controversy, as a result of the lifelong efforts of a very few passionately devoted persons.

In the present subsection we shall discuss briefly the contributions of two of these extraordinary people, namely, Julius Robert Mayer (1814–1878) and James Prescott Joule (1818–1889).

In 1842, at the age of 28, Julius Mayer, then a young physician in his native city of Heilbronn, Germany, published an essay entitled "Remarks on the Energies of Inorganic Nature" in which he proposed, largely on the basis of intuitive, philosophical, and qualitative reasoning, the interchangeability and indestructability of energy. The flavor of Mayer's reasoning can be appreciated from the following quote (to be found on [6, p. 269]) from his 1842 paper:

> Energies are causes: accordingly, we may in relation to them make full application of the principle—*causa aequat effectum*. If the cause c has the effect e, then $e = c$. In a chain of causes and effects, a term or part of a term can never, as plainly appears from the nature of an equation, become equal to nothing. This first property of all causes we call their indestructibility. If after the production of [effect] e, [cause] c still remained in whole or part, there must still be further effects $[f, g, \ldots]$ corresponding to the remaining cause. Accordingly, since c becomes e and e becomes f, etc., we must regard these various magnitudes as different forms under which one and the same entity makes its appearance. This capability of assuming various forms is the second essential property of all causes. Taking both properties together, we may say, causes are quantitatively indestructible and qualitatively convertible entities. Energies are therefore indestructible convertible entities.

The reader will observe the unorthodox nature of Mayer's line of reasoning. His arguments are philosophical, and based on vague verbal structures such as equivalences between "energy" and "causes." This 1842 essay was a remarkable intuitive vision. It was not physics as we know it. It was not a conception stated in the language of mathematics and supported by quantitative measurements. Nevertheless, it was a correct vision. For that vision to be useful it was necessary to formulate it mathematically using the appropriate forms of the potential and kinetic energy applicable to each physical system whether it be mechanical, electrical, chemical, or biologic. This mathematical formulation was, in fact, provided in 1847 by the physiologist, anatomist, and physicist Hermann von Helmholtz (1821–1894) [8].

Despite the speculative and philosophic nature of his arguments on the conservation of heat and energy, Mayer recognized quite clearly the importance of the conversion factor between heat and work. In his essay he says [6], [7].

We will close our disquisition... with a practical deduction. How great is the quantity of heat which corresponds to a given quantity of kinetic or potential energy.

Mayer then went on to estimate the mechanical equivalent of heat in the following way.

Suppose that we endeavor to raise the temperature of a mass (m) of gas $1°$. To do this, we must add an amount of heat ΔQ to the gas. If we measure the amount of heat needed, we will find that the heat needed is different depending upon whether the gas is held at constant volume during the heating or if the gas is held at constant pressure. The two situations are shown schematically in Figure 5.25. In Figure 5.25(a) the gas is heated in a closed container so that its volume remains constant during the heating. In Figure 5.25(b) the gas is heated in a container fitted with a piston that can rise against the atmospheric pressure as the gas is heated. In Figure 5.25(b) the volume of the gas changes during heating, and the pressure of the gas is constant.

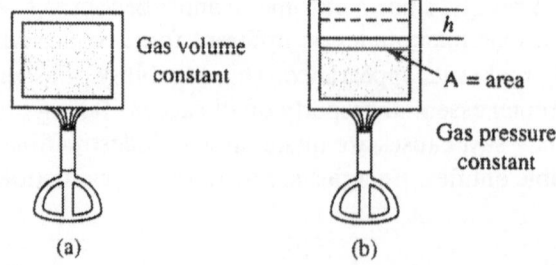

Figure 5.25. The heating of a gas under conditions of (a) constant volume or (b) constant pressure.

The heat needed to raise each gram of gas $1°C$ is called the specific heat C:

$$C = \frac{1}{m}\frac{\Delta Q}{\Delta T}. \tag{5-203}$$

The specific heat as measured in Figure 5.25(a) is called the specific heat at constant volume C_v, while that measured in Figure 5.25(b) is called the specific heat at constant pressure C_p. The crucial point is that these two specific heats are different. According to present-day values for air near room temperature

$$C_p \equiv \frac{1}{m}\left(\frac{\Delta Q}{\Delta T}\right)_p = 0.24 \text{ cal/g} \,^\circ\text{C},$$

$$C_v \equiv \frac{1}{m}\left(\frac{\Delta Q}{\Delta T}\right)_v = 0.17 \text{ cal/g} \,^\circ\text{C}. \tag{5-204}$$

Mayer realized that the difference between C_p and C_v permitted the estimation of the mechanical equivalent of heat. Here is the reasoning needed to make the estimate.

In the process of Figure 5.25(a), if we choose as our system the gas, the inflow of heat goes entirely into a change in the internal energy of the gas. The gas does *no* work. Thus, according to the conservation of heat and energy, we see that if $(\Delta Q)_v = C_v$ is the heat input per gram for a 1° rise in temperature, then the corresponding increase in internal energy is the same as this, i.e.,

$$(\Delta E) = (\Delta Q)_v = C_v. \tag{5-205}$$

On the other hand, in the case of Figure 5.25(b) if $(\Delta Q) = C_p$ is the heat input per gram for a 1° temperature rise, part of the heat input goes into increasing the internal energy and part goes into the mechanical work done by the gas against the atmospheric pressure pressing on the rising piston; i.e.,

$$\Delta E = (\Delta Q)_p - \Delta W = C_p - \Delta W. \tag{5-206}$$

It is important to note that since the temperature rise is the same in the process of Figure 5.25(a) and (b), the change in internal energy is the same in both cases. Thus, by subtracting Equations (5-205) and (5-206), we see that the difference between C_p and C_v is the work done by the piston

$$(C_p - C_v) = \Delta W. \tag{5-207}$$

Since

$$(C_p - C_v) = 0.07 \text{ cal/g} \,^\circ\text{C},$$

all we need to know is the work done per gram by the expanding gas in mechanical units as its temperature increases by 1°. This work is

$$\Delta W = P\Delta V = \left(\frac{F}{A}\right) \times (Ah). \tag{5-208}$$

Here P is atmospheric pressure, and ΔV is the change in volume of 1 g of gas in increasing its temperature $1°$. This volume expansion ΔV is known to be

$$\Delta V = 2.83 \text{ cm}^3/\text{g} °\text{C}. \tag{5-209}$$

Since $P \cong 1 \times 10^6 \text{ dyn/cm}^2$

$$\Delta W \cong 2.83 \times 10^6 \text{ dyn-cm/g}°\text{C} \equiv \frac{0.283 \text{ J}}{\text{g}°\text{C}}$$

Thus we see that an amount of heat equal to 0.07 cal (per gram $°C$) appears as mechanical work amounting to 0.283 J (per gram $°C$). The mechanical equivalent of heat J is defined as

$$J = \frac{\Delta W \text{ in joules}}{\Delta Q \text{ in calories}}. \tag{5-210}$$

Thus

$$J = \frac{0.283 \text{ J/g}°\text{C}}{0.07 \text{ cal/g}°\text{C}}$$

or

$$J \cong 4.0 \text{ J/cal.} \tag{5-211}$$

In fact, using the data known in his day, Mayer found $J = 3.6$ J/cal. The modern value of this mechanical equivalent of heat is $J = 4.184$ J/cal.

　　For a decade or more following the appearance of this essay, Mayer's work was largely ignored despite the fact that much scientific activity was devoted to these questions. He was not a psychologically stable man, and during the long struggle for recognition he did, for a time, lose his mental health. Toward the end of his life he regained himself and was honored for his early visionary insight.

　　James Prescott Joule is the second figure we shall discuss in connection with the law of conservation of energy. Beyond all question, his delicate and increasingly refined experiments, which determined the mechanical equivalent of heat, and which were conducted over a period of about 40 years, were the decisive element in the establishment of this principle. In view of the great merit which attaches to these quantitative measurements, it is altogether proper to characterize

Joule's work as being based upon "measurement, rigorous experiment and observation, and philosophic thought all round the field of physical science" [9].

And yet, if we read Joule's one and only full and clear exposition of his understanding of the law of conservation of energy, we cannot help but be impressed by its hold on him as a basic belief, which he later succeeded in supporting by many detailed experiments. Important historical information, many useful references, and a full account of Joule's exposition of the law of conservation of energy is to be found in a valuable and short article by E. C. Watson in *American Journal of Physics*, **15**, 383 (1947) [10].

Joule's only full and clear exposition of this great principle was presented on April 28, 1847, as a popular lecture at St. Ann's Church Reading Room in Manchester! (His first enunciation on the conversion factor between heat and work was in 1843 in a paper entitled, "On the Caloric Effects of Magneto-Electricity and the Mechanical Value of Heat" which he delivered at the Chemical Section of the British Association Meeting held at Cork. At that time, as Joule remarked later [10], "the subject did not excite much general attention.")

We can do no better than quote directly from his April 28, 1847, paper [10] to show how Joule concluded the validity of the principle of the conservation of energy. (In reading this account one must understand that in Joule's day the term "living force" is what we now call the "kinetic energy" and the term "attraction through space" or "attraction through a given distance" refers to a change in potential energy.) Quoting Joule [10]:

You will at once perceive that the living force of which we have been speaking is one of the most important qualities with which matter can be endowed, and, as such, that it would be absurd to suppose that it can be destroyed, or even lessened, without producing the equivalent attraction through a given distance of which we have been speaking. You will therefore be surprised to hear that until very recently the universal opinion has been that living force could be absolutely and irrevocably destroyed at anyone's option. Thus, when a weight falls to the ground, it has been generally supposed that its living force is absolutely annihilated, and that the labour which may have been expended in raising it to the elevation from which it fell has been entirely thrown away and wasted without the production of any permanent effect whatever. We might reason, a priori, that such absolute destruction of living force cannot possibly take place, because it is manifestly absurd to suppose that the powers with which God has endowed matter can be destroyed any more than that they can be created by man's agency; but we are not left with this argument alone, decisive as it must be to every unprejudiced mind.

At this point we cannot fail to recognize that the notion of the indestructibility of energy is profoundly held *belief*. But Joule then goes on to support this belief more scientifically.

> The common experience of everyone teaches him that living force is not destroyed by the friction or collision of bodies.... How comes it to pass that, though in almost all natural phenomena we witness the arrest of motion and the apparent destruction of living force, we find that no waste or loss of living force has actually occurred? Experiment has enabled us to answer these questions in a satisfactory manner; for it has shown that, wherever living force is *apparently* destroyed, an equivalent is produced which in the process of time may be reconverted into living force. This equivalent is *heat*. Experiment has shown that wherever living force is apparently destroyed or absorbed heat is produced.

Joule then goes on to give examples of the way in which kinetic energy can be transformed into heat, one of which is particularly graphic:

> By fifteen or twenty smart and quick strokes of a hammer on the end of an iron rod of about a quarter inch in diameter placed upon an anvil by an expert blacksmith will render that end of the iron visibly red-hot. Here heat is produced by the absorption of the living force of the descending hammer in the soft iron; which is proved to be the case from the fact that the iron cannot be heated if it be rendered hard and elastic, so as to transfer the living force of the hammer to the anvil.
>
> The general rule then is that wherever living force is apparently destroyed, whether by percussion, friction, or any similar means, an exact equivalent of heat is restored. The converse of this proposition is also true, namely, that heat cannot be lessened or absorbed without the production of living force, or its equivalent attraction through space.

In this part of his discussion we see the hallmark of Joule's character: his commitment to the experimental determination of the mechanical equivalent of heat. This commitment, we could almost call it an obsession, occupied him all his life. He made measurements on all kinds of systems to demonstrate that when mechanical energy was lost, it appeared as heat. Joule's commitment to this work and the profound effect his ideas had on the most perceptive of professional scientists of his day can be appreciated by the following story.

Following his April, 1847, lecture, Joule delivered a short verbal communication at the Oxford meeting of the British Association for the Advancement of Science on June 23, 1847. The section chairman, pressed by the shortness of time, did not ask for comments on the paper and Joule's measurement of the mechanical equivalent of heat would have passed without notice were it not for a young man, William Thomson (later to become Lord Kelvin, one of the foremost men of science in England), who rose and "by his intelligent questions raised a lively interest in the new theory." Joule's results and his insight profoundly impressed William Thomson. Much later he described this first meeting and a further one which demonstrates clearly Joule's overriding commitment to his scientific work. In Lord Kelvin's words [10], written in 1893:

> I can never forget the British Association at Oxford in the year 1847, when in one of the sections I heard a paper read by a very unassuming young man who betrayed no consciousness in his manner that he had a great idea to unfold. I was tremendously struck with the paper. I at first thought it could not be true because it was different from Carnot's theory, and immediately after the reading of the paper I had a few words of conversation with the author James Joule, which was the beginning of our forty years' acquaintance and friendship. On the evening of the same day that very valuable Institution of the British Association, its conversazione, gave us opportunity for a good hour's talk and discussion over all that either of us knew of thermodynamics. I gained ideas which had never entered my mind before, and I thought I too suggested something worthy of Joule's consideration when I told him of Carnot's theory. Then and there in the Radcliffe Library, Oxford, we parted, both of us, I am sure, feeling that we had much more to say to one another and much matter for reflection in what we had talked over that evening. But what was my surprise a fortnight later when, walking down the valley of Chamounix, I saw in the distance a young man walking up the road toward me and carrying in his hand something that looked like a stick, but which he was using neither as an Alpenstock nor as a walking stick. It was Joule with a long thermometer in his hand, which he would not trust by itself in the char-a-banc coming slowly up the hill behind him lest it should get broken. But there comfortably and safely seated on the char-a-banc was his bride—the sympathetic companion and sharer in his work of after years. He had not told in Section A or in the Radcliffe Library that he was going to be married in three days, but now in the valley of Chamounix, he introduced me to his young wife. We appointed to meet again a fortnight later at Martigny to make experiments on the heat of a

waterfall (Sallanches) with that thermometer* and afterwards we met again and again and again, and from that time indeed remained close friends till the end of Joule's life. I had the great pleasure and satisfaction for many years, beginning just forty years ago, of making experiments along with Joule which led to some important results in respect to the theory of thermodynamics. This is one of the most valuable recollections of my life, and is indeed as valuable a recollection as I can conceive in the possession of any man interested in science.

The measurement of the mechanical equivalent of heat fascinated Joule even on his honeymoon. This unswerving, almost obsessive commitment to the demonstration or the understanding of a scientific principle characterizes many of the most important men of science.

Joule used many methods to obtain the mechanical equivalent of heat. One of the first methods was much like the example of Section 5.4.B. He measured the heat produced by an electrical resistance through which current was forced to flow by an electric generator driven by a mechanical device whose energy input could be measured.

His most accurate and effective scheme for measuring the mechanical equivalent of heat consisted in measuring the heat generated by paddlewheels turning in a viscous fluid enclosed in a thermally insulated chamber. The paddlewheel is turned by a falling weight. The energy balance in this system is quite analogous to the case of the weight falling in a viscous fluid which we analyzed in Section 5.2.G(i). After the falling weight (mass m) reaches terminal velocity V_0, all its power is used to produce heat in the fluid

$$\text{Mechanical power} = \frac{dW}{dt} = mgV_0 = \left(\frac{dQ}{dt}\right) = \text{rate of production of heat.}$$

$$(5\text{-}212)$$

Since J, the mechanical equivalent of heat, is the ratio of the mechanical work ΔW done (in joules) to the heat ΔQ produced (in calories), i.e.,

$$\Delta W \text{ (in joules)} = J\Delta Q \text{ (in calories),} \qquad (5\text{-}213)$$

---

*The water at the base of a waterfall must, of course, be warmer than that at the top because the loss in potential energy on falling is converted into heat. In the problems at the end of this chapter we give a problem first stated by Joule in 1845 in which the temperature rise can be related to the height of the waterfall through the mechanical equivalent of heat.

it follows that the mechanical equivalent of heat is given by

$$\frac{mg\,V_0}{(dQ/dt)} = J. \tag{5-214}$$

The mechanical equivalent of heat can be found by measuring the ratio of the mechanical input power and the rate of heat production after the weight has reached terminal velocity. (Before reaching terminal velocity some of the change in potential energy goes into increasing the kinetic energy of the falling weight.)

The heat production dQ/dt could be determined by measuring the rate at which the temperature T rises by the equation

$$\frac{dQ}{dt} = C\left(\frac{dT}{dt}\right). \tag{5-215}$$

Here C is the heat capacity of the fluid. With this system, in which the temperature rise was only a small fraction of a degree, Joule found in 1847 the value $J = 4.20$ J/cal [6]. The modern value is $J = 4.184$ J/cal. In view of his continued experiments, conducted over a period of 40 years, it is fitting and proper that J is called Joules equivalent and that the mks unit of energy, the newton-meter is called the joule.

5.4.D.   Animal Metabolism, Work, and the First Law of Thermodynamics

(i)   Catabolic Rate, Heat and Power Production

The formulation of the first law of thermodynamics had a profound effect on physiology. This effect appears clearly in an 1861 paper by H. von Helmholtz [11]. He describes the change in outlook in this way:

> The majority of the physiologists in the last century, and in the beginning of this century were of the opinion that the processes in living bodies were determined by one principal agent which they termed the "vital principle." The physical forces in the living body, they supposed, could be suspended or again set free at any moment by the influence of the "vital principle," and that by this means, this agent could produce changes in the interior of the body so that the health of the body could thereby be preserved or restored.

The present generation, on the contrary, is hard at work to find out the real causes of the processes which go on in the living body. They do not suppose that there is any other difference between the chemical and mechanical actions *in* the living body and *out of it* than can be explained by the more complicated circumstances and conditions under which these actions take place, and we have seen that the law of conservation of energy legitimizes this supposition. This law, moreover, shows the way in which this fundamental question, which has excited so many theoretical speculations, can be really solved by experiment.

Here Helmholtz is speaking of the experimental studies of the balance of energy between input to the body and its output in mechanical work and heat as predicted by the first law of thermodynamics. This law enables us to place animal metabolism, work, and heat production on a quantitative basis. Writing the first law of thermodynamics as

$$\Delta E = \Delta Q - \Delta W, \tag{5-216}$$

we can characterize the overall energy balances of an entire human or animal body in terms of the changes in each of the quantities in this equation. A human being, whether resting or working is constantly transforming the stored chemical energy in foods into various forms necessary for the maintenance of the function of the various organ systems, tissues, and cells in the body. In this process, called katabolism or catabolism, the internal energy ΔE is constantly decreasing. ΔE is negative, and food must be supplied in order to make up for the catabolic activity. Part of the catabolic activity goes into work ΔW which the body does on systems outside it, and part of it goes into a transfer of heat ΔQ out of the body. ΔQ is negative.

In quantitatively describing animal energetics, it is most useful to measure the rates at which ΔE, ΔQ, and ΔW are changing in time (t). That is, to know the magnitude of each of the terms in the equation

$$\left(\frac{\Delta E}{\Delta t}\right) = \left(\frac{\Delta Q}{\Delta T}\right) - \left(\frac{\Delta W}{\Delta t}\right). \tag{5-217}$$

The term $(\Delta E/\Delta t)$ can be called the catabolic rate, the term $(\Delta Q/\Delta t)$ is the rate of heat production, and the term $(\Delta W/\Delta t)$ is the mechanical power delivered by the body to other systems.

It is of considerable interest to have a quantitative knowledge of the various terms in (5-217) for various forms of human activity and rest, as they give us valuable information on matters such as the efficiency of the human body as a machine, the proper training of an athlete for maximum performance, or the proper conditioning of ordinary persons for the maintenance of physical fitness and a state of well-being. We shall see that the levels of catabolic activity can also be used to detect malfunctioning of the thyroid gland.

(ii) "Calorific Equivalent" of Oxygen

Let us first discuss briefly how each of the terms in (5-217) can be measured. The power output term ($\Delta W/\Delta t$) can be most directly measured. One simply measures the rate at which mechanical work is done in the task involved—whether it be the work done in turning the crank of a bicycle or the work done in pushing a car filled with coal in a mine. The term $-\Delta Q/\Delta t$ is the rate at which heat leaves the body due to evaporation. In principle, this can be measured by studying the subject in an insulated room and measuring the rate at which heat must be removed to keep the room temperature constant. The third term, the catabolic rate, i.e., the rate of change of internal energy of the body, is easier to measure than would appear at first glance. In the process of catabolism, the multitude of organic compounds in our food is transformed ultimately into CO_2, H_2O, urea, and energy. The precise method of this transformation involves many intricate metabolic pathways in which enzymes speed each phase of this transformation. Regardless of the complexity of the details, however, there is but one essential substance whose rate of consumption controls the catabolic rate. That substance is oxygen. Considering the average diet of man, it has been found that the consumption of 1 l of oxygen generates about 4.8 kcal of energy. This relation between oxygen consumption and decrease in internal energy of the body is called the "calorific equivalent of oxygen." If $\Delta O_2/\Delta t$ is the rate of oxygen consumption, the rate as measured by a respirator, then the average catabolic rate $\Delta E/\Delta T$ is given by

$$\left(\frac{\Delta E}{\Delta t}\right) \text{ in kilocalories/second} \simeq 4.8 \left(\frac{dO_2}{dt}\right) \text{ in liters/second.}$$

The biochemical basis for this figure can be seen by considering the energy produced by the complete oxidation of such fundamental food substances as glucose (from carbohydrates), tributyrin (a fat), and ethanol (an alcohol). The heats of reaction and oxygen consumption associated with oxidation of each such substances is shown below [12]:

Glucose (carbohydrate):

$$\underbrace{C_6H_2O_6}_{180\ g} + \underbrace{6O_2}_{134.4\ l} = \underbrace{6CO_2}_{134.4\ l} + \underbrace{6H_2O}_{108\ ml} + 686\ kcal.$$

In this reaction 686 kcal are produced when 134 l of oxygen are consumed. Thus, for this component of the diet the calorific equivalent of oxygen is

$$\frac{686\ kcal}{134.4\ l} = 5.15\ \frac{kcal}{l\ of\ O_2}.$$

Tributyrin (fat):

$$\underbrace{C_3H_5O_3(OC_4H_7)_3}_{302\ g} + \underbrace{18.5O_2}_{414.4\ l} = \underbrace{15CO_2}_{336\ l} + \underbrace{13H_2O}_{234\ ml} + 1941\ kcal.$$

Here the calorific equivalent of oxygen is

$$\frac{1941\ kcal}{414\ l\ O_2} = 4.7\ kcal/l.$$

Ethanol (alcohol):

$$\underbrace{C_2H_5OH}_{46\ g} + \underbrace{3O_2}_{67.2\ l} = \underbrace{2CO_2}_{44.8\ l} + \underbrace{3H_2O}_{54\ ml} + 327\ kcal$$

Here the calorific equivalent of oxygen is

$$\frac{327\ kcal}{67.2\ l} = 4.86\ kcal/l.$$

We list in Table 5.2 the energy production per gram for various food substances in kcal/g and the associated value for the calorific equivalent of heat. Clearly the precise value of the oxygen equivalent depends on the relative proportion of carbohydrate, protein, fat, etc., in the diet. Depending on the character of the food being burned, this calorific equivalent of oxygen varies from ~ 4.5 to ~ 5.05. The actual proportion of food substances oxidized in a given person can be determined from the relative proportion of carbon dioxide produced and oxygen consumed. This ratio is called the respiratory quotient. A respiratory quotient (R.Q.) of 1 cor-

Table 5.2: Energy Content of Food Substance and Calorific Equivalent of Oxygen.

Food	Average energy/g (kcal/g)	Cal. equivalent of oxygen (kcal/l)
Carbohydrate	4.1	5.05
Protein	4.2	4.46
Ethanol	7.1	4.86
Fats	9.3	4.74

responds to about 5.05 kcal/l O_2, while an R.Q. of 0.8 corresponds to an oxygen calorific equivalent of 4.85 kcal/l O_2.

(iii)   Oxygen Consumption and Catabolic Rate for Various Activities—Basal Catabolism

A person's physical fitness, and his ability to perform various types of strenuous athletic activity are determined by his capacity to consume oxygen. As an individual strives to make greater physical efforts, his performance is limited by his ability to consume oxygen. The better the training and fitness, the greater is this maximum oxygen consumption rate. In fact, a current, widely used exercise program classifies the physical fitness of a person in terms of his maximum rate of oxygen consumption [13]. In Table 5.3 we list this classification along with the oxygen consumption in units of ml O_2/(min · kg). This is the oxygen consumption per kilogram of body weight. In Table 5.4 we give a list showing the oxygen consumption rate for various specific activities, along with the equivalent rate of catabolism assuming the calorific equivalent of oxygen as 4.8 kcal/l O_2. There are a number of implications of the results given in this table.

Table 5.3: Physical Fitness Level and Oxygen Consumption Capacity [13].

Fitness level	Rate of oxygen consumption (maximum) (ml/min kg)
Very poor	28
Poor	34
Fair	42
Good	52
Superior	70

Table 5.4: Rate of Oxygen Consumption and Catabolic Rate for Various Levels of Activity [14].

Activity level	Oxygen consumption rate	Catabolic rate for 65 kg man	
	$\frac{1}{m}\left(\frac{\Delta O_2}{\Delta t}\right)$ (ml/min kg)	$-\left(\frac{\Delta E}{\Delta t}\right)$ (kcal/h)	(watts)
Sleep	~ 3.5	~ 70	~ 80
Light Activity (at lecture, slow walking, domestic work)	~ 10	~ 200	~ 230
Moderate Activity (cycling at 10 mph, trotting on horseback, shoveling sand, swim breast stroke at 1 mph)	~ 20	~ 400	~ 465
Heavy Activity (playing soccer, sawing wood)	~ 25	~ 500	~ 580
Quite Heavy Activity (squash, basketball, fast swimming)	~ 30	~ 600	~ 700
Extreme Activity (racing cyclist at 27 mph [15])	~ 70	~ 1400	~ 1600

First of all we observe that when the body is completely at rest the internal energy is still diminishing at a rate of about

$$-\left(\frac{\Delta E}{\Delta t}\right) \cong 70 \text{ kcal/h} = 81 \text{ W}.$$

This is called the basal metabolic rate, or the basal catabolic rate.

Furthermore, when persons are awake and moving about a little, as in a lecture or doing domestic work, their internal energy appears as heat which flows into the room at a rate of between 100–200 W per person. We each generate as much heat as does a bright incandescent light bulb.

The table shows that it is essential that the daily food intake represents an accurate compensation for the loss of internal energy associated with basal catabolism, plus the energy requirements of daily activity. A fairly sedentary man requires a total of about 3000 kcal per day and a sedentary woman requires about 2300 kcal per day. This must be provided by food energy to maintain balance between expenditure and energy gain from food. If he eats more than he expends energetically, he gains weight—which can have important effects on his health. It is important to recognize that it is difficult to reduce weight directly through exercise. In Table 5.2 we see that the loss of 1 g of body fat requires the expenditure of 9 kcal of muscular energy. One pound, i.e., 454 g, requires the expenditure of 4100 kcal. This corresponds to sawing wood at 500 kcal/h for 8 h! It is much easier to control one's weight by resisting the temptation to overeat.

On the other hand, continued strenuous physical activity brings a great need for food. Bicycle "riders in tourist trials covering the standard 100 miles in 8 hours use about 4000 kcal. All these strenuous activities provide a craving for food not satisfied until a day or two's heavy eating has been indulged in by the participants" [15].

(iv)   Mechanical Body Output, and Mechanical Efficiency of the Human Body

In Section 5.4.D(iii) above we presented data on the value of $(\Delta Q/\Delta t)$ associated with various levels of activity. According to the first law of thermodynamics $(\Delta E/\Delta t)$ is related to the heat production rate $(\Delta Q/\Delta t)$ and the external power $(\Delta W/\Delta t)$ delivered to a second system by

$$\left(\frac{\Delta E}{\Delta t}\right) = \left(\frac{\Delta Q}{\Delta t}\right) - \left(\frac{\Delta W}{\Delta t}\right).$$

It is interesting to inquire as to how the energy of catabolism ΔE is divided between ΔQ and ΔW for various situations. From a slightly different point of view, we can look upon the human body as a machine capable of doing work. The gross mechanical efficiency e of the body is defined as the ratio of the useful mechanical power which the body delivers to some external mechanical system, to the rate of change of the internal energy

$$e \equiv \frac{\left(\dfrac{\Delta W}{\Delta t}\right)}{\left(\dfrac{\Delta E}{\Delta t}\right)} = \text{gross mechanical efficiency.} \qquad (5\text{-}218)$$

In this definition $(\Delta E/\Delta T)$ and $(\Delta W/\Delta t)$ are both regarded as being positive numbers.

The magnitude of the efficiency of the human body depends in detail upon the mechanical task involved and also on the level at which the power is delivered. Generally, at low-power output levels, the efficiency is substantially less than at high-power levels.

One of the most efficient schemes for the production of mechanical power from the human body is the bicycle. The power delivered to the foot pedal cranks can be accurately measured using ergometers or dynamometers. The oxygen consumption rate can be measured by respiration devices.

A study of the maximum power delivery rates in bicycling and in rowing shows that for short time periods a well-conditioned person can produce a mechanical power output as large as 2 hp, i.e., $\sim$ 1500 W. This is possible, however, only for about 6 s [16]. The highest output tolerable for a long period of time ($\sim$ 5 h) is about 0.2 hp [15] or $\sim$ 150 W. The "safe" working level for laborers over long periods of time is a power output of about 0.1 hp or $\sim$ 75 W. A summary of these data [15], [16] is shown in Table 5.5.

Table 5.5: Maximum Power Outputs from the Human Body for Various Time Periods ([15], [16].)

Power		Time
hp	watts	
~ 2	~ 1500	6 s
~ 1	~ 750	1 min
~ 0.35	~ 260	35 min
~ 0.2	~ 150	5 h
~ 0.1	~ 75	Laborer's daily work

If the oxygen consumption associated with the mechanical power production is measured, one can measure the efficiency of the body as a machine using the calorific equivalent of oxygen discussed in Section 5.4.D(ii). The efficiency will depend on the rate at which work is done. In Table 5.6 we list (dW/dt) and

Table 5.6: Catabolic Rate, External Power Production and Gross Mechanical Efficiency of Various Tasks (Ref. 17).

Task	hp	Power output (dW/dt) watts	Catabolic rate $\lvert(dE/dt)\rvert$ watts	Gross efficiency e
Cycling	0.15	112	505	0.19
Tramming	0.12	90	525	0.17
Shoveling	0.024	17.5	570	0.03

$\lvert(dE/dt)\rvert$ as measured on mine workers doing external work at a rate of 0.15 hp or less [18]. The forms of work were cycling, shoveling sand and tramming. This last is the pushing of loaded coal cars down a railroad track. Table 5.6 shows that when external work is done at the rate of ~ 0.15 hp, which is about the maximum rate at which a person can steadily deliver power, the gross efficiency is $\sim 18\%$ for such activities as cycling or pushing a heavy car. On the other hand, shoveling is a very inefficient activity. It can be performed only at low output power levels, and it is quite inefficient, e being about 3%.

At very high output power rates the gross efficiency can be rather better than the values listed in Table 5.6. For example, if a racing cyclist is traveling at a speed of 25 mph, the corresponding values of $(\Delta E/\Delta t)$ and $(\Delta W/\Delta t)$ are [15].

$$\left(\frac{\Delta W}{\Delta t}\right) = 0.4 \text{ hp} \cong 300 \text{ W},$$

$$\left(\frac{\Delta E}{\Delta t}\right) = 0.39 \text{ liters/min} = 1360 \text{ W}.$$

$$e \cong 22\%.$$

These data indicate that when the human body is efficiently coupled to a mechanical device it can produce power with an efficiency of about 20%.

It is interesting, in view of these details, and recent measurements to observe how J. P. Joule in 1846 estimated the mechanical efficiency of a horse in doing work as being 25% [18], [19].

A horse, when its power is advantageously applied, is able to raise the weight of 24,000,000 pounds to a height of one foot per day. In the same time (24 hours) he will consume 12 pounds of hay and 12 pounds of corn.

He is therefore able to raise 143 pounds by the consumption of one *grain* of the mixed food. From our own experiments on the combustion of a mixture of hay and corn in oxygen gas, we find that each grain of food, consisting of equal parts of undried corn and hay, is able to give 0.682° to a pound of water, a quantity of heat equivalent to the raising of a weight of 557 pounds to be height of one foot. Whence it appears that one quarter (143/557 ~ 0.256) of the whole amount of vis viva (energy) generated by the combustion of food in the animal frame is capable of being applied in producing a useful mechanical effect—the remaining three quarters being required to keep up the animal heat, etc.

5.4.E.   Basal Metabolism

(i)   Basal Metabolic Rate of Mammals: From the Mouse to the Elephant

The basal metabolic rate is the rate at which the internal energy of an organism is decreasing when it is fasting and completely at rest. Because internal energy is decreasing it is somewhat more correct to call this rate of decrease the basal catabolic rate. A "reference man" weighing 65 kg aged 25 years has a basal catabolic rate of about 1500 kcal/day. A "reference woman" weighing 55 kg aged 25 years has a basal catabolic rate of about 1260 kcal/day. On the other hand, a mouse weighing about 20 g has a basal catabolic rate of about 3 kcal/day, and an elephant of mass 5000 kg has a catabolic rate of about 70,000 kcal/day. It is a remarkable fact that if one makes a graph of the basal catabolic rate of mammals of various sizes as a function of the mass of the animal, one discovers [20], [21] that the catabolic rate (dE/dt) "scales" with mass M in accordance with the formula

$$\left(\frac{dE}{dt}\right) = CM^{3/4}, \tag{5-219}$$

where

$$C \cong 90 \frac{\text{kcal}}{(\text{kg})^{3/4}}.$$

Using this formula one can compute rather accurately the metabolic rate of a sheep, a guinea pig, a man, or a cow, knowing only the mass of the animal. The fact that (dE/dt) varies as $M^{3/4}$ is called Kleiber's law [20]. This is but one example of many such scaling laws in biology. For example, experimental measurements show that the height of trees is proportional to the two-thirds power of the diameter of the tree, or the chest circumference of various primates varies as the mass raised to the $\frac{3}{8}$ power [21]. Other interesting scaling relationships are that the respiratory frequency varies as $M^{-0.25}$, and the heart rate also varies from one mammal to another $M^{-0.25}$. If one compares the heart rate with the lifetime of the animal, one discovers that the ratio of the lifetime to the period of the heart cycle is about the same for all mammals. In fact, then, each mammal seems to live roughly for the same number of heartbeats [21] regardless of its size.

The study of scaling laws, and an effort to find theoretical bases for them, represents an active branch of the sciences of zoology and comparative physiology [22]. To appreciate the kind of reasoning which is employed in explaining such experimentally observed scaling laws, let us discuss briefly the theoretical arguments put forward by T. McMahon [21] to explain Kleiber's law.

The power output of the body's musculature is certainly proportional to the metabolic rate. It is reasonable to presume that the maximum power output P_m of the muscles is proportional to the basal metabolic rate $(dE/dt)_b$, i.e.,

$$\left(\frac{dE}{dt}\right)_b \propto P_m. \tag{5-220}$$

If we now examine how the maximum power P_m depends on the size of the muscle mass, we can deduce how the basal metabolic rate changes with the size of the animal.

The power output of a muscle is given by

$$P_m = \sigma A \left(\frac{dl}{dt}\right). \tag{5-221}$$

Here σ is the force per unit area exerted by the muscle: it is the stress exerted by the muscle. A is the cross-sectional area of the muscle, and dl/dt is the speed with which the muscle shortens on contraction. Studies of muscles show that regardless of species and size, both (dl/dt), the speed of shortening, and stress σ are constant. The stress which can be generated by muscle is always of the order of a few kilograms per cm^2, i.e., $\sigma \sim 2$–4 atm. Since σ and (dl/dt) are independent of the size of the muscle, we see that the maximum power exerted by an individual muscle mass is proportional to its cross-sectional area A. Thus, apart from constants independent of species, we can expect that the basal metabolic rate is proportional to the *cross-sectional area* of the animal's muscles

$$\left(\frac{dE}{dt}\right)_b \propto A. \tag{5-222}$$

The question we now must answer is how this cross section depends upon the mass of the animal.

In the crude figure (Figure 5.26) we characterize the size of a body in terms of l, the length of the limbs and trunk, and d, their width: The cross-sectional area of muscle is clearly proportional to d^2.

$$A \propto d^2. \tag{5-223}$$

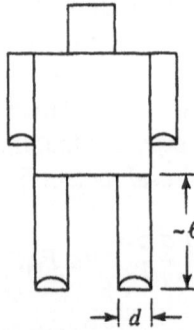

Figure 5.26. Characterization of the dimensions of the mammalian body in terms of limb length l, and limb width d.

If we could determine how the width of the limbs varies with the mass of the animal, we could predict how the basal metabolic rate depends on mass. In fact, one can deduce the dependence of d on mass by a line of reasoning proposed by McMahon [21]. Just as the height of a tree is related to the width of the trunk by the relation

$$l \propto d^{2/3}, \tag{5-224}$$

so the length of the limbs of animals is related to their thickness in the same way. Experimental data on primates support this relationship between length and width of limbs [21]. (The relationship $l^3 = c^1 d^2$ is also known theoretically [21] as the criterion for the point at which a long column will buckle under its own weight when displaced from its vertical orientation.) With this relationship in mind we can immediately deduce the connection between the mass M and the limb thickness d. Since M is proportional to volume and the volume of the animal is proportional to

ld^2, we have

$$M \propto ld^2. \tag{5-225}$$

But according to (5-224):

$$1 \propto d^{2/3}.$$

Therefore we conclude that

$$M \propto d^{8/3} = (d^2)^{4/3}. \tag{5-226}$$

But since $A \propto d^2$, we find that

$$M \propto A^{4/3}. \tag{5-227}$$

Thus the cross-sectional area of muscle is related to the mass M of the body as a whole by the relation

$$A \propto M^{3/4}. \tag{5-228}$$

Combining this result with (5-222), we see finally that one expects the basal metabolic rate to vary as $M^{3/4}$, i.e.,

$$\left(\frac{dE}{dt}\right)_b \propto A \propto M^{3/4}, \tag{5-229}$$

which is in agreement with Kleiber's law.

These arguments "explain" Kleiber's law by relating the basal metabolic rate to the cross-sectional area of muscle mass in the animal. This area in turn is shown to vary as mass raised to the power $\frac{3}{4}$.

The line of reasoning employed in analyzing Kleiber's law is typical of "dimensional analysis" which is used in many branches of engineering and science, as well as the present example which comes from comparative physiology.

It should be noted that ofttimes physiologists treat the basal metabolic rate as if it scaled as the *surface* area of the animal [24]. As a result of this, metabolic rates are sometimes expressed units of (dE/dt) per unit of surface area. This type of characterization of basal metabolic rate would be useful if (dE/dt) varied as $M^{2/3}$. However, it does not, and as a result the specification of $[(dE/dt)$ surface area] for various mammals shows that this quantity varies by a factor of about a factor 2 as one goes from mouse to cow [24]. The ratio $(dE/dt)_b/M^{3/4}$, on the other hand, is about the same for all mammals from mouse to elephant.

(ii)   Energy Requirements of Various Body Organs

The basal metabolic rate $(dE/dt)_b$ is a reflection of the rate at which each organ and part of the body does work. Part of the basal rate $(dE/dt)_b$ goes into the work of the heart in pumping blood around the body. Part goes into the work of breathing. Part goes into chemical transformations which take place in the liver. The liver plays an important role in the synthesis, storage, and conversion of proteins, carbohydrates, and fats. These chemical processes require chemical work to be done. Part of the basal metabolic rate goes into the work of the kidney which separates the waste products of catabolism from the blood by the performance of osmotic work. The neurologic activity of the brain requires energy and the work of the brain is an important contribution to (dE/dt). In Table 5.7 we list the contribution to $(dE/dt)_b$ from various organs. In columns 3 and 4 the experimental values of these contributions are presented as the result of a measurement of the oxygen

Table 5.7: Contribution of Various Organs to the Basal Metabolic Rate of a Healthy Man of Mass $\sim$ 65 kg [24].

(1) Organ	Rate of energy consumption (expt.)		(4) Percent of basal metabolic rate	(5) Rate of Energy consumption (theoretical) (kcal/day)
	(2) $\left(\dfrac{dO_2}{dt}\right)$ (ml/min)	(3) (kcal/day)		
Heart	17	117	$\sim$ 7%	140
Lung				35
Kidney	26	180	$\sim$ 10%	$(0.74/e) \simeq 180$
Liver and spleen	67	470	$\sim$ 27%	
Brain	47	325	$\sim$ 19%	
Skeletal muscle	45	310	$\sim$ 18%	
(Remainder by difference)	48	328	$\sim$ 19%	
Total	$\sim$ 250	1730	100%	

consumption of each organ [24]. In column 5 we list theoretical calculations of the catabolic rate associated with each organ. In the following paragraphs we show how these theoretical calculations are made.

It is interesting to observe that though the brain constitutes only 2.5% of the body weight it is responsible for 20% of the basal metabolic rate. The process of hard thinking produced only a very small increase in the oxygen consumption by the brain. The liver is the most important single organ insofar as basal metabolism is concerned.

Let us now compute theoretically the contribution $(dE/dt)_b$ from the heart, the lungs, and the kidney. In each case we will do this by looking at each organ as a system which performs external work at a rate (dW/dt). We can calculate (dW/dt) for each organ. In each case then we have

$$\left(\frac{dE}{dt}\right) = \left(\frac{dQ}{dt}\right) - \left(\frac{dW}{dt}\right)$$

To compute dE/dt, we need to know dQ/dt for each organ, but we don't know this. To overcome this difficulty, we will estimate the gross efficiency e of each organ, and then compute (dE/dt) using the definition

$$e = \frac{(dW/dt)}{(dE/dt)}$$

The Work of the Heart.   Mechanically the heart is a pump. Schematically it functions as shown in Figure 5.27. The blood volume enters through the mitral valve. The heart muscle contracts, thereby decreasing the volume of the heart. This increases the pressure on the blood closing the mitral valves and forcing blood out the aortic valve.

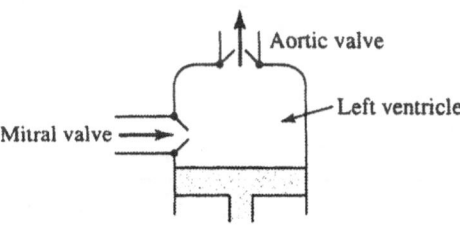

Figure 5.27. Schematic diagram of the heart as a pump.

The work done in a single compression of the heart is

$$\Delta W = \int \vec{F} \cdot d\vec{r}.$$

Since the force on the blood is a hydrostatic pressure, it is more appropriate to write ΔW in terms of the area of the heart wall (A), as

$$\Delta W = \int \left(\frac{F}{A} \right) \vec{A} \cdot d\vec{r} = \int \left(\frac{F}{A} \right) d\Omega \qquad (5\text{-}230)$$

or

$$\Delta W = \int_{\Omega_1}^{\Omega_2} P \, d\Omega. \qquad (5\text{-}231)$$

Here P is the pressure that the heart wall exerts on the blood during the contraction of the heart muscle. $d\Omega$ is a differential volume of blood, and $\Omega_2 - \Omega_1$ is the volume of blood which is forced out of the heart during each compression. The pressure exerted by the heart muscle increases as the blood flows out, and then decreases back to the diastolic pressure when the stroke volume is fully expelled. If we plot pressure as a function of the volume of blood in the heart, we get schematically the result shown in Figure 5.28. According to this figure we can express the pressure as a function of the outflow volume Ω as

$$P(\Omega) \cong P_0 + \left(\frac{P_m - P_0}{\Delta \Omega} \right) (\Omega - \Omega_1), \qquad (5\text{-}232)$$

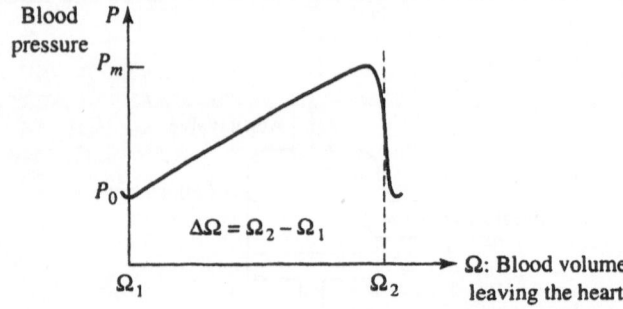

Figure 5.28. Pressure on blood during contraction of the heart muscle (systole).

$\Delta\Omega$ is the volume of blood forced out of the heart in a single contraction. From the ballistocardiograph record we know $\Delta\Omega \cong 75$ cm^3. Also from measurements of blood pressure we know that

$$P_m \cong 120 \text{ mmHg} = \text{systolic pressure,}$$

$$P_0 \cong 80 \text{ mmHg} = \text{diastolic pressure.}$$

We may now integrate (5-231)

$$\Delta W = \int_{\Omega_1}^{\Omega_2} \left\{ P_0 + \left(\frac{P_m - P_0}{\Delta\Omega} \right) (\Omega - \Omega_1) \right\} d\Omega, \qquad (5\text{-}233)$$

$$\Delta W = \left[P_0 + \frac{1}{2} \left(\frac{P_m - P_0}{\Delta\Omega} \right) \Delta\Omega \right] \Delta\Omega, \qquad (5\text{-}234)$$

where $\Delta\Omega = \Omega_2 - \Omega_1 \cong 75$ cm^3. The quantity inside the square brackets in (5-234) is nothing other than the average blood pressure during systole, i.e., 100 mmHg $= 1.33 \times 10^5$ (dyn/cm^2). Thus we find numerically that

$$\Delta W = 1.33 \times 10^5 \text{ dyn/cm}^2 \times 75 \text{ cm}^3 = 1 \times 10^7 \text{ dyne cm} = 1 \text{ J.}$$

Thus in each heartbeat the work done is about 1J. Since the time for one heartbeat is about 1 s, we have

$$\left(\frac{\Delta W}{\Delta t} \right) \cong 1 \times 10^7 \text{ dyn-cm/s} = 1 \text{ J/s} = \frac{1}{4.19} \text{ cal/s} = 0.24 \text{ cal/s.} \qquad (5\text{-}235)$$

In a single day there are 8.64×10^4 s. Thus in units of kcal/day we find that the work of the heart is

$$\left(\frac{\Delta W}{\Delta t} \right) = (0.24)(86.4) \text{ kcal/day} \cong 21 \text{ kcal/day.}$$

The energy expenditure for this work is

$$\left(\frac{dE}{dt} \right) = \frac{1}{e_{\text{heart}}} \left(\frac{dW}{dt} \right). \qquad (5\text{-}236)$$

The efficiency of a human heart ranges from 12–30% depending on the output rate [25]. If we take the efficiency of a human heart near the resting state as 15%, we find

$$\left(\frac{dE}{dt}\right) \cong 140 \text{ kcal/day}. \tag{5-237}$$

This is the value that is listed in column 5 of Table 5.7. You will observe that (dE/dt), as measured experimentally for the heart from the oxygen consumption of this organ, is ~ 117 kcal/day. Considering the crudeness of the present calculation, we may regard this calculation as in good agreement with the measured value.

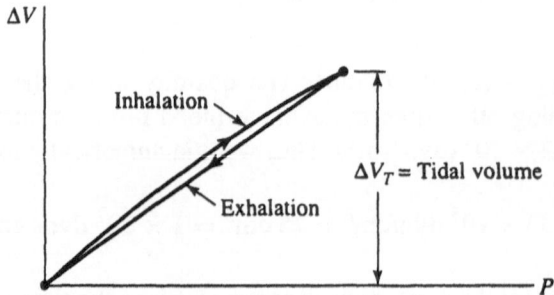

Figure 5.29. Pressure versus volume change for the lung.

The Work of Breathing. We now estimate the work done by the thorax and abdomen to expand and contract the lungs during breathing. In Figure 5.29 we plot the relationship between the volume ΔV of air expelled or inhaled into the lung as a function of the pressure the thorax and diaphragm exert upon the lung. In this figure P' is the pressure of the air in the lung minus atmospheric pressure. ΔV is the change in the volume of air in the lung. The slope of this graph is called the "compliance" of the lung–thorax–abdomen system

$$C \equiv \left(\frac{d(\Delta V)}{dP'}\right). \tag{5-238}$$

The work done in expelling the tidal volume (ΔV_T)—the volume of air expelled in one breath—is just

$$\Delta W = \int P' d(\Delta V).\tag{5-239}$$

But P' itself is a function of ΔV as seen in Figure 5.29. In fact

$$P' = \left(\frac{1}{C}\right)\Delta V.\tag{5-240}$$

Thus

$$\Delta W = \int_{\Delta V=0}^{\Delta V_T} \frac{\Delta V}{C} d(\Delta V) = \frac{1}{2C}(\Delta V_T)^2.\tag{5-241}$$

This is the mechanical work done in a single inspiration or expiration

$$\Delta W = \frac{1}{2C}(\Delta V_T)^2.\tag{5-242}$$

The rate at which work is done is ΔW times the number of breaths (in plus out) per unit time. Calling this (dn/dt), we have that the rate at which the work of breathing is done is

$$\frac{dw}{dt} = \Delta W\left(\frac{dn}{dt}\right) = \frac{1}{2C}(\Delta V_T)^2 \frac{dn}{dt}.\tag{5-243}$$

We give below the actual numerical values for the quantities that enter into this formula

$$V_T = \text{tidal volume} = 500 \text{ cm}^3,$$

$$C = 0.1 \text{ l/cmH}_2\text{O} \cong 0.1 \text{ cm}^3/\text{dyn/cm}^2,$$

$$\frac{dn}{dt} = \text{breathing rate} = 40 \text{ breathes (in plus out)/min}$$

Thus

$$\left(\frac{dW}{dt}\right) = \frac{(5 \times 10^2)^2 40}{2(0.1)} = 5 \times 10^7 \text{ egs/min} = 5 \text{ J/min},\tag{5-244}$$

or converting to kcal/day, we have

$$\left(\frac{dW}{dt}\right) = \left(\frac{5}{4.19}\right)\frac{60 \times 24}{1000} \text{ kcal/day} = 1.7 \text{ kcal/day.} \tag{5-245}$$

To compute the associated rate of change of internal energy (dE/dt), we need the efficiency with which the thorax and abdominal muscles do work. Mead and Martin estimate that respiratory muscle efficiency is about 5% [26]. Thus we have

$$\left(\frac{dE}{dt}\right) = \left(\frac{1}{0.05}\right)\left(\frac{dW}{dt}\right) \cong 35 \text{ kcal/day.} \tag{5-246}$$

We also give this value in column 5 of Table 5.7. Comparing this with the observed total basal metabolic rate of 1730 kcal/day, we see that the work of breathing contributes only about 2% to the total basal rate.

The Work of the Kidney.   The kidneys serve the function of regulating the composition and volume of the extracellular fluids of the body. It removes from the blood the waste products of metabolism and maintains the concentration of the blood constituents within very narrow limits. Each day the kidney filters out of the blood a volume of fluid fifteen times greater than the total volume of blood plasma. 99% of this filtrate is returned to the blood, after processing, at the renal tubules. The ultimate filtrate retained by the kidneys and passed in the urine consists of small ions, such as Na^+, K^+, Ca^{2+}, NH_3, Cl^-, SO_4^-, and HCO_3^-. The urine also contains, in addition to water, larger molecules, such as urea, creatinine, uric acid, and other amino acids. These molecules are the products of protein catabolism. If the uric acid molecules precipitate, they form crystals which, on growing, produce painful kidney stones.

The concentration of all the various substances in the urine is roughly two to three times the concentration in the blood plasma. Thus the kidney concentrates the substances flowing in the plasma and work must be done to do this. The work is called osmotic work because the kidney functions against a concentration gradient and hence against a gradient of osmotic pressure.

In Volume II of this series we will show that the osmotic work done per unit time is given by the following formula:

$$\left(\frac{dW}{dt}\right)_{\text{osmotic}} = \left(\frac{dn}{dt}\right) RT \ln\left(\frac{C_2}{C_1}\right). \tag{5-247}$$

Here

$$\left(\frac{dn}{dt}\right) = \text{number of moles of urine solutes secreted per unit of time,}$$

$C_2 = $ total concentration of filtrates in urine,

$C_1 = $ total concentration of filtratable molecules in plasma,

$R = $ gas constant per mole $= 1.987 \text{ cal/}^\circ\text{K}$,

$T = $ body temperature $(310\,^\circ\text{K})$.

We may now compute (dW/dt) using the following values [27], the various parameters entering (5-247).

$$\left(\frac{C_2}{C_1}\right) = 3,$$

$$\left(\frac{dn}{dt}\right) = C_0 \frac{d\Omega}{dt}.$$

where

$$C_0 = \text{concentration of solutes in urine}$$

$$= 0.8 \text{ mol/l,}$$

$$\left(\frac{d\Omega}{dt}\right) = \text{volume of urine produced per day,}$$

$$\left(\frac{d\Omega}{dt}\right) = 1.5 \text{ l/day,}$$

therefore $\quad \left(\frac{dn}{dt}\right) = 1.2 \text{ mol/day,}$

$$\left(\frac{dW}{dt}\right) \cong 1.2(2)(310)\ln 3 = 740 \text{ cal/day.}$$

Thus the osmotic work of the kidney is about 740 cal/day. To compute the contribution of the kidney to the basal metabolic rate, we need to know the efficiency with which the kidney does osmotic work. The efficiency of the kidney is low, and not well known [27]. In fact, if we take (dE/dt) as measured experimentally from the oxygen consumption of the kidney as 180 kcal/day from Table 5.7, we see that $e = (dW/dt)/(dE/dt) \cong 0.740/180 \sim 0.4\%$. If we adopt this value for efficiency, we get, of course, the figure of $(dE/dt) = 180$ kcal/day. The work of the kidney is responsible for $\sim 10\%$ of the total basal metabolic rate.

This is about as far as we can go now in estimating theoretically the daily work and energy expenditures of various body organs.

To conclude this discussion of the basal metabolic rate, it is interesting to note that the measurement of the basal metabolic rate has been used to detect abnormalities in the secretion of thyroid hormones. In excess secretion (hyperthyroidism) the basal rate is between 30% and 70% higher than that of the normal. In deficient secretion (hypothyroidism) the basal rate is 20% to 40% lower. In Figure 5.30 we show the probability with which a particular metabolic rate is measured in the normal, the hyperthyroid, and the hypothyroid person [28].

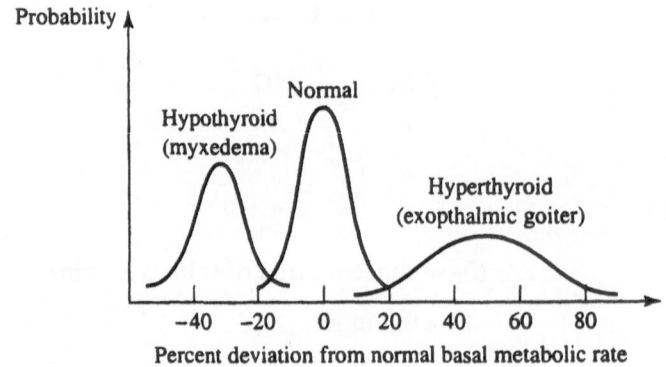

Figure 5.30. Probability of measuring a deviation from the basal metabolic rate for various thyroid diseases.

5.5   Rotational Motion

There is one elaboration of Newton's laws of motion that is beyond the scope of this text: the field generally described as "Rigid Body Dynamics." A thorough treatment of this topic, on an elementary level, is to be found in the text by Kleppner and Kolenkow [1].

We shall give here only a very brief treatment of a restricted part of the general subject of the rotational motion of a rigid body: We confine ourselves to the case of rigid bodies with an axis of rotation that is fixed in space. This fixed axis may be an actual axis, as for instance in the case of a rotating piece of machinery; more generally, any rigid body pinned down at *two* fixed points is free to rotate only about an axis defined by the straight line passing through these two points.

This restriction leads to an enormous simplification of the problems, mainly because now the motion of such a body is completely specified by just knowing the value of a single angle θ as a function of time (see Figure 5.31).

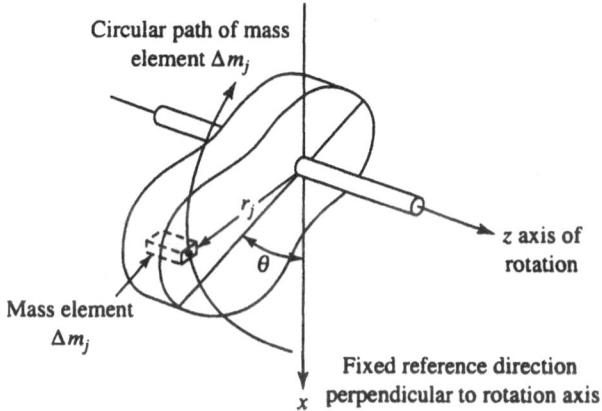

Figure 5.31. Definition of the angle θ.

The kinematic constraint of a fixed axis forces every part of the body to move in a circle whose center is on the axis of rotation. As described in Section 1.3.A, the velocity of such a part of the body, at a distance r from the axis, is given by

$$v = r\frac{d\theta}{dt} = r\omega. \tag{5-248}$$

The dynamics of this restricted motion deals with the question of how the angular position θ and the angular velocity ω change with time, given the forces that act on the body. In answering this question, the previously developed work–energy relation will again be extremely helpful. Recall that in Section 5.2.A it was shown how this relation emerged from Newton's laws. In order to apply it in the present context, we shall find an expression for the kinetic energy of the rotating body in terms of the angular velocity $\omega = d\theta/dt$. We shall also find an expression for the work done on the body by the applied forces, as the body changes its angular position from θ_0 to θ_1. By combining these results, we shall have at hand an expression for the work energy relation in terms of the angular position θ and angular velocity ω.

(i)   Kinetic Energy of the Rotating Body

Dividing the body into small elements Δm, a distance r away from the axis, we can write the kinetic energy as

$$T = \frac{1}{2} \sum_j \Delta m_j v_j^2. \tag{5-249}$$

By using (5-248), this may be rewritten as

$$T = \frac{1}{2} \sum_j \Delta m_j r_j^2 \left(\frac{d\theta}{dt} \right)^2. \tag{5-250}$$

Now $d\theta/dr$ has a common value for all mass elements Δm_j, and may therefore be factored out of the sum

$$T = \frac{1}{2} \left(\sum_j \Delta m_j r_j^2 \right) \left(\frac{d\theta}{dt} \right)^2. \tag{5-251}$$

This important result contains the statement that the kinetic energy T of a body rotating about a fixed axis can be written as

$$T = \frac{1}{2} I \left(\frac{d\theta}{dt} \right)^2 = \frac{1}{2} I \omega^2, \tag{5-252}$$

that is, it is proportional to the square of the angular velocity ω, and to a constant I, defined as

$$I = \sum_j \Delta m_j r_j^2 = \int dm r^2, \tag{5-253}$$

whose value depends only on the shape and mass of the rigid body, and on the position of the axis of rotation, but does not depend on either θ or ω. I is called the *moment of inertia* of the body with respect to the z-axis. It has the dimension of mass $\times$ (length)^2. The determination of the numerical value of I for a given body is a problem of calculus, and is not our main concern here.

(ii)   Work Done on the Rotating Body

The forces acting on the body considered here are certainly in part forces of constraint (e.g., the elastic forces inside the body, which maintain its rigidity, plus the constraints that maintain the axis of rotation in its fixed position in space). These forces do no work. Figure 5.32(a) gives an example of an external force that does work, and changes the state of motion of rotating piece: a water jet driving a turbine wheel.

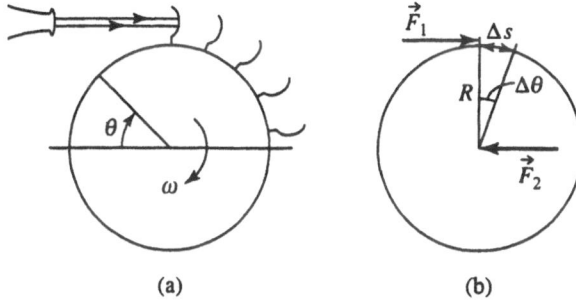

(a)   (b)

Figure 5.32. (a) Water jet hitting a turbine blade. (b) Forces on a blade and turbine axle.

In Figure 5.32(b) the force $\vec{F}_2$ exerted by the bearings on the turbine axle is equal and opposite to $\vec{F}_1$, the force exerted by the jet on the blades: $\vec{F}_1 + \vec{F}_2 = 0$; this is indeed required since the center of mass of the turbine stays at rest. So $\vec{F}_1$ and $\vec{F}_2$ form a "couple," to which we assign a *torque*

$$\tau = RF_1. \tag{5-254}$$

(A torque may have either a positive or negative sign. We will adhere to the following sign convention: The definition of the angle θ, as in Figure 5.31, defines a positive sense of rotation, in which θ increases, that is, $\omega = d\theta/dt$ is positive. A torque τ is considered positive if it induces a positive rotation of the body, negative if it induces a rotation in the opposite sense.) We are now in a position to calculate the work done by the jet on the moving blade. This work is (see Figure 5.32(b))

$$\Delta W = \vec{F}_1 \cdot \vec{\Delta}S = F_1 \Delta S = F_1 R \Delta \theta \tag{5-255}$$

for a displacement $\vec{\Delta}S$ of the blade position. (Notice that F_2, being a force of constraint, does not contribute to the work ΔW!) In the last part of this equation we have used the fact that if the blade is hit by the jet at a point a distance R from the axis, then the displacement of that point can be written as

$$\Delta S = R \Delta \theta.$$

Now the product $F_1 R$ in (5-255) defines the torque τ of the couple $(\vec{F}_1, \vec{F}_2)$, so that we may write

$$\Delta W = \tau \Delta \theta. \tag{5-256}$$

This then states that the work done by the forces acting on the rotating body, as its angular position changes by $\Delta \theta$, is given by the product of applied torque and angular change.

Equations (5-252) and (5-256) will now be used to formulate the work–energy relation: Consider a rigid body, allowed to rotate about a fixed axis; let τ be the torque exerted by the external forces on the body. Consider the state of the body at two times t_0 and t_1. At the earlier time, t_0, the body has angular position θ_0 and angular velocity ω_0. The corresponding values at the later time t_1 are θ_1 and ω_1. As the body moves from θ_0 to θ_1 in the time interval $t_0 \rightarrow t_1$, the amount of work done on the body is

$$W_{\theta_1 \leftarrow \theta_0} = \int_{\theta_0}^{\theta_1} dW = \int_{\theta_0}^{\theta_1} \tau \, d\theta. \tag{5-257}$$

This must then be equal to the change in kinetic energy of the body

$$\Delta T = (T)_{\theta_1} - (T)_{\theta_0}.$$

Using (5-252) for T, we have thus the statement

$$\tfrac{1}{2} I \omega_1^2 - \tfrac{1}{2} I \omega_0^2 = \int_{\theta_0}^{\theta_1} \tau \, d\theta. \tag{5-258}$$

This is our fundamental result. We can exploit this result in several ways, all of which are in close analogy to the application of the work–energy relation for translational motion.

First, let us again introduce the concept of power, or the rate, at which work is done. In analogy to the procedure in Sections 5.2 and 5.3, we apply (5-258) to a small increment of angle $\Delta\theta$: $\theta_1 = \theta_0 + \Delta\theta (\Delta\theta \ll 1)$. The work–energy relation may then be written as

$$\Delta T = (T)_{\theta_0+\Delta\theta} - (T)_{\theta_0} = \tau\,\Delta\theta$$

and, upon dividing by the corresponding time interval $\Delta t = t_{\theta_0+\Delta\theta} - t_{\theta_0}$, we have

$$\frac{\Delta T}{\Delta t} = \tau\frac{\Delta\theta}{\Delta t}.$$

In the limit of infinitesimally small $\Delta\theta$ and Δt, the left-hand side of this equation represents the rate of change of kinetic energy, the right-hand side the rate of work, or power:

$$\frac{dT}{dt} = \tau\frac{d\theta}{dt} = \tau\omega \equiv P. \tag{5-259}$$

We saw that in the case of rigid-body rotation about a fixed axis, the kinetic energy has the form

$$T = \tfrac{1}{2}I\omega^2.$$

A change of kinetic energy thus requires a change in angular velocity, and hence an angular acceleration $d\omega/dt$. With the above expression for T, we have indeed

$$\frac{dT}{dt} = \tfrac{1}{2}I\frac{d\omega^2}{dt} = I\omega\frac{d\omega}{dt}. \tag{5-260}$$

Inserting this into (5-259) gives us another important result

$$I\frac{d\omega}{dt} = \frac{d}{dt}(I\omega) = \tau. \tag{5-261}$$

The quantity $(I\omega)$ is called the *angular momentum* of the rigid body about its axis of rotation. Equation (5-261) can then be put in words as follows: The rate of change of the angular momentum of a rigid body is equal to the applied torque. In making this verbal statement, one must be aware of the necessity of defining both the angular momentum and the torque in reference to the actual axis about which the body is allowed to rotate. Equation 5-261 is the rotational analog of the relation developed in Chapter 4 (Eqn. (4-7)), stating that the rate of change of the linear momentum (generally just called momentum) of a body is equal to the applied net force.

(iii)   The Physical Pendulum

An illustration of the angular momentum–torque relation is provided by the motion of the so-called "physical pendulum" (see Figure 5.33).

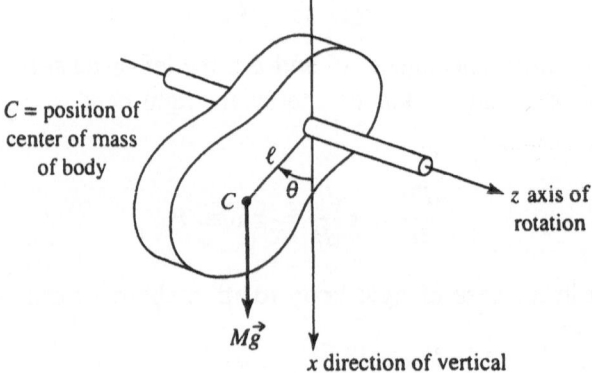

Figure 5.33. Physical pendulum.

Of all the forces acting on the body, only gravity, $M\vec{g}$, has a nonzero torque about the z-axis

$$\tau_{\text{grav}} = -Mg\ell \sin \theta. \tag{5-262}$$

(Notice the − sign, in agreement with the sign convention previously established: τ tends to decrease the angle θ.) Inserting this expression for τ into (5-261) gives the equation of motion

$$I \frac{d\omega}{dt} = -Mg\ell \sin \theta, \tag{5-263}$$

$dω/dt$ is the angular acceleration

$$\frac{d\omega}{dt} = \frac{d}{dt}\left(\frac{d\theta}{dt}\right) = \frac{d^2\theta}{dt^2}.$$

At $\theta = 0$, the rigid body is in equilibrium. If displaced from this position, and then released, it will perform an oscillation about $\theta = 0$. For a small amplitude oscillation (that is, small angles θ), we can approximate $\sin\theta$ by θ, and simplify (5-263) to

$$I\frac{d^2\theta}{dt^2} \cong -Mg\ell\theta. \tag{5-264}$$

The solutions to this equation are harmonic oscillations about $\theta = 0$:

$$\theta(t) = \theta_M \sin(\Omega t + \phi)$$

with

$$\Omega^2 = \frac{Mg\ell}{I}$$

Hence, the period of oscillation is

$$T = \frac{2\pi}{\Omega} = 2\pi\sqrt{\frac{I}{Mg\ell}}. \tag{5-265}$$

It is possible, in this case, to use the energy method to describe this oscillation. Gravity is a conservative force, and the work done by such a force can be expressed in terms of a change of potential energy. The present problem may serve as an illustration of this point

$$W_{\theta_1,\theta_0} = \int_{\theta_0}^{\theta_1} \tau\, d\theta = -Mg\ell \int_{\theta_0}^{\theta_1} \sin\theta\, d\theta$$
$$= +(Mg\ell\cos\theta_1 - Mg\ell\cos\theta_0).$$

We see that the work done in moving the body from position θ_0 to position θ_1 takes the form

$$W_{\theta_1, \theta_0} = U(\theta_1) + U(\theta_0)$$

and thus defines a potential energy

$$U(\theta) = -Mg\ell \cos\theta + \text{constant}.$$

We will choose the constant in such a way as to have a zero value of the potential energy at the equilibrium point $\theta = 0$: $U(0) = -Mg\ell + \text{constant} = 0$. Thus, the constant has the value $+Mg\ell$, and $U(\theta)$ may be written as

$$U(\theta) = Mg\ell(1 - \cos\theta). \tag{5-266}$$

With this, the work–energy relation [(5-258)] can thus be cast into the format of a statement of conserved total energy

$$(T)_{\theta_1} - (T)_{\theta_0} = W_{\theta_1, \theta_0} = -U(\theta_1) + U(\theta_0)$$

or

$$(T)_{\theta_1} + U(\theta_1) = (T)_{\theta_0} + U(\theta_0). \tag{5-267}$$

At this point, all considerations pertaining to the application of the energy methods, as discussed in this chapter, are applicable to (5-267). The only differences is that here position and motion of the body are described by means of an angle θ and an angular velocity $\omega = d\theta/dt$ rather than by a coordinate x and linear velocity $v = dx/dt$. Again using the present example of the oscillation of the physical pendulum, we have

$$\tfrac{1}{2}I\omega^2 + Mg\ell(1 - \cos\theta) = E_{\text{tot}}.$$

At $\theta = 0$, the potential energy is zero, whereas at the maximum amplitude, θ_M, the kinetic energy is zero, and the energy is all potential:

$$0 + Mg\ell(1 - \cos\theta_M) = E_{\text{tot}}.$$

This gives us E_{tot} in terms of θ_M, and we can now write $1/2I\omega^2 = Mg\ell(\cos\theta - \cos\theta_M)$ and in this way establish the dependence of ω on θ:

$$\omega(\theta) = \sqrt{\frac{2Mg}{I}(\cos\theta - \cos\theta_\mathrm{M})}.$$

The further pursuit of this topic now follows exactly the procedure illustrated in Section 5.2.E.

5.6   References and Supplementary Reading

[1] *An Introduction to Mechanics* by D. Kleppner and R. Kolenkow. McGraw-Hill, New York (1973).

[2] *The Techniques of Springboard Diving* by C. Batterman. MIT Press, Cambridge, MA (1968).

[3] "Physics and the Vertical Jump," by E. J. Offenbacher. *American Journal of Physics* **38**, 829 (1969).

[4] "A Dynamical Analysis of the Standing Vertical Jump," by P. H. Gerrish. PhD Thesis, Teachers College, Columbia University (1934).

[5] "The Dimensions of Animals and Their Muscular Dynamics," A. V. Hill, *Sci. Prog. London* **38** (No. 150), 209 (1950).

[6] *Introduction to Concepts and Theories in Physical Science*, 2nd ed., by Gerald Holton (revised by S. G. Brush). Addison Wesley, Boston, MA (1973).

[7] "Bemerkungen über die Kräfte der Unbelebten Natur," by R. J. Mayer, *Ann. Chem. Pharm.* **42**, 233 (1842). A translation by G. C. Foster, first published in *Phil. Mag.* **24**, 371 (1862), is also reproduced in W. F. Magie's *A Source Book in Physics* (McGraw-Hill, New York, pp. 197–203) and *Isis* **13**, 27–33 (1929).

[8] *Popular Scientific Lectures* by H. von Helmholtz. Dover, New York (1962). Paperback reprint of 1881 edition.

[9] *Lord Kelvin Popular Lectures and Addresses*, Vol. 2, MacMillan (1894), p.564.

[10] "Joules Only General Exposition of the Principle of Conservation of Energy," by . C. Watson, *Amer. J. Phys.* **15**, 383 (1947). *Joule and the Study of Energy* by Alex Wood. G. Bell and Sons, London (1925 and 1934); *Nature* **26**, 618 (1882); and *The Life of William Thomson, Baron Kelvin of Largs* by Silvanus P. Thompson. Macmillan, London (1910).

[11] "On the Application of the Law of the Conservation of Force to Organic Nature," H. von Helmholtz, *Proc. of the Royal Institution* **3**, 356 (1861).

[12] *Physiology and Biophysics* by Ruch and Patton, 19th ed., p. 1037, Saunders.

[13] *Aerobics* by K. H. Cooper. Bantam Books and M. Evans, New York (1968).

[14] *A Companion to Medical Studies* by Passmore and Robson. F. A. Davis, Philadelphia (1968), p. 4.11. Also private communication with Professor D. Wilson, Department of Mechanical Engineering, MIT.

[15] "A Note on Estimation of Energy Expenditure of Sporting Cyclists," by F. R. Whitt, *Ergonomics* **14**, 419 (1971).

[16] "Maximizing Human Power Output by Suitable Selection of Motion Cycle and Load," by J. W. Harrison, *Human Factors* **12**(3), 315 (1970).

[17] "Inter- and Intra-Individual Differences in Energy Expenditure and Mechanical Efficiency," by C. H. Wyndham et al., *Ergonomics* **9**, 17 (1966).

[18] *Muscular Work. A Metabolic Study with Special Reference to the Efficiency of the Human Body as a Machine* by F. G. Benedict and E. P. Cathcart. Carnegie Institute of Washington Publ. #187, Washington, DC (1913).

[19] "Experiment and Observation on the Mechanical Powers of Electro-Magnetism," Scoresby and Joule, *Phil. Mag.* **28**, 454 (1846).

[20] *The Fire of Life* by M. Kleiber. Wiley, New York (1961).

[21] "Size and Shape in Biology," by T. McMahon, *Science* **179**, 1201 (1973).

[22] *How Animals Work* by K. Schmidt-Nielson. Cambridge University Press, Cambridge (1972).

[23] "The Dimensions of Animals and Their Muscular Dynamics," A. V. Hill, *Sci. Prog. London* **38** (No. 150), 209 (1950).

[24] *A Companion to Medical Studies*, Vol. I, by R. Passmore and J. S. Robson. F. A. Davis, Philadelphia (1968), pp. 4.8, 4.9.

[25] *Starlings Human Physiology* by H. Davson and M. G. Eggleston, (14th ed.), Lea and Febriger, Philadelphia (1968), p. 188.

[26] *Principles of Respiratory Mechanics: Course in Respiratory Physiology*, by J. Mead and H. Martin, Harvard Medical School.

[27] "The Work of the Kidney in the Production of Urine," by H. Borsook and H. M. Winegarden, *Proc. National Acad. Sci.* **17**, 3 (1931).

[28] Ref. [12, p. 1047].

5.7 Problems

(The problems are grouped in sections and arranged roughly in order of increasing difficulty.)

Work, Potential, and Kinetic Energy

1. A jack, of the kind used to change a tire, is built as illustrated. The pitch of the worm gear is $\frac{1}{4}$ in per turn. L is 1 ft. What force F must be applied to the handle to lift a weight of 500 lb? Assume no friction, so that work done on the jack equals work delivered by jack.

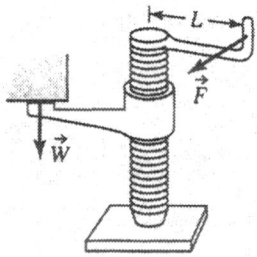

2. The town of Yverdon in Switzerland has installed a trolleybus, which runs on the energy stored in a flywheel. The kinetic energy of a cylindrical flywheel is

$$E_{\text{kin}} = \pi^2 M R^2 f_0^2,$$

where

$$M = \text{mass of wheel} = 10^3 \text{ kg},$$
$$R = \text{radius of wheel} = 1 \text{ m},$$
$$f_0 = \text{number of revolutions per second at full speed}$$
$$= 10^2/\text{s}.$$

(a) Find the value of the stored energy, in joules.

(b) Reexpress this energy in kilowatt hours.

(c) Assume that the average power consumption of the running bus is 30 hp $(1 \text{ hp} = \frac{3}{4}\text{kW})$. For how long a time can the bus run on the stored energy?

3. During a one-legged jump, such as in the hurdle of a springboard dive, or the high jumper's takeoff, part of the "lift" is provided by the lifting of the arms and one leg-knee. Suppose both arms represent 15% of the body weight, and one leg 20%. Suppose also that the (upward) velocity of each of these is about 50% larger than the extension velocity of the grounded leg.

   (a) Compute the contribution of the arm and leg motion to the total kinetic energy of the body at takeoff.

   (b) Compute the contribution of the arm and leg raising motion to the total rise ΔH of the center of mass of the body.

4. In Chapter 4 we showed that if a body of mass m, area A, moving with velocity v, is brought to a halt in collision with an absorbing medium, such as snow, the average pressure p, acting on the body during impact, is

$$p = mv^2/2Ah,$$

   h being the distance the body sinks into the absorbing medium.

   (a) Derive this result from a work–energy consideration.

   (b) Suppose a person jumps from a window of a burning building into a life net held by firemen 30 ft below. How deeply should the net sink for the person to avoid injury. The critical pressure, above which injury occurs, is $p_0 \cong 3.0$ lb/in^2.

Energy Conservation

5. As a tourist attraction, the city of Geneva has built a fountain in the lake, which sends a jet of water 300 ft high into the air. What is the velocity of the water at the nozzle? (Give your answer in ft/s; use $g = 32$ ft/s^2.)

6. Pole vaulting. In this sport the athlete picks up speed by running toward the hurdle, and then lifts himself up and over the hurdle by means of a long pole. A skillful athlete manages to keep his center of mass below the top of the hurdle, by an estimated 25 cm. The world record for this sport is $H = 5.3$ m. h, the height of the athlete's center of mass while running is $h = 1$ m.

   (a) Assuming the athlete converts all his kinetic energy into potential energy, how fast must he run to achieve the world record?

   (b) How does this speed compare with that of a sprinter, who covers 100 m in 10 s?

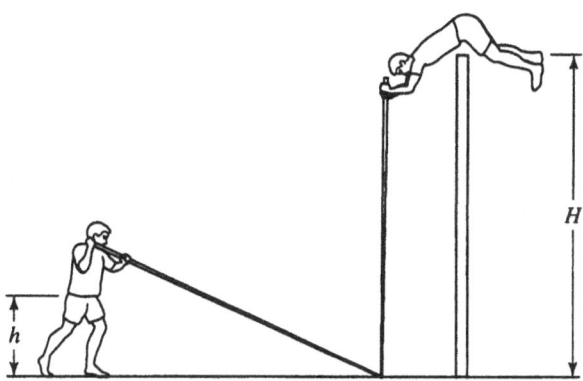

7. A toy car runs without friction on a track which forms a circular loop of radius R:

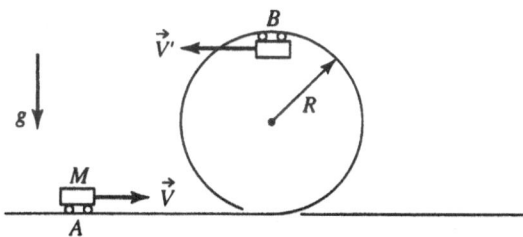

(a) Assuming the car has speed v at point A, what is its speed v' at point B?

(b) What is the minimum speed it must have at B so as not to fall off the track?

8. The kinetic energy of a *rolling* cylinder of mass M is:

(i) $E_{\text{kin}} = Mv^2$ for a solid cylinder;

(ii) $E_{\text{kin}} = \frac{3}{2}Mv^2$ for a thin-walled, hollow cylinder.

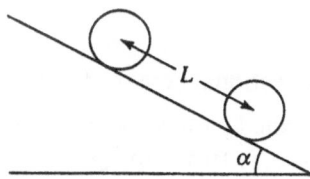

Find the speeds attained by the two types of cylinders after rolling down a distance L on an inclined plane, starting with zero velocity. Compare this speed with that attained by a block, *sliding* down without friction.

Power

9. An airplane flying at speed v must overcome a drag resistance

$$F_{\text{drag}} = \tfrac{1}{2}C_D\rho v^2,$$

C_D being a constant, ρ the density of air.

(a) What *power* must the airplane engines deliver to enable the plane to fly at *constant speed* v?

(b) Assuming *constant* engine power, how much could the plane *gain* in cruising speed v, if it flew at a higher altitude, where the density ρ of air would be lower by a factor 2?

10. In the Pelton-turbine of a hydroelectric power plant, a water jet of velocity v_0 hits the turbine blade, is split in two, and has its direction reversed.

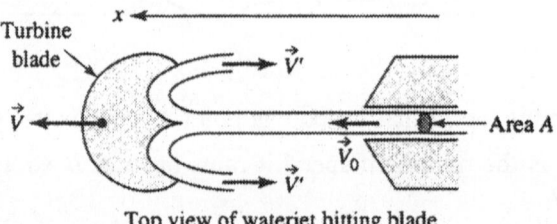

Top view of waterjet hitting blade

(a) Show that the rate at which kinetic energy emerges from the nozzle is

$$\frac{d E_{\text{kin}}}{dt} = \tfrac{1}{2}\rho A v_0^3,$$

A being the cross-sectional area of the jet, ρ the density of water.

(b) Show that the deflected jet has a velocity $\vec{v}' = 2\vec{V} - \vec{v}_0$, as long as $\vec{V}$, the velocity of the blade, is less than $\vec{v}_0$.

(c) Show that the rate-of-momentum change of the jet is

$$\frac{d\vec{p}}{dt} = \rho A v_0 2(\vec{V} - \vec{v}_0).$$

Determine the force $\vec{F}$ on the turbine blade.

(d) The power delivered to the turbine is

$$P = \vec{F} \cdot \vec{V} = FV.$$

Show that P has its maximum value when $V = \frac{1}{2}v_0$, and that in this case, P is equal to the rate of kinetic energy delivered by the jet.

(e) Show that this result may also be inferred from (b) and the work energy relation.

Motion of System by the Energy Method

11. The universal oscillator. Consider some system, whose motion can be described by means of a parameter $q(t)$ (q may be a displacement, an angle, a pressure, a current, etc.). Suppose that the kinetic and potential energies of this system can be written as

$$E_{\text{kin}} = \frac{B}{2}\left(\frac{dq}{dt}\right)^2, \qquad E_{\text{pot}} = \frac{C}{2}q^2,$$

and that the *total* energy $e = E_{\text{kin}} + E_{\text{pot}}$ is *conserved*.

(a) Show that the condition $E = $ constant, or $dE/dt = 0$, leads to an oscillator equation for $q(t)$.

(b) Give the expression for the period of oscillation in terms of the two constants B and C.

12. A wooden disc of radius R, mass M, has a peg of mass m stuck into its side, a distance r from the center. If the disc is allowed to *roll* on a flat table, it is found to oscillate about its equilibrium position ($\theta = 0$). The kinetic energy of the rolling wheel is

$$E_{\text{K}} = MR^2\left(\frac{d\theta}{dt}\right)^2$$

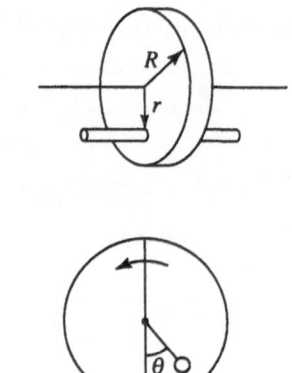

and that of the peg is (for small θ)

$$E_K \cong \frac{m}{2}(R-r)^2 \left(\frac{d\theta}{dt}\right)^2$$

(a) What is the potential energy of the wheel and peg, as a function of θ for small θ?

(b) From this energy expression, find the *period* of oscillation of this system. *Hint:* Consult Problem 11.

13. A U-shaped glass tube is filled with liquid of density ρ. p_0 is atmospheric pressure. Let a be the inner cross-sectional area of the tube, and let L be the total length of the liquid column.

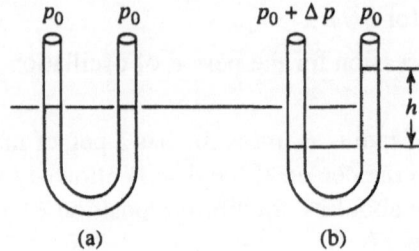

(a)   (b)

(a) What is the pressure difference Δp necessary to produce a difference h between parts 1 and 2 of the tube?

(b) Starting from equilibrium (a), assume Δp being increased gradually so as to bring about situation (b). Show that the total *work* done by the pressure difference on the liquid is

$$W = \tfrac{1}{4}\rho g a h^2$$

(c) If you have both ends of the U-tube open (atmospheric pressure p_0), the liquid may oscillate about the equilibrium position (a). Show that the total energy of this oscillation is

$$E = \tfrac{1}{2}\rho a L \left(\tfrac{dh}{dt}\right)^2 + \tfrac{1}{4}\rho g a h^2.$$

(d) From this expression, and *using the analogy* with the expression for the energy of the "universal oscillator" (see Problem 11), find the natural frequency ω_0 of oscillation of the liquid column.

14. Air drag damping of a pendulum. Let the motion of a pendulum be described by the displacement s of the bob from its equilibrium position.

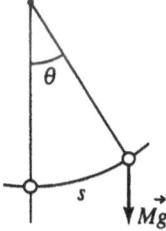

(a) Show that the energy of the pendulum is, for $|\theta| \ll 1$:

$$E = \frac{M}{2}\left(\frac{ds}{dt}\right)^2 + \frac{Mg}{2\ell}s^2.$$

(b) Apply the energy conservation condition $dE/dt = 0$ to determine the equation of motion, and the angular frequency ω of the pendulum.

(c) Show that the average kinetic and potential energies, (averaged over one period of motion), are the *same*.

(d) Let a be the amplitude of the oscillation that is described by

$$s(t) = a \cos \omega t.$$

Show that the total energy may be written as

$$E = \frac{M}{2}\omega^2 a^2.$$

(e) Assume now that this oscillation is weakly damped by a drag force $F_D = \alpha v^2 = \alpha(ds/dt)^2$, always opposed to the motion. The rate of energy loss of the pendulum is then

$$\frac{dE}{dt} = -F_D v = -\alpha v^3,$$

and the average rate of energy loss is

$$-\alpha \langle v^3 \rangle_{av} = -\alpha \left(\frac{4}{3\pi}\right) a^3 \omega^3.$$

(You may try to prove this—if not, at least convince yourself that from a *dimensional* consideration, this term must have the form $-\text{const.}\alpha(a\omega)^3$.) We conclude that the amplitude a must slowly decrease with time. Establish the equation

$$\frac{da}{dt} = -\left(\frac{4}{3\pi}\right)\left(\frac{\alpha\omega}{M}\right) a^2(t).$$

(f) Let a_0 be the initial amplitude of the pendulum. Verify that the solution of the equation for $a(t)$ is

$$a(t) = \frac{a_0}{1 + t/t_0}$$

and express t_0 in terms of a_0, M, α, ω.

(g) A big pendulum can swing for a long time. Consider a steel ball with mass $M = 30$ kg, suspended on a wire of length $L = 25$ m. The air drag coefficient α is then $\alpha \cong 10^{-2}$ kg/m. Show that for $a_0 = 1$ m, one finds for the time constant t_0 the value $t_0 \simeq 3$ h!

Energy Dissipation and Heat

15. A block of lead, sliding without friction on a horizontal track with speed v, collides head-on with another lead block, of the same mass M, initially at rest.

The collision is totally inelastic; after the collision, the two blocks move together with speed $v/2$.

(a) What fraction of the initial kinetic energy is lost in the collision?

(b) The specific heat of lead is $c_v = 0.03$ cal/g. If $v = 20$ m/s, what is the temperature increase of the lead blocks? (1 cal = 4.2 J).

16. Niagara Falls are about 176 ft high. The average flow of water is given as 15×10^6 ft^3/min.

(a) Using the specific heat of water, $c_v = 1$ cal/g, find the temperature difference between the water upstream and downstream of the Falls.

(b) Find the total amount of heat generated by the Falls in kcals/s.

(c) About 20% of the water is diverted to generate hydroelectric power. Using the conversion factor 1 kcal/s = 4.2 kW; what is the power output of this station?

17. A soft lead bullet is fired against a steel armor plate. Given the specific heat of lead as 0.03 cal/g, and the melting temperature of lead as 327 °C, at what speed must it be fired so as to reach melting temperature on impact?

18. In the great Arizona crater, created by the impact of a meteorite, large amounts of fused silica have been found, which indicated that on impact, rock heated to a temperature of at least 3000 °F. The impact speed of the meteorite must have been at least equal to the escape velocity $v_c = \sqrt{2gR}$, R being the Earth's radius.

(a) The specific heat of silica is 0.3 cal/g °C. Calculate the minimum impact speed of the meteorite, and check whether a temperature of 3000 °F could in fact have been reached.

(b) The total mass of the meteorite is estimated to have been 5×10^8 kg. If 50% of the meteorite's energy was used to vaporize ground water, compute the total mass of steam generated. (Assume the heat of vaporization of water is 540 cal/g °C.)

Energy and Metabolism

19. Mountain climbing. A hiker climbs a mountain, whose summit is 1200 m (4000 ft) above base camp. He reaches it in 4 h. His mass is 70 kg.

(a) Calculate the total work W_0 needed to overcome the altitude difference of 1200 m. Express it in joules, kilocalories, and kilowatt hours.

(b) This work is done, at a steady rate, in 4 h. State the rate of work (power), in watts, and in kilocalories per hour.

(c) For plain walking on a horizontal track, the body expends power at about three times the basal metabolic rate of 80 W or 70 kcal/h. Add to this the power required to overcome the altitude difference, assuming that only 15% of the energy liberated in catabolism is converted into mechanical energy, the remainder into heat. Find the *total* body power requirement for the hike, in watts, kilocalories per hour, and the total energy used up in the 4 h climb, in kilocalories and kilowatt hours.

(d) Calculate the rate of heat production of the body in kilocalories per second.

(e) Find the amount of heat generated per kilogram of body weight in kilocalories per second. At what rate would the body temperature rise if this heat were not dissipated?

(f) One donut has an energy content of about 300 kcal. How many donuts must the climber eat at the summit to replace his lost energy?

20. Salmon that spawn in Stuart Lake, British Columbia, must swim 640 miles up the Fraser River from the Pacific Ocean. If you catch a salmon that has swum this distance, it is not very good to eat; this problem shows why.

    The fish makes the journey in about 20 days without eating, at an average speed of 1.3 mph. However, because a fish must swim against the current, its effective swimming speed is 2.6 mph, twice as great! Studies show that a migrating salmon swimming at this speed uses oxygen at a rate of 0.5 g/h for each kilogram of body weight. The energy released by oxidizing either fat or protein is 3.3 kcal for each gram of oxygen used.

(a) What is the total metabolic energy used by a 3 kg salmon to complete its journey?

(b) Suppose the salmon oxidizes 2 g of body fat for each 1 g of protein as it swims. The energy released for 1 g of fat and 1 g of protein is 9 kcal and 4 kcal, respectively. How many grams of fat and protein does the fish use up during its journey? (A salmon is about 15% fat, by weight, just prior to starting the journey. What percentage of fat has it used up?)

CHAPTER 6

Feedback, Control, and Stability in Physical and Biological Systems

6.1 Introduction

6.1.A. Heat Engines, Power Production, and Automatic Control

In Chapter 5 we developed the work–energy formulation of mechanics. Within that framework we included the recognition that heat was a form of energy, and we were able to obtain the general law of the conservation of energy, namely, the first law of thermodynamics. The first law of thermodynamics relates quantitatively the heat flow into a system, the work done by the system, and the change in internal energy of the system. The first law of thermodynamics makes it clear that mechanical work can be converted into heat and, conversely, that heat can be converted into mechanical work. This law does not, of course, tell us *how* to make the conversion between heat and mechanical work. In fact, the first important heat engine—the steam engine—was invented by Matthew Boulton and James Watt long before the enunciation of the first law of thermodynamics. The first steam engine was sold in 1783, while the first law of thermodynamics was first enunciated about 1842. Most of the prime movers of modern technology are heat engines. They obtain heat from some fuel: wood, coal, oil, or atomic nuclei. This fuel is "burned" either by combustion with oxygen, or by nuclear fission, to produce heat. The heat may then be converted directly to mechanical work as in the case of a gasoline engine or a steam engine. Or, alternatively, it may be used to generate electric energy by turning the shafts of electric generators. These generators produce electric power which in turn can do mechanical work, or generate heat, or cool, using electrically driven motors, heaters, or refrigerators.

441

The heat engine is the major prime mover of modern technological societies.* These engines have produced profound social, economic, military, and political changes and have brought us to the point of encroachment on the ecological balance between industrial man and his natural environment. The enormity of the change wrought by these engines can be appreciated, at least in part, by the following consideration.

Prior to 1783 the scale of power source available to man was that which he could himself produce or that which could be produced by an animal. A farmer with a horse or two to help till his fields or pull his wagons and carriage was fortunate. We may estimate that at the time of the founding of the United States near the end of the eighteenth century, the average power available to each person was about 0.1 hp or about 0.1 kW. At present, at the turn of the twentieth century the power consumption of the United States is about 25×10^{11} W. With a population of 200×10^6 people, this corresponds to a power consumption of about 10 kW for each man, woman, and child in the United States: a one-hundredfold increase. This staggering increase in power consumption has revolutionized the standard of living in this and other technological societies.

The utilization of these enormous power sources in such fields as transportation (air, ship, railway, automobile); communication (radio, telephone, television); power distribution; and industrial processing (steel mills, metal rolling plants, chemical industries, textiles) requires, above all, a means to accurately *measure* and *control* the output of these devices. The need to regulate accurately the output of the myriad of devices powered by heat engines has led to the development of an entire intellectual discipline—the theory of automatic control [1], [2]. By automatic control we mean that the operating device automatically maintains its output very accurately despite variations in the input, or variations in the "load" on the device.

The hallmarks of automatic control are: first, an element that *measures* the output; second, a means of comparing that output with the desired output; and third, a means of *feeding back* this information to the input in such a way as to minimize the deviation of the output from the desired level. In fact, feedback and control systems are now as essential to the modern technological society as are the prime movers themselves.

---

*An exception is Switzerland. In that country of glaciers and high mountains, hydroelectric power, power generated from falling water, is a major source of energy.

6.1.B.   Brief History of the Field of Automatic Control

Prior to the industrial revolution only a very few automatic control devices were known. An interesting and detailed history of these devices is found in the works of O. Mayr [3], [4]. Until the beginning of the nineteenth century, the automatic control devices were limited to three different types. The first was a float valve invented by Ktesibos, a Greek mechanician in the third century B.C. This float value was used to provide a steady rate of rise of water in a tank. This steady rise in level was used to indicate the passage of time in the water clocks of antiquity. Modifications of Ktesibos' float value were used to regulate water level in troughs and fluid containers until the thirteenth century A.D. Somehow the device was lost, only to be reinvented in England about 1750. It was used then to regulate the water level in domestic cisterns. Independently, Ivan Polzunov in Russia, about 1765, invented a float valve regulator to maintain the water level in a steam boiler. Also in 1758 James Brindley patented another float valve device for the same purpose. In Figure 6.1 we show Polzunov's scheme.

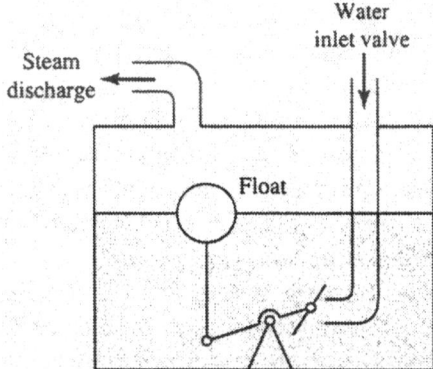

Figure 6.1. I. Polzunov's float valve for control of the water level in a steamboiler.

Clearly when the level of water falls below a level preset by the length of the vertical rod beneath the float, the valve opens and water flows in. This continues until the float rises to the preset level and then the inlet valve closes.

The second type of regulator was the thermostat. This was invented early in the seventeenth century by Cornelius Drebbel, a Dutch engineer [3], [4]. Drebbel's device was markedly improved by a Parisian inventor named Bonnemain who patented his invention in 1783. Bonnemain's device controlled the temperature of an incubator for the hatching and raising of chickens. The device consisted of a temperature "feeler" consisting of an iron rod surrounded by a lead tube. The temperature was "felt" by the expansion or contraction of the length of the lead tube. This motion was coupled by the lever to the air inlet to the furnace in such a way that an increase in furnace temperature reduced the air input to the fire and vice versa. This temperature control device permitted Bonnemain to operate, between ca. 1778 and 1794, a large progressive chicken farm outside Paris which provided chickens to the royal court and market in Paris.

The third type of control device, introduced about 1787, was the centrifugal or conical pendulum. We have already discussed the mechanics of this system in Section 2.6 as an example of the application of Newton's law of motion. We showed there that the height of the balls (h) above the vertical position of the stationary balls is related to the angular velocity ω by

$$h = \ell - \frac{g}{\omega^2}.$$

Here ℓ is the length of the pendulum, and g is the acceleration of gravity. In order that the balls fly outward at all, the angular frequency must exceed $\omega_0 = \sqrt{g/\ell}$. The conical pendulum can be regarded as a device which transforms angular velocity to a height reading. In the centrifugal governor the up and/or down motion of the balls is converted into the up–down motion of a sleeve on the axis of the governor. This sleeve is attached to an actuating lever that controls the input into the device being controlled in such a way that an increase in angular velocity of the device acts to reduce the input and vice versa. We show in Figure 6.2, a diagram of the centrifugal or conical pendulum as a governor.

The conical pendulum was first used as a speed control device in *windmills*. (It is interesting to note that many of the great British mechanical and civil engineers of the eighteenth and nineteenth centuries began their careers as millwrights.) In 1787 Thomas Mead patented the use of a centrifugal pendulum to control the speed of grindstones in a mill driven by wind. The millstones in a windmill had the tendency to move apart from one another as the angular velocity of the stones increased. This resulted in coarse grinding of the grain. To correct this, Mead arranged the output lever of the centrifugal pendulum to control the pitch of the sails on the windmill in such a way that as the speed of the millstones increased the pitch of the blades decreased; thus keeping the output (angular velocity of the grinding wheels) constant despite changes in the input (wind speed).

The centrifugal pendulum really came into its own, however, when used in conjunction with the steam engine of Boulton and Watt. The adaptation of the centrifugal pendulum to control the output angular velocity of the steam engine came

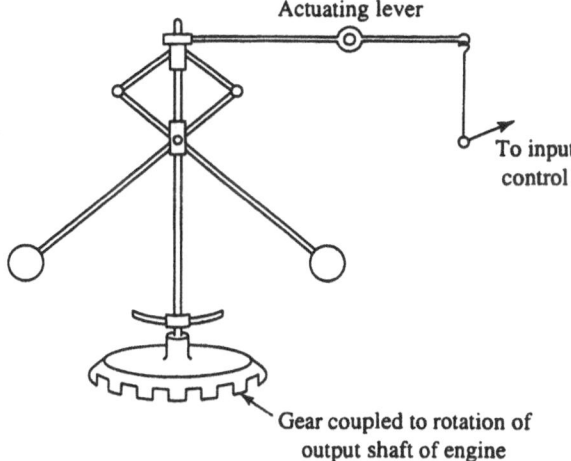

Figure 6.2. The centrifugal pendulum as a controller or governor.

about in the following way [3], [4]: In 1784 Boulton and Watt became involved in the construction of a mill driven by a steam engine. To supervise the construction of the mill, they hired a 23-year-old millwright, John Rennie, who had just completed his apprenticeship under a famous Scots millwright, Andrew Meikle. In May 1788 Matthew Boulton visited this mill, the Albion Mill, and discovered that Rennie had installed a centrifugal pendulum to detect the speed of the grindstones. The output of the pendulum was then applied to a "lift-tenter"—a device that pressed down on the wheels so as to maintain the spacing of the grindstones despite changes in their speed. Boulton and Watt realized that the speed-sensing centrifugal pendulum could serve as a means of directly controlling the speed of the steam engine. All that was needed was to *apply the output* of the centrifugal pendulum through the activating lever *to the input* pressure to the pistons of the engine through the input throttling valve. In fact, Boulton and Watt had just improved the design of this throttling valve in such a way that it was simple to couple it to the activating lever of the centrifugal pendulum. The actual detailed design of the "centrifugal speed governor" was placed in the hands of John Southern, a 30-year-old assistant of Watt's. By the end of 1788 the governor was designed and soon installed on the steam engine. Thus, the speed control scheme of the millwright found its way to become the speed governor of the prime mover of the industrial revolution. The centrifugal governor, spinning dramatically at the top of each steam engine, automatically controlling the speed, regardless of changes in the load or steam pressure, captured the imagination of all who saw it. The flying balls became, in fact, the very symbol of automatic control.

In the following section we shall analyze the controlled and uncontrolled operation of the steam engine as a detailed example that illustrates quantitatively several important principles of the theory of feedback and control. In this analysis we will find theoretically, as engineers in the nineteenth century found experimentally, that as the "feedback" of the governor is increased (that is, the reduction ΔP in steam pressure is made greater and greater for a fixed increase in output speed $\Delta \omega$), a point is reached where the governor fails to control, and starts to oscillate or "hunt." The governor becomes unstable! The governor, designed to *abate* the effect of external disturbances, can, under certain circumstances, actually *abet* these disturbances. We shall see that because of the finite "response time" of the governor its action can change from *negative feedback*, in which the disturbances of the output are reduced, to *positive feedback*, in which the disturbances of the output are actually amplified.

The first mathematical analysis of instability in control system governors was made by James Clerk Maxwell [5], [6], the founder of the mathematical theory of electromagnetism. In his 1868 paper, Maxwell laid the foundation for the mathematical theory of feedback and control, and reduced the problem of stability to the location of the roots of the characteristic of the differential equation that describes the motion of the governor. The theory showed how "hunting" could be avoided by a proper choice of the design parameters of the governor.

Following Maxwell's work, the subject of the automatic regulation of prime movers developed steadily but slowly. The culmination of this phase of development of control theory was marked by the publication in 1921 of M. Tolle's book [7], giving the differential equations that describe the motion of several different types of prime movers and their governors. In the 1920s and early 1930s there were only a few hundred control engineers in the United States, and these were mechanical engineers.

In the late 1920s and early 1930s the subject of feedback and control received important impetus from the solution of problems involved in the electrical engineering of communications systems. The problem of transcontinental telephone communication was the impetus for this development [8]. In order that a voice signal be transmitted over transcontinental distances, it was necessary to employ in the line several hundreds of electronic vacuum tube amplifiers, or "repeaters" as they were called. Fluctuations in the output gain of these amplifiers produced a signal distortion and degradation in direct proportion to the number of repeaters used. It became absolutely essential to control automatically the gain of these amplifiers to a very high degree. The practical means for this gain control was the invention of the "negative feedback amplifier" by H. S. Black. This amplifier was, in fact, the electronic expression of the feedback and control principle previously used in mechanical systems. Black's invention of this negative feedback amplifier in 1927 almost at once raised further problems. The question was whether or not such an amplifier, which has a great "loop gain," would be stable over the range of frequencies represented in voice communication. The answer to this question required a more sophisticated mathematical analysis of stability than had previously been needed for mechanical systems. Important contributions to the theory of stability of oscillators and amplifiers were made by H. Nyquist [9], H. Bode [8], and others [8].

And then came World War II. This dramatically accelerated the field of feedback and control theory. The pressing needs of war resulted in control systems associated with radar, with the guidance of antiaircraft guns, with automatic pilots for aircraft, and for the guidance of missiles, to say nothing of the growth of control devices for the acceleration of industrial production. Even the atomic reactors developed during the war had their famous boron control rod which regulated the rate of nuclear fission in the pile.

Following the war, scientists and engineers began to appreciate the versatility and the power of the ideas of feedback, control, and stability in illuminating vital problems in fields as diverse as psychology, economics, physiology, and sociology. This broadened concept of the applicability of control theory required a new word to describe the field. This word, cybernetics, from the Greek word for "steersman," was coined by the famous mathematician, N. Wiener, in 1947. His book, *Cybernetics: or Control and Communication in the Animal and the Machine* [10], ushered in the application of the theory of automatic controls to a wide gamut of human concerns. Physiology, particularly neurophysiology, was one of his principle interests. Here is a sample of his outlook [10]:

> To sum up: the many automata of the present age are coupled to the outside world both for the reception of impressions and for the performance of actions. They contain sense organs, effectors, and the equivalent of a nervous system to integrate the transfer of information from the one to the other. They (the automata) lend themselves very well to description in physiological terms. It is scarcely a miracle that they can be subsumed under one theory with the mechanisms of physiology.

Claude Bernard, the physiologist, in 1878 recognized the remarkable stability of the internal environment of the living organism. W. B. Cannon, the famous physiologist of the Harvard Medical School, in fact, coined the now widely used term "homeostasis" to describe the extraordinary stable maintenance of such variables as glucose concentration, hydrogen ion concentration, ionic strength, and erythrocyte concentration in the blood. Other examples of physiological control systems that clearly operate using the principles of feedback and control are the rate of breathing and the rate of work of the heart under normal and working conditions. On the microscopic level the production of protein by ribosomes and messenger RNA proceeds until the protein needed for the cell is produced. Then the production stops. In cancer the growth of cells does not stop. The feedback and control system fails. In fact, the very existence of living organisms is based upon the existence of networks of sensing, feedback, and control systems. The failure of these systems can lead to disease and death. It is interesting to observe that Wiener's book on cybernetics was, in fact, the outgrowth of a series of lectures on quantitative physiology organized at the Harvard Medical School by W. B. Cannon and attended by physiologists, mathematicians, physicists, and engineers [10].

6.1.C.   Mechanics and Feedback and Control Theory

It is proper to inquire why we include a chapter on feedback and control theory in this textbook on mechanics. We do so because we believe that this field, like thermodynamics, while rooted in mechanics, provides important organizing principles for physiology and molecular biology. Just as thermodynamics provides a basis for the quantitative analysis of energy exchanges in living systems, so does automatic control theory provide a basis for the quantitative understanding of biological regulatory systems. Nearly every macroscopic and microscopic process in living organisms requires the action of the principles of feedback and control. Examples are legion—to name just a few which come immediately to mind, we have: the on–off control of protein production from its RNA template; the viral bacteriophage control of bacterial enzymes and DNA; and the neuromuscular control of the eye in tracking and focusing. Furthermore, the development of prosthetic replacements for missing limbs or the medical use of natural or built-in control systems in the human body to combat such ailments as angina pectoris requires an understanding of the principles of feedback and control.

Mechanics provides us with a means of developing the principles of the theory of feedback, control, and stability. In Section 6.2 we analyze quantitatively the mechanical control of that first prime mover, the steam engine, and develop some of the major ideas in control theory. We develop these even further in Section 6.3 in the discussion of the control of temperature in a house or in a water bath. Following this we shall apply these principles to some physiological examples, e.g., control of body temperature in man and control of glucose concentration in the blood.

6.2   A Mechanical System Under Automatic Control: The Steam Engine and its Centrifugal Governor

6.2.A.   Description of the Overall Operation of the Steam Engine

In Figure 6.3 we give a schematic drawing that contains the major operating elements of the steam engine without the governor. Fuel burned in a furnace heats water enclosed in the steam boiler. The steam pressure produced depends on the strength of the boiler and the effectiveness of the heat transfer from the furnace to the water. In the early days the steam pressure was about 50 psi. By 1862, the development of steel production through the Bessemer process permitted stronger boilers and resulted in steam pressures as high as 250 psi. By the 1920s superheating of steam produced steam pressures as high as 1500 psi.

The steam pressure (p') produced by the steam boiler then passes through a throttling valve which reduces the pressure by an amount determined by the an-

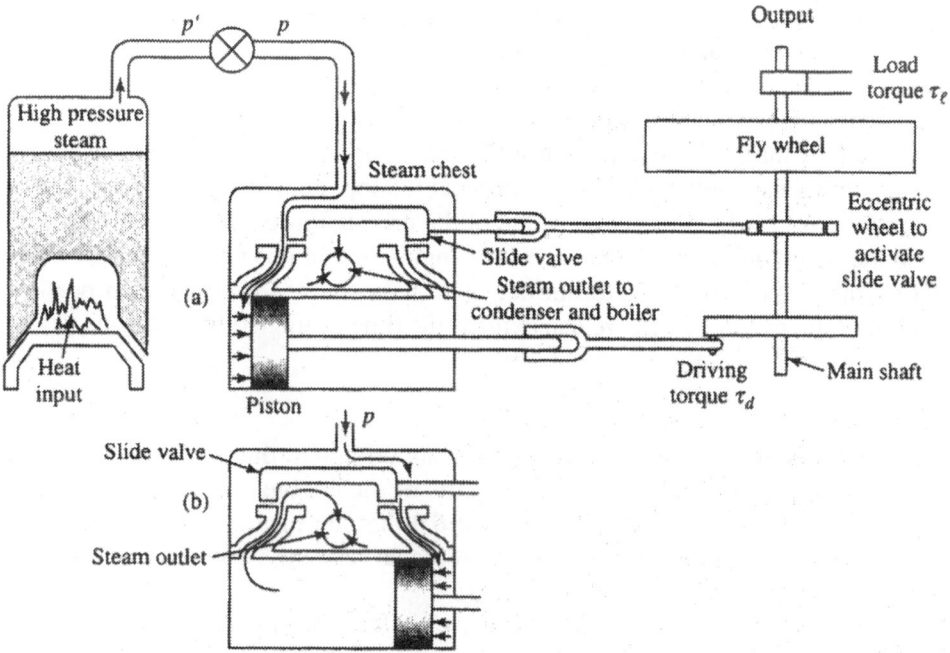

Figure 6.3. Schematic drawing showing the principal elements in the operation of a steam engine—without the action of the governor.

gular setting of the valve. When the governor is in place, the activating lever of the centrifugal governor controls the angular setting of this throttle valve, in such a way as to reduce the output pressure from the valve when the engine speed increases and vice versa.

The output pressure p from the throttle valve passes into the steam chest that contains the driving piston of the engine. The steam chest is, in fact, an ingenious device for transforming the one-way push of the steam against the piston into a reciprocating driving motion. The transformation is accomplished with the aid of the slide valve shown in Figure 6.3. When the slide valve is in the position shown in (a), the steam pressure pushes the piston to the right. The steam behind the piston is at low pressure and is pushed out of the chest through the steam outlet which connects to a cooling condenser and eventually returns as water to the boiler by a route not shown in the drawing. As the piston is driven to the right-hand end of its travel, the slide valve, which is activated by an eccentric wheel fitted with a slipping collar attached to the main shaft of the flywheel, moves to the position shown in block (b). In this position the steam pressure p is now introduced *behind* the piston and it drives the piston to the left. In stage (b), the steam to the left of the piston is at low pressure since the steam outlet is now accessible. The piston thus is driven to the left. In this way, the piston is constantly driven in a reciprocating motion back and forth from right to left.

The piston's reciprocating motion is transformed into the rotary motion of the drive shaft by means of a connection to the drive wheel. In essence, then, the piston exerts, during each full cycle of its reciprocal motion, a fluctuating driving torque upon the drive shaft. The average value of the driving torque depends, in general, upon the steam pressure p and also on the angular speed (ω) with which the main shaft turns. (Equivalently, the torque depends on the frequency of the reciprocating motion of the piston.) Thus, we can express the time average driving torque as

$$\tau_a = \tau_a(p, \omega). \tag{6-1}$$

The precise functional dependence of the mean driving torque will depend on the detailed construction and motion of the parts of the engine. In general, however, we can expect from the drawing in Figure 6.3 that the mean torque τ_d increases with driving steam pressure and will decrease at very high engine speed, as the piston speeds away from the expanding gas.

The driving torque is applied to the drive shaft of the engine. The load torque, as well, is applied to the drive shaft. In general, the time average load torque τ_ℓ will depend on the angular velocity of the main shaft. For example, if the load is the propeller of a ship, the resistance of the water to the rotation of the propeller, and hence the loading torque, will increase very rapidly with increasing propeller angular speed. Thus, we may write

$$\tau_\ell = \tau_\ell(\omega). \tag{6-2}$$

The drive shaft is usually equipped with a flywheel with a large moment of inertia (I). The purpose of the flywheel is essentially to provide the drive shaft with a great rotational inertia, so as to help reduce changes in the angular velocity of the drive shaft when the load or driving torques fluctuate for a short time. In effect, we may look upon the flywheel as a reservoir in which rotational kinetic energy $\frac{1}{2}I\omega^2$ is stored. If there should be a change in the load torque $\Delta\tau_\ell$, which requires more energy from the engine, this can be provided from the flywheel with a relatively small change in angular velocity if the flywheel has a large moment of inertia.

6.2.B.   Mathematical Model for the Operation of the Steam Engine—Without Automatic Control

(i)   Equation of Motion for the Engine Output Shaft

In Figure 6.4 we show a block diagram representation of the various components of the steam engine. The input to the steam engine may be regarded as the steam pressure p entering the steam chest. The output of the engine is the angular velocity of the drive shaft. We seek now to make a quantitative analysis of the uncontrolled steam engine. We shall examine how the engine comes up to speed; what determines its steady state operating speed; and how the output operating speed changes

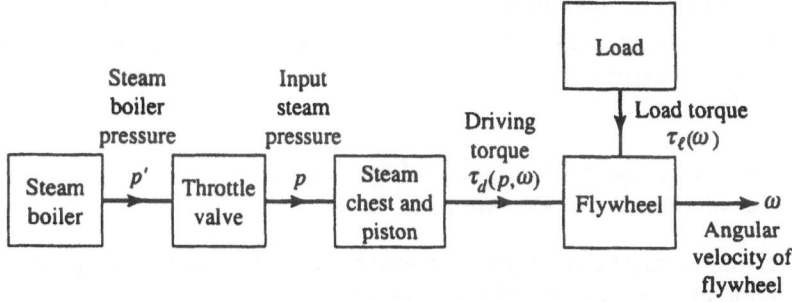

Figure 6.4. Block diagram of a steam engine without governor (open loop operation).

with changes either in load or input pressure. Finally, we examine how rapidly the engine adjusts itself, in the absence of control, to changes in input or load. The angular velocity of the main shaft of a steam engine is typically about 240 rev/min or $\omega \simeq 8\pi$ rad/s. The frequency then is 4 cycle/s. During this short period we may regard the load torque τ_ℓ as being constant in time. Also we shall regard the driving torque as being a constant equal to its average value over a cycle. The equation that describes the angular velocity (ω) of the main shaft is then given by

$$I\frac{d\omega}{dt} = (\tau_d(p, \omega) - \tau_\ell(\omega)) \,. \tag{6-3}$$

The rate of change of the angular momentum of the flywheel and shaft is equal to the net applied torque. (See Section 5.5 for the equation of motion of a rotating rigid body.) When the engine is turned on, $\omega = 0$. The driving torque τ_d has some value determined by the pressure, i.e., $\tau_d(\omega = 0) = \tau_d(p, 0)$ and the load torque has some value $\tau_\ell(0)$. The rate of change of the engine speed is $\tau_d - \tau_\ell$, and the engine speed will increase linearly with time. If τ_d and τ_ℓ both were independent of ω, the speed would simply keep increasing and the engine would never reach a steady speed. The reason that the engine speed actually does stop changing and reaches some operating value is that the driving torque in general falls with frequency and the load torque increases with frequency until a point is reached that they become equal to one another. At this point $\tau_d = \tau_\ell$ and $d\omega/dt = 0$. In Figure 6.5 we give one possible frequency dependence for τ_d and τ_ℓ. We shall assume, for simplicity, that over the operating region the driving torque is independent of the frequency, and that the load torque increases in the manner shown in Figure 6.5. To determine the precise variation of ω with time, it is necessary to know the dependence of τ_ℓ on ω. In actual engine operation, this buildup of engine speed to its steady value is not so important. What is more interesting is the response of the engine to *changes* in load or input pressure.

(ii)   Steady State Operating Speeds of the Engine.
Effect of Changes in Pressure and Load

The operating point of the engine is determined by the condition that the driving torque and load torques be equal to one another, i.e.,

$$\tau_d(p) = \tau_\ell(\omega). \tag{6-4}$$

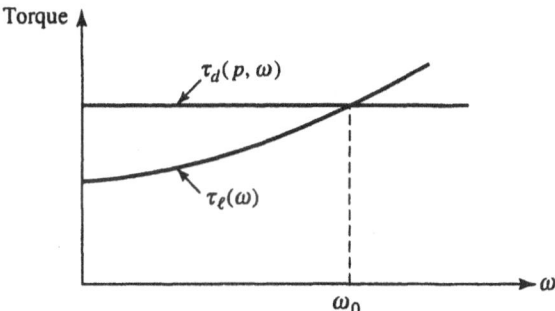

Figure 6.5. Assumed variation of the driving torque and load torque with a frequency at fixed steam pressure.

Let us suppose that for some value of steam pressure $p = p_0$, and the corresponding angular velocity is ω_0. This means that

$$\tau_d(p_0) = \tau_\ell(\omega_0). \tag{6-5}$$

Now we consider what happens to the engine speed if there is a change in the steam pressure. For example, as the fuel in the boiler burns away, the steam pressure will fall and as a result the engine speed will fall. This can produce marked changes in engine speed, and it was to correct such undesirable effects that the centrifugal governor was added to the steam engine.

In Figure 6.6 we see graphically how to determine the steady state change ($\Delta\omega_0$) in engine speed associated with a rise Δp_0 in steam pressure. We can ex-

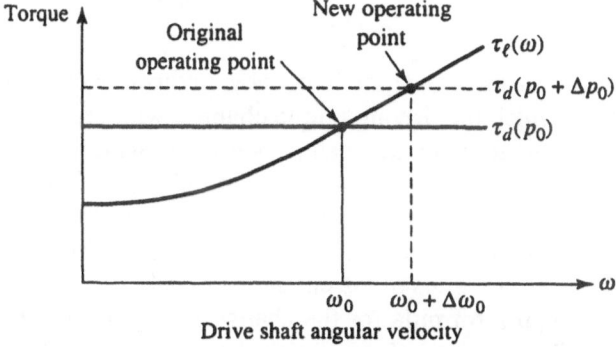

Figure 6.6. Graphical determination of a change in engine speed upon the change in driving pressure.

press the change in engine speed analytically when the changes Δp and $\Delta \omega$ are not too big. This is done by relating $\tau_d(p_0 + \Delta p)$ to $\tau_d(p_0)$ by a Taylor's expansion; and retaining only terms linear in Δp, namely

$$\tau_d(p_0 + \Delta p) = \tau_d(p_0) + \left(\frac{d\tau_d}{dp} \right)_{p_0} \Delta p + \cdots . \tag{6-6}$$

We can also relate the values of load torque τ_ℓ near ω_0 to those near $\omega_0 + \Delta\omega$, if $\Delta\omega$ is not too big, in a similar way:

$$\tau_\ell(\omega_0 + \Delta\omega) = \tau_\ell(\omega_0) + \left(\frac{d\tau_\ell}{d\omega} \right)_{\omega_0} \Delta\omega + \cdots . \tag{6-7}$$

When the steam pressure then changes to $p_0 + \Delta p_0$, we must have that in the final state the net torque is zero, i.e.,

$$\tau_d(p_0 + \Delta p_0) = \tau_\ell(\omega_0 + \Delta\omega_0). \tag{6-8}$$

This implies, on using (6-7), (6-6), and (6-5) in (6-8), that

$$\left(\frac{d\tau_d}{dp} \right)_{p_0} \Delta p_0 = \left(\frac{d\tau_\ell}{d\omega} \right)_{\omega_0} \Delta\omega_0 \tag{6-9}$$

or

$$\Delta\omega_0 = \left[\frac{(d\tau_d/dp)_{p_0}}{(d\tau_\ell/d\omega)_{\omega_0}} \right] \Delta p_0. \tag{6-10}$$

We see then that the change in engine frequency, in this model, is directly proportional to the rate at which the driving torque changes with pressure, and inversely proportional to the rate at which the load torque varies with frequency. In the following section we shall see quantitatively how this change in engine speed can be markedly reduced when the system is under the automatic control of the governor. In (6-10) the numerator and denominator are both assumed to be known from a knowledge of the characteristics of the engine and load.

We can also obtain a formula for the change in engine speed when the load torque fluctuates. The response of the engine depends on how the change in load depends on the frequency ω. We shall assume that the additional load is like an additional weight applied to the output shaft and is therefore independent of fre-

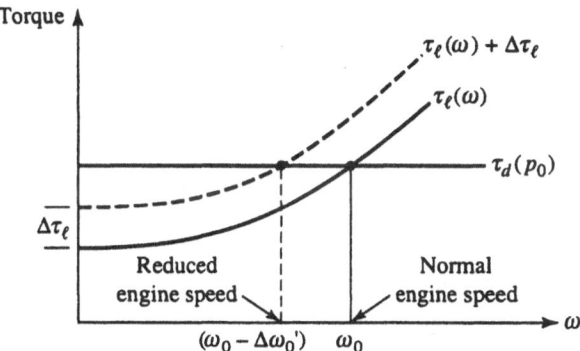

Figure 6.7. Graphical determination of the change in engine speed ($\Delta\omega_0'$) on change in the loading torque.

quency. In this case, the graphical representation of the change in operating point of the engine is shown in Figure 6.7. The change $\Delta\omega_0'$ in speed produced by such a change in load is computed analytically as follows. In the steady state, after an increase in the load, we have

$$\tau_d(p_0) = \tau_\ell(\omega) + \Delta\tau_\ell. \tag{6-11}$$

Using the expansion of (6-7) for $\tau_\ell(\omega)$ in the vicinity of ω_0, we have

$$\tau_d(p_0) = \tau_\ell(\omega_0) + \left(\frac{d\tau_\ell}{d\omega}\right)_{\omega_0} \Delta\omega_0' + \Delta\tau_\ell. \tag{6-12}$$

But since

$$\tau_d(p_0) = \tau_\ell(\omega_0)$$

from (6-5), the condition that, before the change in load, the steam pressure was p_0 and the output angular velocity was ω_0, we find from (6-12) that the decrease in engine speed indicated graphically in Figure 6.7 is

$$\Delta\omega_0' = -\frac{\Delta\tau_\ell}{(d\tau_\ell/d\omega)_{\omega_0}}. \tag{6-13}$$

Thus, for this type change in load torque, the change in output angular velocity is again inversely proportional to the derivative of the load torque with frequency.

(iii)   Temporal Response of the Steam Engine to Changes in Steam Pressure and Load Torque

In Section 6.2.B(ii) above, we computed the long-term change ($\Delta\omega$) in engine speed that results when the steam pressure or the load torque changes. In the present section we compute the time response of the uncontrolled steam engine to such changes. That is, we examine how the angular velocity ω changes in time in its approach to the steady state velocity $\omega_0 + \Delta\omega$.

To carry out this objective, we return to the basic equation (6-3) for the rate of change of the angular velocity of the flywheel

$$I\frac{d\omega}{dt} = \tau_d(p) - \tau_\ell(\omega),\tag{6-14}$$

where I is the moment of inertia of the output drive shaft. We consider the situation in which for $\tau < 0$ the engine is running steadily at speed ω_0 under a steam pressure into the steam chamber of p_0. Thus for $\tau < 0$, we have

$$I\frac{d\omega}{dt} = 0,$$

$$\tau_d(p_0) = \tau_\ell(\omega_0),\tag{6-15}$$

and

$$\omega = \omega_0.$$

After $t = 0$ let us assume that the input steam pressure rises suddenly to $p_0 + \Delta p_0$. We know from our previous analysis that the engine speed will rise from ω_0 to $\omega_0 + \Delta\omega_0$ when the system has settled into its new operating state. This is indicated in the schematic graph of ω versus time shown in Figure 6.8.

To obtain the precise way in which ω approaches $\omega_0 + \Delta\omega$, we make use of the Taylor's series expansions in (6-6) and (6-7) and the condition that $\tau_d(p_0) = \tau_\ell(\omega_0)$. This gives the following equation of motion for ω:

$$I\left(\frac{d\omega}{dt}\right) = \left(\frac{d\tau_p}{dp}\right)_{p_0}\Delta p_0 - \left(\frac{d\tau_\ell}{d\omega}\right)_{\omega_0}(\omega - \omega_0)\tag{6-16}$$

with the conditions

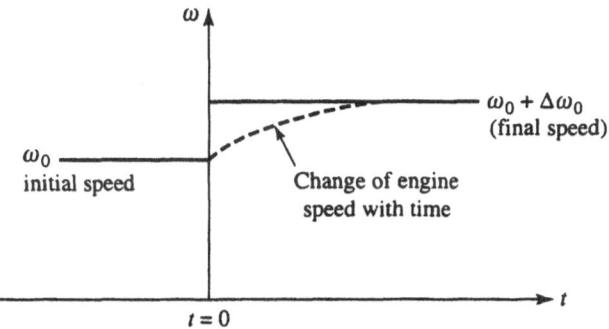

Figure 6.8. Schematic drawing of time dependence of the engine speed upon changing input pressure.

$$\text{at } t = 0 \ , \quad \omega = \omega_0,$$

$$\text{at } t = \infty, \quad \omega \equiv \omega_f = \omega_0 + \Delta\omega_0 = \omega_0 + \frac{(d\tau_d/dp)_{p_0}}{(d\tau_\ell/dw)_{\omega_0}}\Delta p_0. \quad (6\text{-}17)$$

To solve (6-16) subject to the conditions of (6-17), we rearrange (6-14) as follows:

$$I\left(\frac{d\omega}{dt}\right) = \left\{\left[\frac{(d\tau_p/dp)_{p_0}}{(d\tau_\ell/dw)_{\omega_0}}\Delta p_0 + \omega_0\right] - \omega\right\}\left(\frac{d\tau_\ell}{dw}\right)_{\omega_0}. \quad (6\text{-}18)$$

The quantity inside the square brackets is the final operating frequency $\omega_f = \omega_0 + \Delta\omega_0$ (see (6-8)), so that we can write (6-18) as

$$\left(\frac{d\omega}{dt}\right) = (\omega_f - \omega)\frac{1}{I}\left(\frac{d\tau_\ell}{d\omega}\right)_{\omega_0}$$

or

$$\frac{d\omega}{\omega_f - \omega} = \frac{dt}{(I/(d\tau_\ell/d\omega)_{\omega_0})}. \quad (6\text{-}19)$$

Defining the characteristic time t_0 as

$$t_0 \equiv \frac{I}{(d\tau_\ell/d\omega)_{\omega_0}}. \quad (6\text{-}20)$$

(One can easily confirm that this combination does in fact have the units of time.)

We find that the equation of motion of ω can be put in the form

$$\frac{d\omega}{\omega_f - \omega} = \frac{dt}{t_0}. \tag{6-21}$$

The solution for this equation, when $\omega = \omega_0$ at $t = 0$, is found by direct integration of both sides. This yields

$$\omega(t) - \omega_0 = (\omega_f - \omega_0)\left(1 - e^{-t/t_0}\right). \tag{6-22}$$

The engine speed approaches the final value exponentially with a time constant t_0 given by (6-20). The time variation of ω is shown in Figure 6.9: the quantity t_0 is the characteristic time of response of the engine. This time is proportional to the moment of inertia of the flywheel and inversely proportional to $(d\tau_\ell/d\omega)_{\omega_0}$. It is clear now why it is advantageous to have a flywheel with a large moment of inertia. In this case, if the steam pressure or load torque fluctuates rapidly in comparison to t_0, the engine will not be able to respond to these fast fluctuations and the steam engine output speed will remain relatively stable. It is also interesting to note that, in the present model of the engine operation, one can measure the important load characteristic $(d\tau_\ell/d\omega)_{\omega_0}$ indirectly by observing the time constant t_0 for the response of the engine to an abrupt change in steam pressure. (Of course, since t_0 also depends on I, we presume this is easily determined from the dimensions of the flywheel.)

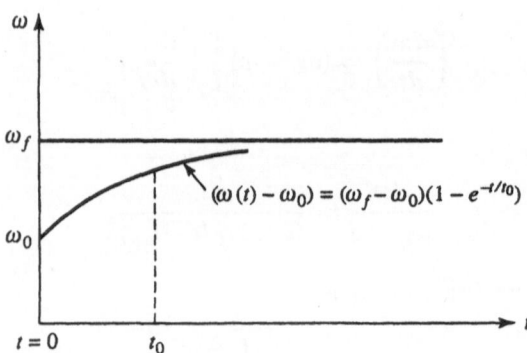

Figure 6.9. Angular velocity of the output shaft of the steam engine as a function of time after a sudden increase in steam pressure.

We can show finally that the time constant t_0 also determines the response of the engine to sudden changes in engine load. Suppose then that at $t \geq 0$ instead of changing steam pressure the load torque changes from $\tau_\ell(\omega_0)$ to $\tau_\ell(\omega_0) + \Delta\tau_\ell$. Let $\Delta\tau_\ell$, as in Section 6.2.B(ii), not depend upon engine speed. Then the differential equation for the time dependence of ω is just

$$I\left(\frac{d\omega}{dt}\right) = \tau_d(p_0) - \left(\tau_\ell(\omega_0) + \Delta\tau_\ell + \left(\frac{d\tau_\ell}{d\omega}\right)_{\omega_0}(\omega - \omega_0) + \cdots\right) \qquad (6\text{-}23)$$

or, rearranging, we have

$$I\left(\frac{d\omega}{dt}\right) = \left[\left(\omega_0 - \frac{\Delta\tau_\ell}{(d\tau_\ell/d\omega)_{\omega_0}}\right) - \omega\right]\left(\frac{d\tau_\ell}{d\omega}\right)_{\omega_0}. \qquad (6\text{-}24)$$

Since from (6-11), the final engine speed ω_f' on change of the load is

$$\omega_f' = \left(\omega_0 - \frac{\Delta\tau_\ell}{(d\tau_\ell/d\omega)_{\omega_0}}\right), \qquad (6\text{-}25)$$

we see that the equation of motion of the output shaft is

$$\left(\frac{d\omega}{dt}\right) = (\omega_f' - \omega)\frac{1}{I}\left(\frac{d\tau_\ell}{d\omega}\right)_{\omega_0}. \qquad (6\text{-}26)$$

This is of just the same form as (6-17). Thus the engine speed will exponentially coast downward to $\omega_0 - \Delta\omega$ on increasing load torque with a characteristic time $t_0 = I/(d\tau_\ell/d\omega)_{\omega_0}$ which is the same as that found when the pressure is increased.

6.2.C.   Quantitative Analysis of the Operation of the Steam Engine with Feedback Control from a Centrifugal Pendulum

In the previous sections we saw that changes in steam pressure or load torque produce changes in the steady state operating speed of the steam engine. Such changes or "offsets" are generally undesirable in practice. We also saw that the use of a flywheel with a large moment of inertia was useful in suppressing the effects of changes in pressure or load when these changes occur rapidly compared to the engine response time t_0. However, for changes of load or pressure that occur over long periods compared to t_0, the full "offsets" in engine speed will result.

In the present section we will examine how the feedback of deviations of the engine speed from ω_0 to the input pressure can dramatically reduce the steady state changes in engine speed. We shall also see how the feedback decreases the response time of the controlled engine. Finally, we shall examine how the feedback and control system can become unstable and produce large fluctuations in engine speed when the feedback becomes too great, i.e., when the offsets are too tightly controlled.

(i)   Differential Equation of the Steam Engine—With the Centrifugal Governor Included

In Figure 6.10 we show diagrammatically how the centrifugal pendulum mechanically couples the output speed of the drive shaft ω to the pressure applied to the pistons. The centrifugal pendulum is coupled to the drive shaft of the engine by a belt which drives the gear at the base of the pendulum. The speed of the conical pendulum (ω') is directly proportional to the speed of the output shaft ω. The constant of proportionality, that is fixed by the coupling geometry, should have a size such that when the engine shaft has angular velocity near its normal value ω_0, the centrifugal pendulum should rotate at a frequency ω' somewhat larger than $\sqrt{g/\ell}$, where ℓ is the length of the pendulum. (When $\omega' \gg \sqrt{g/\ell}$ the pendulum balls are flying nearly horizontally, and the height does not change much when ω' changes. Also when $\omega' < \sqrt{g/\ell}$ the balls won't fly out at all. (See Chapter 2, Section 2.7 and the problems at the end of this chapter.))

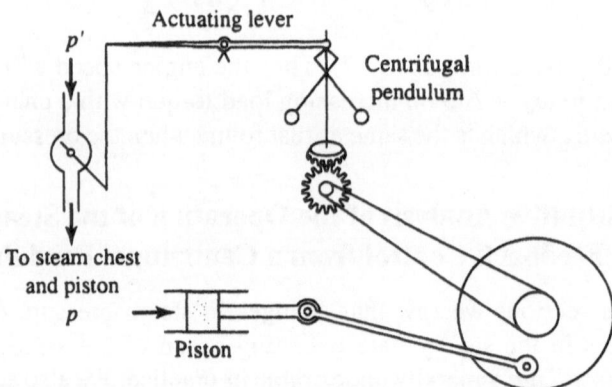

Figure 6.10. Mechanical means of coupling the output speed back into the input pressure p to the steam chest. An increase in engine speed works to decrease the input pressure.

As the balls move up due to an increase in engine angular velocity, the actuating lever twists the throttling valve in such a way so as to reduce the steam pressure to the pistons. We then see that there is a *feedback* between the output (ω) and the input (p). The feedback is *negative feedback* in the sense that an increase in engine speed is fed back to *reduce* the steam pressure.

In Figure 6.11 we show the block diagram of the various components of the controlled steam engine. We note, on comparing this block diagram with that of Figure 6.4, that the controlled steam engine is a "closed loop system." The fact that the output is connected to the input results in the block diagram which contains a "closed loop." (The uncontrolled system is a so-called "open loop" system.) The feedback through the governor closes the loop. With the aid of this diagram and our discussion of this system, we can now write the equations of motion of the governed steam engine.

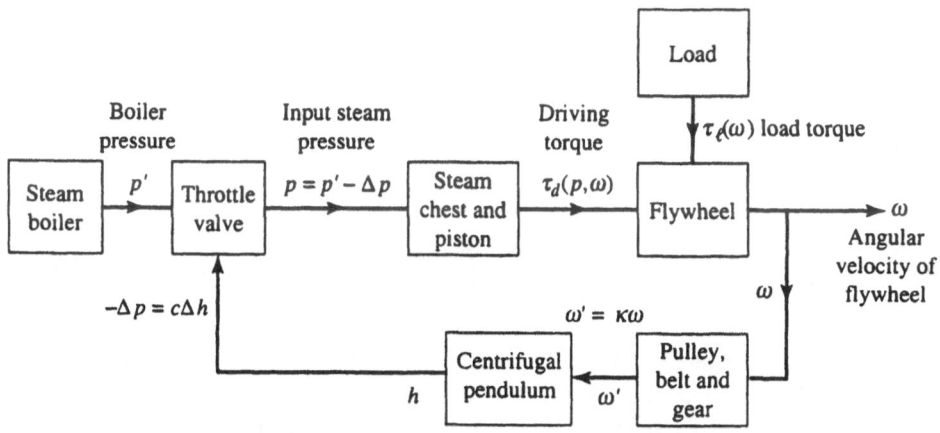

Figure 6.11. Block diagram of a steam engine with governor (closed loop operation).

As before, we have that the change in angular momentum of the flywheel is equal to the difference between driving torque $\tau_d(p)$ and load torque $\tau_\ell(\omega)$:

$$I\frac{d\omega}{dt} = \tau_d(p) - \tau_\ell(\omega). \tag{6-27}$$

In the feedback case the crucial point is that the steam pressure (p) into the pistons is now a function of the output angular velocity ω. $p = p(\omega)$ and

$$\tau_d(p) = \tau_d(p(\omega)).\tag{6-28}$$

It is important to recognize that in writing $p = p(\omega)$ we are assuming essentially that a change in output frequency ω is felt *immediately* at the input pressure. There will, of course, have to be some time lag in this response because the governor itself cannot change the height of its balls instantaneously when its rotation frequency changes. (In Section 2.7 we estimated the time delay required for the response of the balls.) We shall see later that the delay in response of the governor can lead to "instability" in the control of the engine. At present, however, we shall assume that the response of the controlled system, as a whole, to changes in output load or steam pressure is so slow that in comparison with this time the governor response is essentially instantaneous.

To examine both the steady state and the transient response of the steam engine to changes in load or steam pressure, we must examine how τ_d and τ_ℓ change in response to changes near the operating point. By expanding $\tau_d(p)$ and $\tau_\ell(\omega)$ in a Taylor's expansion around the operating points p_0 and ω_0, and retaining only terms linear in Δp and $\Delta\omega$, we find

$$\tau_d(p_0 + \Delta p) = \tau_d(p_0) + \left(\frac{d\tau_d}{dp}\right)_{p_0} (p - p_0) + \cdots,\tag{6-29}$$

$$\tau_\ell(\omega_0 + \Delta\omega) = \tau_\ell(\omega_0) + \left(\frac{d\tau_\ell}{d\omega}\right)_{\omega_0} (\omega - \omega_0) + \cdots.\tag{6-30}$$

These are identical in form to (6-6) and (6-7). In the present case, however, we must keep in mind that changes in the steam pressure p can be produced either by changes in the boiler pressure (we call this change Δp_0), *or* by changes in the output frequency $\Delta\omega$. We express this as follows:

$$p - p_0 = \Delta p_0 - k(\omega - \omega_0).\tag{6-31}$$

This equation represents the crucial difference between the controlled and uncontrolled system. We observe in this equation that changes in steam pressure are now proportional to changes in output speed. The change in input variable is *directly proportional* to the change in the output variable. This type of control system is therefore called a "proportional control system." We shall see later that there are devices in which the input is proportional to the time derivative of the output variable. This is a "derivative control system." Also there are "integral control systems" in which the input is proportional to the time integral of the output variable.

Finally, the input may not be related linearly to the output. Such systems are called nonlinear control systems. The mathematics of such devices is generally quite difficult.

In (6-31) the quantity k which has the dimensions of (pressures × seconds) is the constant of proportionality relating change in engine speed to change in piston pressure. The negative sign before the k represents the fact that the system is a "negative feedback" device. On placing (6-31) into (6-30), and putting the result and (6-31) into (6-27), we obtain the basic differential equation that describes the changes in the system about the steady state operating point. This equation is

$$I\left(\frac{d\omega}{dt}\right) = \left(\frac{d\tau_d}{dp}\right)_{p_0} \Delta p_0 - k\left(\frac{d\tau_d}{dp}\right)_{p_0}(\omega - \omega_0) - \left(\frac{d\tau_\ell}{d\omega}\right)_{\omega_0}(\omega - \omega_0). \quad (6\text{-}32)$$

Here we have used $\tau_d(p_0) = \tau_\ell(\omega_0)$. The first two terms on the right-hand side of this equation represent the change in driving torque associated, respectively, with a change in steam pressure (Δp_0) and with a change in the output frequency (through the feedback). The final term on the right-hand side of (6-32) is the change in load torque associated with a change in the angular velocity of the output shaft. If we wish to include the effect of a constant change in the magnitude of the load torque, i.e., $\tau_\ell(\omega_0)$ becomes $\tau_\ell(\omega_0) + \Delta\tau_\ell$, we need only add a constant term $(-\Delta\tau_\ell)$ to the right-hand side of (6-32), just as we did in the discussion above (6-11) and (6-22).

By recombining terms in (6-32) we find that the differential equation for the change in output speed (ω) of the engine, including feedback, is

$$I\left(\frac{d\omega}{dt}\right) = \left(\frac{d\tau_d}{dp}\right)_{p_0} \Delta p_0 - \left[\left(\frac{d\tau_\ell}{d\omega}\right)_{\omega_0} + k\left(\frac{d\tau_p}{dp}\right)_{p_0}\right](\omega - \omega_0). \quad (6\text{-}33)$$

This equation has the same form as (6-16), which describes the response of the uncontrolled steam engine. In the present case, however, the coefficient of the $(\omega - \omega_0)$ term now includes the negative feedback term $-k(d\tau_p/dp)_{p_0}$.

(ii)   Steady State Changes in Engine Output Angular Velocity Produced by Changes in Steam Pressure or Load Torque: The "Open Loop Gain"

When the steam pressure changes the engine angular velocity will change until the net torque on the flywheel becomes zero, i.e., the right-hand side will be zero in the new steady state. This condition is achieved when the output angular velocity changes to the value

$$(\omega - \omega_0) \equiv \Delta\omega_1 = \frac{(d\tau_d/dp)_{p_0}\,\Delta p_0}{\left[(d\tau_\ell/d\omega)_{\omega_0} + k\left(d\tau_p/dp\right)_{p_0}\right]}. \tag{6-34}$$

This can be written as

$$\Delta\omega_1 = \left[\frac{(d\tau_d/dp)_{p_0}}{(d\tau_\ell/d\omega)_{\omega_0}}\Delta p_0\right] \times \frac{1}{\left(1 + k(d\tau/dp)_{p_0}/(d\tau_\ell/d\omega)_{\omega_0}\right)}. \tag{6-35}$$

From our results for the uncontrolled steam engine [(6-10)] we see that the expression in square brackets is just $\Delta\omega_0$, the frequency change produced by a pressure change Δp_0 without feedback. We should also observe that the second term in the denominator of (6-35) is a dimensionless quantity which we will define as G, and give it the name "open loop gain"

$$G \equiv k\frac{(d\tau_p/dp)_{p_0}}{(d\tau_\ell/d\omega)_{\omega_0}} = \text{open loop gain.} \tag{6-36}$$

With these relations in we see that the frequency offset ($\Delta\omega_1$) with feedback is related to the frequency offset ($\Delta\omega_0$) without feedback by the relation

$$\Delta\omega_1 = \frac{\Delta\omega_0}{(1+G)}. \tag{6-37}$$

The factor $1/(1+G)$ is called, in automatic control parlance, the "minification" of the system. It shows how the feedback reduces or "minifies" the response of the system to changes in the input variable.

Generally, the open loop gain G is a number very much larger than unity. We can understand the significance of this quantity physically in the following way. Let us imagine that we open the closed feedback loop shown in Figure 6.11 between the flywheel and the input to the pulley, belt, and gear system which drives the centrifugal pendulum. We show this point in the simplified diagram in Figure 6.12. Imagine that we put a small angular velocity change $\delta\omega$ into a system at the opened position A. Suppose further that $\Delta\omega$ is the change this produces in the flywheel output speed. Then, we shall show that the ratio of $-\Delta\omega/\delta\omega$ is exactly equal to G. *Proof:* A change $\delta\omega$ in output engine speed produces a change δp_0 in steam pressure given by

$$\delta p_0 = -k\delta\omega, \tag{6-38}$$

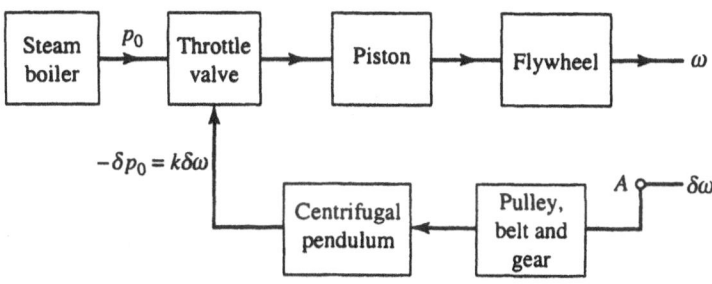

Figure 6.12. Opening the loop to obtain the open loop gain G.

from the very definition of the meaning of k given in (6-31). The response of the uncontrolled, open loop engine to such a change in input pressure is, according to the argument leading to (6-10), given by

$$\Delta\omega = \frac{(d\tau_d/dp)_{p_0}}{(d\tau_\ell/d\omega)_{\omega_0}}(-k\delta\omega).$$

Therefore, we see that

$$-\left(\frac{\Delta\omega}{\delta\omega}\right) = k\frac{(d\tau_d/dp)_{p_0}}{(d\tau_\ell/d\omega)_{\omega_0}} \equiv G. \tag{6-39}$$

Thus the complicated coefficient $G = k(d\tau_p/dp)_{p_0}/(d\tau_\ell/d\omega)_{\omega_0}$ is a dimensionless measure of the effectiveness of the governor in changing the steam engine frequency a great deal $(-\Delta\omega)$, when its input to the governor is changed a small amount $(\delta\omega)$. The gain G, in principle, can be increased to very large values. Equation (6-37) shows that as a result of these large values of G, the engine speed offsets can be very greatly reduced by the feedback control. Values of $G \sim 100$ are possible.

If one considers the effect of changes in load torque $(\Delta\tau_\ell)$ instead of changes in steam pressure, one finds the following change in engine speed

$$(\Delta\omega_1)' = \frac{(\Delta\omega_0)'}{(1+G)}. \tag{6-40}$$

Here $(\Delta\omega_0)'$ is the change in engine speed produced in the uncontrolled steam engine by a change $\Delta\tau_\ell$ in load torque $(\Delta\omega_0' = -\Delta\tau_\ell/(d\tau_\ell/d\omega)_{\omega_0}$, (6-13)). The proof of this result [(6-40)] is given as an exercise for the reader at the end of this chapter. The result of (6-40) shows that the feedback control suppresses, with equal minification, changes in steam pressure and load torque in the present model.

It is worth noting that while this feedback system can markedly diminish changes or "offsets" in the engine speed, that there is always some offset. This is characteristic of a proportional type controller.

(iii) Temporal Response of the Steam Engine Under Proportional Feedback Control to Changes in Steam Pressure

The feedback control of the centrifugal pendulum not only reduces fluctuations in the engine speed, it also reduces the response time of the engine to sudden changes in steam pressure (or load). We can see precisely how the engine response gets quicker from the equation of motion (6-33). Suppose that for $t < 0$, $(\Delta p_0) = 0$ and $\omega = \omega_0$. Suddenly at $t = 0$ and afterward the steam pressure rises to a value (Δp_0). We know from our discussion above that the engine speed will, in the steady state, finally come to the final value ω_{f_1}:

$$\omega_{f_1} = \omega_0 + \Delta\omega_1 = \omega_0 + \frac{(d\tau_d/dp)_{p_0}\Delta p_0}{(d\tau_\ell/d\omega)_{\omega_0}(1 + G)}. \tag{6-41}$$

This is shown in the graph (Figure 6.13). We can now compute how ω approaches ω_{f_1} by starting from the equation of motion of the engine [(6-33)]:

$$I\frac{d\omega}{dt} = \left(\frac{d\tau_d}{dp}\right)_{p_0}\Delta p_0 - \left[\left(\frac{d\tau_\ell}{d\omega}\right)_{\omega_0} + k\frac{d\tau_p}{dp}\right](\omega - \omega_0),$$

$$I\left(\frac{d\omega}{dt}\right) = \left\{\left[\frac{(d\tau_d/dp)_0\Delta p_0}{(d\tau_\ell/d\omega)_{\omega_0}(1 + G)} + \omega_0\right] - \omega\right\}\left(\frac{d\tau_\ell}{d\omega}\right)_{\omega_0}(1 + G), \tag{6-42}$$

where $G = k(d\tau_p/dp)/(d\tau_\ell/d\omega)_{\omega_0}$ as in (6-36). The quantity in square brackets in (6-42) is just ω_{f_1}, the final engine speed with feedback. We can express (6-42) in the same form as the (6-18) which applies in the case of an uncontrolled engine, by substituting ω_{f_1} for the quantity in square brackets and on dividing by I, the moment of inertia of the engine shaft and flywheel. This gives

$$\left(\frac{d\omega}{dt}\right) = (\omega_{f_1} - \omega)\frac{1}{I}\left(\frac{d\tau_\ell}{d\omega}\right)_{\omega_0}(1 + G), \tag{6-43}$$

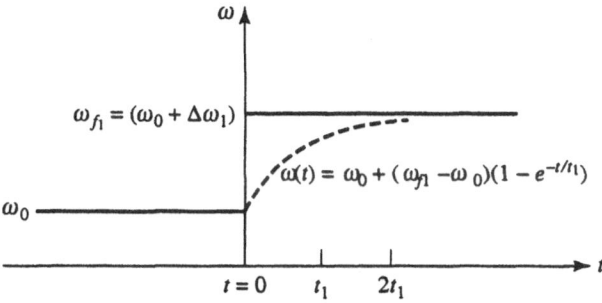

Figure 6.13. Temporal response of engine speed (ω) to changes in steam pressure with the governor acting.

which can also be written as

$$\frac{d\omega}{(\omega_{f_1} - \omega)} = \frac{dt}{t_1},\tag{6-44}$$

where t_1 is the new characteristic time for the exponential relaxation of the engine speed to its final value. t_0 is given by

$$t_1 = \frac{I}{(d\tau_\ell/d\omega)_{\omega_0} (1 + G)}.\tag{6-45}$$

Since the relaxation time for the uncontrolled engine is (from (6-20)):

$$t_0 = \frac{I}{(d\tau_\ell/d\omega)_{\omega_0}},$$

we see that

$$t_1 = \frac{t_0}{1 + G}.\tag{6-46}$$

Thus, the characteristic time t_1 for the response of the controlled steam engine is smaller, by a factor $1/(1 + G)$, than the response time of the uncontrolled engine. The minification factor $1/(1 + G)$ describes both the reduction in the steady state offsets and the response time of the engine. Equation (6-44) may be integrated directly to give the following result for the time dependence of the engine speed

$$(\omega(t) - \omega_0) = (\omega_f - \omega_0)(1 - e^{-t/t_1}). \tag{6-47}$$

The steam engine, under control, approaches its final speed exponentially in a characteristic time t_1 which is smaller, by a factor $1/(1 + G)$, than the characteristic time of the uncontrolled engine. A graph of (6-47) is shown in Figure 6.13.

To show graphically the remarkable effect of feedback control in reducing both the steady state offsets and the response time of the engine, we show in Figure 6.14 a comparison of the engine response to a change in pressure in both the controlled and uncontrolled cases. In this graph we have chosen $G = 4$. Even when the open loop gain is as small as 4, as is the case in this drawing, the effect of feedback in reducing fluctuations in engine speed is very dramatic. It can be shown quite easily that the time response of the engine to changes in load is also exponential with the same characteristic time t_1 as applies for changes in steam pressure.

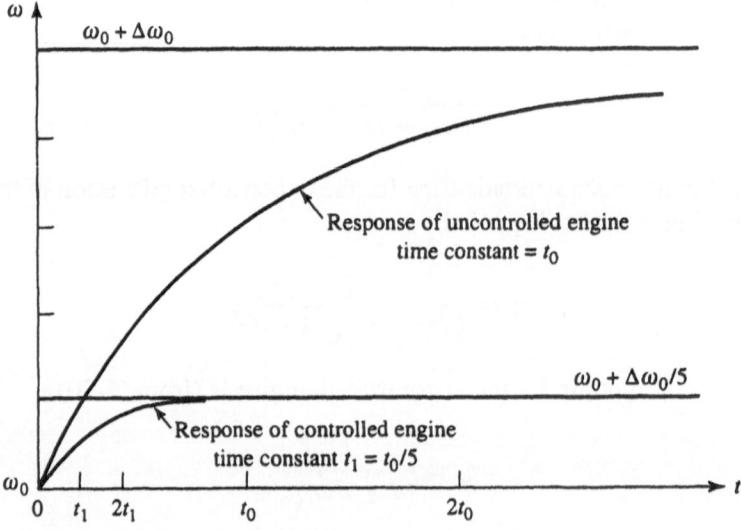

Figure 6.14. Comparison of the steady state and temporal response of uncontrolled and controlled steam engines for the case that the open loop gain $G = 4$.

It is understandable then that designers should wish to increase the open loop gain to very high values. When they built systems with such very high loop gains they found, to their surprise, that the centrifugal governor no longer governed. In fact, the feedback system became *unstable* and instead of reducing fluctuations in engine speed, it actually produced large fluctuations in engine speed. The fly ball governor was observed to "hunt" up and down as if it were searching for the position at which it should operate, but was always doomed to miss and overshoot the correct position.

Instability is, in fact, a characteristic feature of feedback and control systems which always occurs when the feedback becomes too large. There is a very large and important mathematical literature on the problem of stability and instability in feedback and control systems. The first paper on the subject was provided by James Clark Maxwell in 1858. [5]. In the following section we present an elementary analysis of the instability of the proportional control system as exemplified by the centrifugal governor. The theory of instabilities is in general quite complex mathematically. In the following section we shall concentrate on the essential physical basis for the instability and provide a simple mathematical treatment for the precise point at which instability occurs in the centrifugal governor system.

6.2.D.   Instability in the Feedback and Control System of the Steam Engine

In all our previous discussions of the controlled steam engine, we assumed that the centrifugal governor is capable of detecting changes in a time (t_g) which is short in comparison to the characteristic time during which the engine speed actually can change. If this fundamental assumption is not satisfied, the automatic control system, instead of diminishing fluctuations in input or load, can actually *amplify* such fluctuations. This amplification leads quickly to very large oscillations in the output speed of the engine. The control system then becomes *unstable* and the governor hunts up and down in a vain search for the stable operating point. We now examine semiquantitatively the conditions under which this fundamental assumption fails. We shall also discuss how the failure of this assumption produces instability in the feedback system.

(i)   Response Time of the Centrifugal Pendulum (t_g) and the Characteristic Time for Engine Speed Changes

In our discussion in Section 6.2.B(iii) we saw that in the absence of control the inertia of the flywheel prevents the engine speed from rapid changes. We saw that the engine, in fact, approaches a new steady state speed exponentially with a time constant t_0 where $t_0 = I/(d\tau_\ell/d\omega)_{\omega_0}$. In practice, this time will be of the order of several minutes. When the system is under automatic control, we saw in Section 6.2.C(iii) that the response time t_1 of the engine can be dramatically smaller than t_0. Indeed we showed that t_1 is given by

$$t_1 = t_0/(1 + G), \tag{6-48}$$

where G is the "open loop gain" of the control system. By making G very large t_1 becomes very small. As a result, it is quite possible that t_1 can be made equal to or smaller than the response time of the centrifugal pendulum itself.

In Chapter 2, Section 2.7, we discussed the motion of the centrifugal pendulum. We showed that if the angular velocity of the pendulum suddenly changes then the balls will rise up to the new height and oscillate about it with a period of oscillation T given roughly by $T \cong 2\pi \sqrt{(\ell/g)}$, where ℓ is the length of the pendulum. If the pendulum is well designed, there will be sufficient friction in it so that instead of oscillating about the new height it will coast up to the new height exponentially with the time $t_g \simeq T \simeq 2\pi \sqrt{\ell/g}$. Thus we see that the characteristic response time of the centrifugal pendulum governor is about

$$t_g \sim 2\pi \sqrt{\ell/g}. \tag{6-49}$$

It is important to recognize that this characteristic response time of the governor is quite fundamental and is determined essentially by the mechanical inertia of the pendulum.

Suppose the pendulum is about 12 in long, i.e., $\ell \sim 1$ ft. Then since $g = 32$ ft/s^2, we find that $t_g \sim 1$ s. We now can estimate the value at which the open loop gain G is so large that the response time t_1 for the engine speed is equal to the response time of the governor, i.e.,

$$t_g = \frac{t_0}{1 + G} \tag{6-50}$$

or

$$(1 + G) = \frac{t_0}{t_g} \sim \frac{3\text{–}4 \text{ min}}{1\text{–}2 \text{ s}} \sim 150. \tag{6-51}$$

Thus we can expect that t_1 is about equal to t_g when the gain in the feedback loop is of the order of 100 or so. Increase of the gain beyond this level will decrease the engine response time to the point that the centrifugal governor will simply be unable to respond quickly enough to control charges in the engine speed.

(ii) Qualitative Discussion of the Origin of Instability

One can understand qualitatively how instability results when the governor response time t_g is about the same size as the characteristic time $(t_0/(1 + G) = t_1)$

for the controlled engine. Suppose, after operating at pressure p_0 and speed ω_0, the steam pressure changes suddenly to $p_0 + \Delta p_0$. If the governor response is sluggish, in effect it does not govern, at first, and the engine speed starts changing toward the uncontrolled offset $\Delta\omega_0$ with the time constant t_0. The engine speed crosses the desired frequency offset ($\Delta\omega_1 = \Delta\omega_0/(1 + G)$) at a time $t = t_0/(1 + G)$. At this point the governor starts to respond because its characteristic time t_g is about equal to $t_0/(1 + G)$. By the time the governor is able to respond fully, the engine speed exceeds, or "overshoots," the control level $\omega_0 + \Delta\omega_1$. The governor finally calls for a decrease in steam pressure. The engine responds and the frequency is driven downward. However, once again, before the governor can respond to the decrease in speed, the engine "undershoots" the control level. When the governor eventually reacts to the fall in speed, it calls for an increase in steam pressure and another overshoot occurs. In this way, the overshoot–undershoot cycle is perpetuated, and the engine speed oscillates or "hunts." Instead of stable control the feedback system produces an unstable oscillation.

In the following section we present a careful mathematical analysis of the transition between stable control and unstable oscillations. This analysis is quite fascinating from the mathematical point of view—particularly if you enjoy mathematics. The techniques employed are elementary, but not easy. A student who finds the next section heavy-going can pass over it. Its omission will not affect your ability to understand subsequent sections.

(iii)   Quantitative Analysis of Instability in a Feedback-Control System

We now present a mathematical analysis of the instability in the steam engine control system which we discussed qualitatively in the preceding section. The equation of motion of the angular velocity (ω) of the flywheel shaft is, as always, given by

$$I\left(\frac{d\omega}{dt}\right) = \tau_d(p) - \tau_\ell(\omega). \qquad (6\text{-}52)$$

(See (6-3).)

The steam pressure into the steam chest at time t is $p(t)$. In the controlled steam engine the steam pressure is determined in part by the difference between the output speed and the steady state operating speed ω_0. When the centrifugal pendulum has a delay in its response, the feedback part of the steam pressure is determined by the engine speed not at the present time t, but at some previous time $t - t_g$ where t_g is the "lag time" of the centrifugal governor. With this thought in mind we can alter (6-31), which described the steam pressure in the case of instantaneous governor response, by using in its place the equation

$$p(t) - p_0 = \Delta p_0(t) - k(\omega(t - t_g) - \omega_0). \qquad (6\text{-}53)$$

Here $\Delta p_0(t)$ is the change in steam pressure produced by changes in the boiler. We shall assume that $\Delta p_0(t) = 0$ for $t < 0$ and $\Delta p(t) = \Delta p_0 = $ constant for ≥ 0. The term $k(\omega(t - t_g) - \omega_0)$ represents the feedback part of the steam pressure change. It states that this feedback depends on the engine speed, not at the present time (as was the case in (6-31)), but at some previous time $t - t_g$, where t_g is the response time of the governor. In a completely rigorous treatment one would write the equation of motion of the centrifugal pendulum and its coupling to that of the engine to examine the instabilities in this system. In fact, that is what J. C. Maxwell did in 1868. The essential features of the instability, however, are adequately represented by (6-53).

a) Equation of Motion Including Delay in Governor.   As before, we represent the dependence of load torque $\tau_\ell(\omega)$ on frequency by the Taylor expansion

$$\tau_\ell(\omega(t)) = \tau_\ell(\omega_0) + \left(\frac{d\tau_\ell}{d\omega}\right)_{\omega_0} (\omega(t) - \omega_0) + \cdots, \qquad (6\text{-}54)$$

and retain only the linear terms in the expansion. To include the effect of a change in the magnitude of the load torque, we need only add a term $\Delta\tau_\ell$ to the right-hand side of (6-54). To avoid unnecessary clutter, we will not consider such changes in this section.

From (6-53) we see that we can write the driving torque $\tau_d(p)$ as

$$\tau_d(p(t)) = \tau_d(p_0) + \left(\frac{d\tau_d}{dp}\right)_{p_0} (p(t) - p_0) + \cdots$$

or

$$\tau_d(p(T)) \cong \tau_d(p_0) + \left(\frac{d\tau_d}{dp}\right)_{p_0} \Delta p_0 - k\left(\frac{d\tau_d}{dp}\right)_{p_0} (\omega(t - t_g) - \omega_0). \qquad (6\text{-}55)$$

Putting (6-55) and (6-54) into (6-52) gives the following equation of motion for $\omega(t)$:

$$I\left(\frac{d\omega}{dt}\right) = \left(\frac{d\tau_d}{dp}\right)_{p_0} \Delta p_0 + \omega_0 \left(k \left(\frac{d\tau_d}{dp}\right)_p + \left(\frac{d\tau_\ell}{d\omega}\right)_{\omega_0}\right)$$

$$- k \left(\frac{d\tau_d}{dp}\right)_{p_0} \omega(t - t_g) - \left(\frac{d\tau_\ell}{d\omega}\right)_{\omega_0} \omega(t). \tag{6-56}$$

By factoring out of the right-hand side of this equation the term $(d\tau_\ell/d\omega)(1+G)$, where, as before

$$G = k(d\tau_d/dp)/(d\tau_\ell/d\omega)_{\omega_0}$$

is the open loop gain, we find after elementary algebraic manipulation that (6-56) can be expressed as

$$I\left(\frac{d\omega}{dt}\right) = \left\{\left(\omega_0 + \frac{\Delta\omega_0}{1+G}\right) - \left(\frac{G\omega(t - t_g) + \omega(t)}{(1+G)}\right)\right\} \left(\frac{d\tau_\ell}{d\omega}\right)_{\omega_0} (1 + G). \tag{6-57}$$

Noting now that the final frequency of the instantaneously controlled steam engine is just

$$\omega_{f_1} = \omega_0 + \frac{\Delta\omega_0}{(1+G)} = \omega_0 + \Delta\omega_1 \tag{6-58}$$

from (6-33) and (6-41). We also note that the response time t_1 of the controlled steam engine, as defined in (6-45), is just

$$\frac{1}{t_1} = \left(\frac{d\tau_\ell}{d\omega}\right)_{\omega_0} \frac{(1+G)}{I}. \tag{6-59}$$

Thus (6-57) can be written as

$$\left(\frac{d\omega}{dt}\right) = \left(\omega_{f_1} - \left(\left(\frac{G}{1+G}\right)\omega(t - t_g) + \frac{1}{1+G}\omega(t)\right)\right)\frac{1}{t_1}. \tag{6-60}$$

In the case that $G \gg 1$, which is the condition under which instabilities do occur, we can neglect the term $\omega(t)/(1+G)$ in comparison to the term $G\omega(t-t_g)/1+G$, and then we finally find the equation

$$\left(\frac{d\omega}{dt}\right) \cong \left(\omega_{f_1} - \omega(t - t_g)\right)\frac{1}{t_1}. \tag{6-61}$$

This equation is similar to (6-43) for the instantaneous controlled steam engine except for the very important delay $t - t_g$ in the angular velocity term $\omega(t - t_g)$.

   This differential equation for ω has the boundary condition that for $t \leq 0$, $\omega = \omega_0$. Furthermore, in the case that $t_g = 0$, we saw that the final angular velocity is ω_{f_1}. It is thus reasonable to change the variables of the problem slightly and to use instead of the variable $\omega(t)$, the variable

$$\Omega(t) = (\omega_{f_1} - \omega(t)), \tag{6-62}$$

$\Omega(t)$ has the property that at $t = 0$:

$$\Omega = \Omega(0) - \quad _{f_1} - \omega_0), \tag{6-63}$$

and when $t_g = 0$, and $t \to \infty$:

$$\Omega(t \to \infty) = 0.$$

With $\Omega(t)$ as the variable of the problem we have that (6-61) becomes

$$\frac{d\Omega(t)}{dt} = -\frac{\Omega(t - t_g)}{t_1}. \tag{6-64}$$

This equation is the basis of the mathematical analysis of the response of the control system when delay is present. In the limiting case where no delay exists, i.e., $t_g = 0$, the solution of (6-64) is, of course,

$$\Omega(t) = \Omega(0)e^{-t/t_1}.$$

In this case the system relaxes exponentially to the final engine speed.

   In the following parts of this section we obtain the solution of this basic equation in the general case $t_g \neq 0$. The mathematics employed in obtaining the solution of (6-64) is really quite interesting—even fascinating—in its own right. In what follows we shall investigate (6-64) using two methods: (i) the method of piecewise integration, and (ii) the method of the trial solution.

b) Solution Using Method of Piecewise Integration.   Equation (6-64) states that the derivative $(d\Omega/dt)$ at the present time (t) is equal to the negative of the value of Ω/t_1 at the previous time $t - t_g$. Since it is assumed, as the initial condition of the equation, that for $t < 0$, $\Omega = \Omega(0)$, we see that during the time interval $0 \le t \le t_g$ the equation for $\Omega(t)$ is

$$d\Omega(t') = \frac{-\Omega(0)}{t_1} dt'.$$

Integrating this equation between $t' = 0$ and $t' = t$, we find

$$\Omega(t) = \Omega(0) \left(1 - \frac{t}{t_1} \right) \quad \text{for} \quad 0 \le t \le t_g. \tag{6-65}$$

We now proceed to the interval $t_g \le t \le 2t_g$. Equation (6-64) takes the form

$$\frac{d\Omega(t')}{dt'} = -\frac{\Omega(0)}{t_1} \left(1 - \frac{(t' - t_g)}{t_1} \right),$$

where t' ranges between $t_g < t' < 2t_g$. Integrating this equation between the limits $t_g < t' < t$, we have

$$\left(\Omega(t) - \Omega(t_g) \right) = -\frac{\Omega(0)}{t_1} \int_{t'=t_g}^{t'=t} \left(1 - \frac{(t' - t_g)}{t_1} \right) dt'$$

$$= -\Omega(0) \left\{ \left(\frac{t - t_g}{t_1} \right) - \frac{(t - t_g)^2}{2t_1^2} \right\}. \tag{6-66}$$

Since $\Omega(t_g) = \Omega(0)(1 - t_g/t_1)$, we find finally that

$$\Omega(t) = \Omega(0) \left(1 - \frac{t}{t_1} + \frac{1}{2} \left(\frac{t}{t_1} - \left(\frac{t_g}{t_1} \right) \right)^2 \right). \tag{6-67}$$

If we now define the ratio (t_g/t_1) as

$$\alpha = (t_g/t_1), \tag{6-68}$$

we see that (6-67) has the form

$$\Omega(t) = \Omega(0)\left(1 - \frac{t}{t_1} + \frac{1}{2}\left(\frac{t}{t_1} - \alpha\right)^2\right) \tag{6-69}$$

for $t_g < t < 2t_g$.

We may now obtain $\Omega(t)$ for the interval between $2t_g$ and $3t_g$ in a perfectly similar manner. In this interval (6-64) has the form

$$\frac{d\Omega(t')}{dt'} = -\frac{\Omega(0)}{t_1}\left(1 - \frac{(t' - t_g)}{t_1} + \frac{1}{2}\left(\left(\frac{t' - t_g}{t_1}\right) - \alpha\right)^2\right) \tag{6-70}$$

and $2t_g < t' < 3t_g$. Integrating this equation between $t' - 2t_g$ and $t' = t$, we have

$$\Omega(t) - \Omega(2t_g) = -\frac{\Omega(0)}{t_1}\int_{t'=2t_g}\left\{1 - \frac{(t' - t_g)}{t_1} + \frac{1}{2}\left(\left(\frac{t' - t_g}{t_1}\right) - \alpha\right)^2\right\}dt'$$

on performing the integration, and using the fact that

$$\Omega(2t_g) = \Omega(0)\left(1 - 2 + \tfrac{1}{2}\alpha^2\right),$$

we obtain

$$\Omega(t) = \Omega(0)\left\{1 - \frac{t}{t_1} + \frac{1}{2!}\left(\frac{t}{t_1} - \alpha\right)^2 - \frac{1}{3!}\left(\frac{t}{t_1} - 2\alpha\right)^2\right\}$$

for $2t_g \leq t \leq 3t_g$.

Thus, we see that by breaking the time domain into intervals such that the first $n = 1$ is $0 \leq t \leq t_g$; the second $n = 2$ is $t_g \leq t \leq 2t_g$; and $n = 3$ is $2t_g \leq t \leq 3t_g$, etc., we can successively compute the value of Ω in interval $n + 1$ simply by knowing its value in the previous interval. The result for $\Omega(t)$ in the nth interval, i.e., $(n - 1)t_g \leq t \leq nt_g$ was given to us by Mr. Norman Mazer (a student of physics at MIT and of medicine at the Harvard Medical School). The result is

$$\Omega(t) = \Omega(0)\left\{1 - \left(\frac{t}{t_1}\right) + \frac{1}{2}\left(\frac{t}{t_1} - \alpha\right)^2 - \frac{1}{3!}\left(\frac{t}{t_1} - 2\alpha\right)^3\right.$$
$$\left. + \cdots + \frac{(-1)^n}{n!}\left(\frac{t}{t_1} - (n - 1)\alpha\right)^n\right\}. \tag{6-71}$$

For $(n - 1)t_g < t < nt_g$. This series is equal to the exponential function e^{-t/t_1} when the time delay $\alpha t_1 = t_g$ is equal to zero.

The equations above for each interval are ideal for computer calculations. The precise form of $\Omega(t)$ depends on the value of the parameter $\alpha = t_g/t_1$, the ratio of the governor delay to the response time of the controlled system. Mr. Mazer has kindly provided the graphs shown in Figure 6.15, giving $\Omega(t)$ for various values of α. The graphs show that in the region in which $\alpha < (\pi/2)$, the angular velocity of the output shaft eventually stabilizes, i.e., $\Omega = \omega_f - \omega(t)$ does finally go to zero. However, when $\alpha = (t_g/t_1) > \pi/2$, the engine speed undergoes ever larger amplitude fluctuations as time goes on. For such values of α the feedback control system becomes unstable. (In practice, of course, the increasing amplitude of oscillations of the engine speed will be limited by the nonlinearities in the frequency variation of the torque.) Figure 6.15 clearly shows how the oscillations develop as the response time of the governor gets larger and larger.

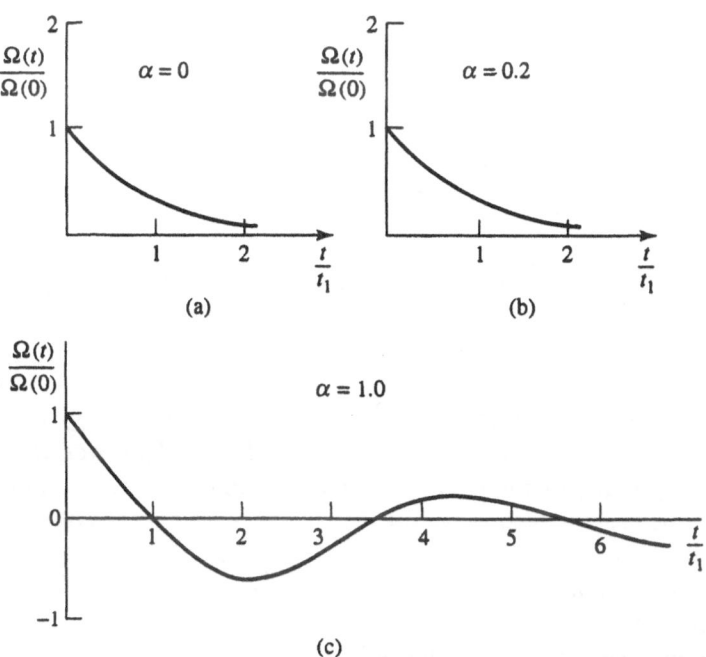

Figure 6.15. (a), (b) and (c): $\Omega(t)$ for $\alpha = 0$, $\alpha = 0.2$, and $\alpha = 1$ as computed using the solution given in (6-71).

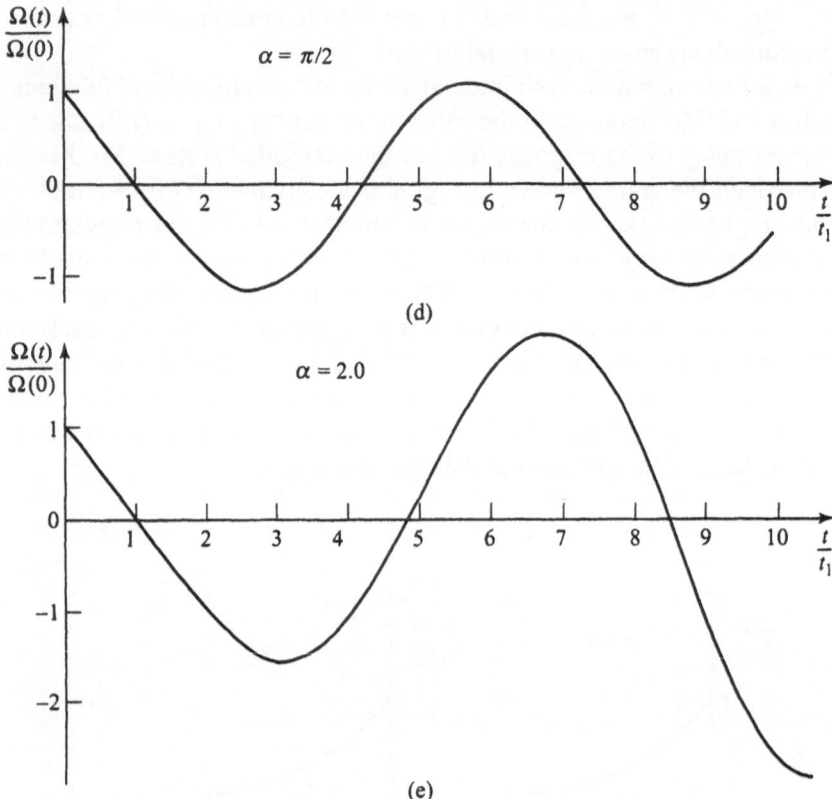

Figure 6.15. (d) and (e): $\Omega(t)$ for $\alpha = \pi/2$ and $\alpha = 2.0$ showing the onset of instability in the feedback control system computed from (6-71).

c) Solution Using the Method of Trial Solution.   The analysis given above does indeed provide a closed form for the solution of (6-64). However, to actually investigate the features of the solution, a considerable amount of computation is required. In the present section we present an alternative means of solving (6-64) which has the merit that it permits a precise determination of the point at which instabilities develop, and enables us to estimate the characteristic frequency of the unstable oscillations.

We examine whether a trial solution

$$\Omega(t) = \Omega(0)e^{-\gamma t} \cos \nu t \qquad\qquad (6\text{-}72)$$

satisfies (6-64). Substituting (6-72) into each side of (6-64) in turn, we find:

$$\frac{d\Omega}{dt} = -\Omega(0)\nu e^{-\gamma t} \sin \nu t - \Omega(0)\gamma e^{-\gamma t} \cos \nu t, \qquad (6\text{-}73)$$

$$-\frac{\Omega(t - t_g)}{t_1} = -\frac{\Omega(0)}{t_1} e^{-\gamma t} e^{\gamma t_g} \cos \nu t \cos \nu t_g$$

$$-\frac{\Omega(0)}{t_1} e^{-\gamma t} e^{\gamma t_g} \sin \nu t \sin \nu t_g. \qquad (6\text{-}74)$$

This second equation is obtained by using the trigonometric formula for $\cos \nu(t - t_g)$. On comparing (6-73) and (6-74) we see that the trial solution will, in fact, satisfy (6-64) provided that the parameters ν and γ jointly satisfy the conditions

$$\nu = \frac{e^{\gamma t_g}}{t_1} \sin \nu t_g, \qquad (6\text{-}75)$$

$$\gamma = \frac{e^{\gamma t_g}}{t_1} \cos \nu t_g. \qquad (6\text{-}76)$$

These transcendental equations permit the determination of ν and γ once t_1 and t_g are specified. Thus, the numerical values of the parameters (ν, γ) entering the solution for $\Omega(t)$:

$$\Omega(t) = \Omega(0)e^{-\gamma t} \cos \nu t$$

are fixed once t_g and t_1 are fixed.

It is important to observe that there may be values of t_g and t_1 for which γ will be a negative number. In this case, the engine speed does not settle down to a final steady state, but instead becomes unstable, i.e., the amplitude of the oscillations grow in time. We shall presently show that the threshold for this instability occurs when

$$\alpha = \frac{t_g}{t_1} = \frac{\pi}{2}, \qquad (6\text{-}77)$$

when $\alpha < \pi/2$, the parameter γ is positive, and the system is stable. That is, after a sudden change in steam pressure the system finally settles down to the steady state frequency $\Omega \to 0$ and $\omega(t) \to \omega_{f_1}$. When $\alpha > \pi/2$, the parameter γ is negative. As time goes on the amplitude of oscillation in the engine speed gets larger and larger. The system is unstable. When $\alpha = \pi/2$ the parameter γ is zero. We are on the threshold of instability.

This dependence of γ on α, and also the corresponding dependence of the oscillation frequency ν on α and t_1, can be obtained by examining the values of ν and γ predicted by (6-75) and (6-76). For this purpose it is convenient to define two new dependent variables x and y instead of ν and γ, and two new independent variables α and t_1 instead of t_1 and t_g. The definitions are as follows. Let

$$x = \nu t_g,$$
$$y = \gamma t_g. \tag{6-78}$$

Under this definition (6-75) and (6-76) now become

$$x = \alpha e^y \sin x,$$
$$y = \alpha e^y \cos x. \tag{6-79}$$

For specified values of $\alpha = t_g/t_1$ the allowed values of x and y are those which simultaneously satisfy both of the following equations:

$$y = x \cot n\, x \tag{6-80}$$

and

$$x^2 = \alpha^2 e^{2y} - y^2. \tag{6-81}$$

If we make a plot of each of these two equations in the x–y plane, the point of intersection (P) of the two curves corresponding to the actual given value of α will determine x and y for this value of α. This in turn will permit the determination of ν and γ from (6-78) since both t_1 and t_g are presumed known.

In Figure 6.16 we show the graph of (6-80) and (6-81). The form of the curve of (6-80) is independent of the ratio $\alpha = t_g/t_1$. On the other hand, the shape and position of the second equation (6-81) does depend on the exact magnitude of α.

It is interesting to notice that the curve of (6-81) crosses the $y = 0$ axis at the points $x = \pm\alpha$. We plot (6-81) for $\alpha = \pi/2$ and for $\alpha = 1.0$ and $\alpha = 2.0$. We see from the location of the intersection point P between the curves $x^2 + y^2 = \alpha^2 e^{2y}$ and $y = x \cot n\, x$, that when $\alpha < \pi/2$ the damping factor $y = \gamma t_g$ is indeed positive and the oscillations die out. When $\alpha > \pi/2$, the solution point occurs for values of y which are negative and the oscillations *grow* in time.

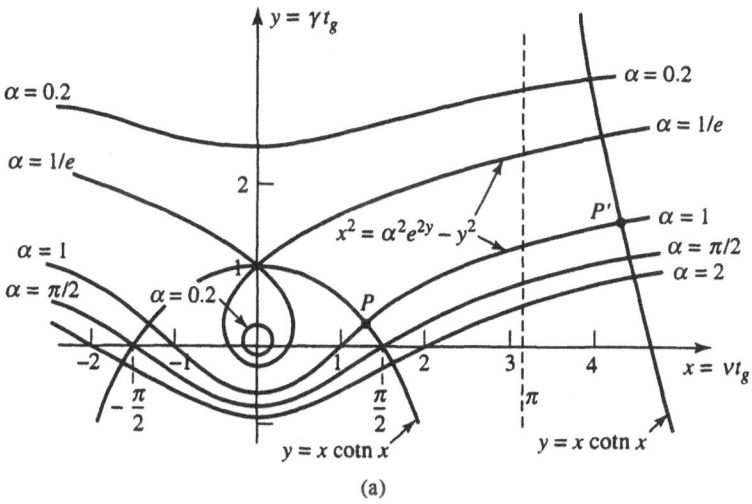

Figure 6.16. (a) A graph of equations (6-80) and (6-81).

We also note that for each value of α the curve of $x^2 + y^2 = \alpha^2 e^{2y}$ intersects $y = x$ cotn x at an infinite number of positions as x ranges from $\pm\pi$ to $\pm\infty$. These correspond to various high-frequency "modes of oscillation" of the system. We concern ourselves here only with the region $-\pi < x < \pi$ in which the lowest frequency modes of the system occur. The general solution of (6-64) consists of a sum over all the modes P_i:

$$\Omega(t) = \Omega(0) \sum_{P_i} a(P_i) e^{-\gamma(P_i)t} \cos \nu(P_i)t, \qquad (6\text{-}82)$$

P_i represents the ith solution point of (6-80) and (6-81). The $a(P_i)$ are coefficients chosen to satisfy the initial conditions on (6-64). In the analysis in (i) the initial condition was that for $-t_g < t < 0$, $\Omega(t) = \Omega(0)$. Similarly, in the present form of solution we obtain the magnitude of each of the $a(P_i)$ by requiring that $\Omega(t) =$ constant *between* $-t_g < t < 0$. It is interesting to observe that the inclusion of the time delay in the simple first-order differential equation $d\Omega/dt = -(\Omega_f - \Omega)$ results in the need to specify not just a single initial condition, but requires the specification of the function over an entire time interval t_g before the full solution can be constructed. In general, then, we must expect that many terms will be needed in the sum of functions given in (6-82). In the present discussion we will only consider the behavior of the *lowest* frequency solution, i.e., we consider only the region $-\pi < \nu t_g < \pi$.

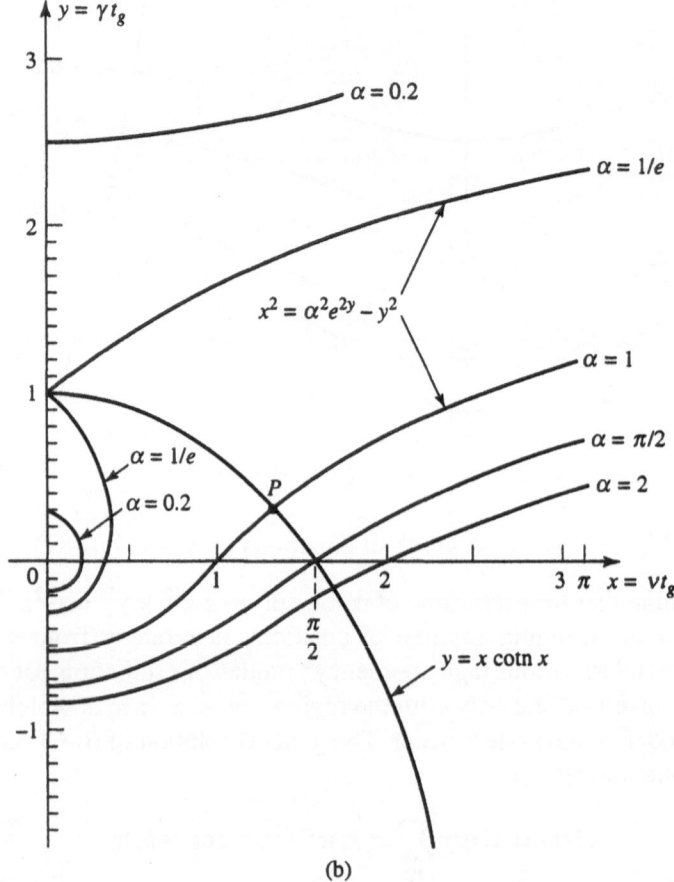

(b)

Figure 6.16. (b) A graph of $y = x \cotn x$ and $x^2 = \alpha^2 e^{2y} - y^2$ showing the detail between $x = 0$ and $x = \pi$, for various values of α. The graph for negative values of x is the mirror image of that for positive values of x.

Figure 6.16 permits us to determine not only the onset of instability, but also to estimate the magnitude of the oscillation period and decay or divergence rate γ for the lowest frequency mode. The oscillation period T is given by

$$T = \frac{2\pi}{v}. \tag{6-83}$$

Consider now $\alpha = \pi/2$. At this value of t_g/t_1 the decay rate $\gamma = 0$. The period of the threshold oscillation is found from the fact that here $v = \pi/2t_g$

(for $-\pi < x < \pi$):

$$T = 4t_{\mathrm{g}} = 2\pi t_1, \tag{6-84}$$

since $t_{\mathrm{g}} = (\pi/2)t_1$. It is interesting to observe from Figure 6.15 that the complete solution is seen from its graph to have a period of $2\pi t_1$ as predicted here.

We also observe that when α increases from $\pi/2$ to very large values the frequency of oscillation does not change very strongly. The shortest period of oscillation occurs when

$$\nu t_{\mathrm{g}} = \pi,$$

i.e.,

$$T_{\min_{\alpha \to \infty}} = 2t_{\mathrm{g}}. \tag{6-85}$$

Thus the period ranges from

$$4t_{\mathrm{g}} < T < 2t_{\mathrm{g}} \tag{6-86}$$

as α ranges from $\pi/2 < \alpha < \infty$. The divergence rate increases markedly as α grows in comparison to $\pi/2$. For $\alpha = 2$, the divergence rate is $-\gamma t_{\mathrm{g}} \cong 0.2$, or since $t_{\mathrm{g}} = 2t_1$, $\sim \frac{1}{10}t_1$. Since the period of oscillation is about $T \sim 2\pi t_1$, we see that the amplitude of oscillation increases an amount $e^{0.63}$ in one period for $\alpha = 2$. In Figures 6.17a and 6.17b we give graphs of the form of the trial solution for values of $\alpha = 1, 2$, and $\pi/2$, using only the values of ν and γ corresponding to the "lowest mode," i.e., $-\pi < x < \pi$.

By careful comparison of these results, shown in Figure 6.17 with those of Figure 6.15, we find that the single form of solution agrees quite well with the complete solution shown in Figure 6.15 when $\alpha = 1$. When $\alpha = \pi/2$ the single mode solution is shifted slightly in phase relative to the rigorous solution. When $\alpha = 2$ the single mode solution is even more shifted relative to the rigorous solution. The phase shift is $\phi \sim 2\pi/10$ for $\alpha = 2$. However, the increase in amplitude and the frequency of the single mode solution is in rather good agreement with the rigorous result for this value of α ($\alpha = 2$).

The great virtue of the approach based on the assumed form ($e^{-\gamma t} \cos \nu t$) is that it gives us a good "feel" for how the oscillation frequency and the damping rate depends upon α. It also gives us an exact criterion for the onset of instability. The closed form of solution, while it gives us the complete rigorous solution, provides

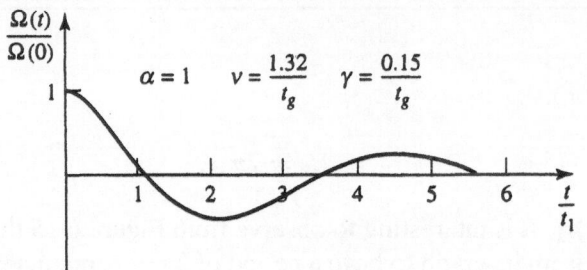

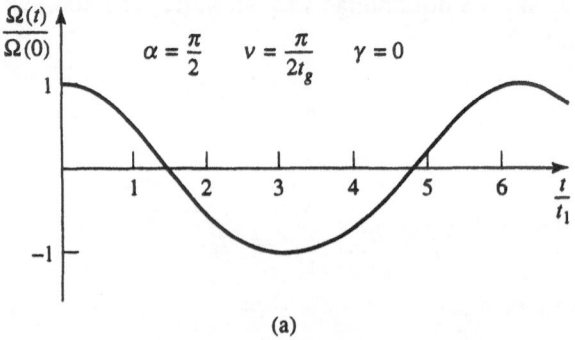

(a)

Figure 6.17. (a) $\Omega(t)$ as calculated using the form $e^{-\gamma t} \cos \nu t$ and including only the lowest frequency mode.

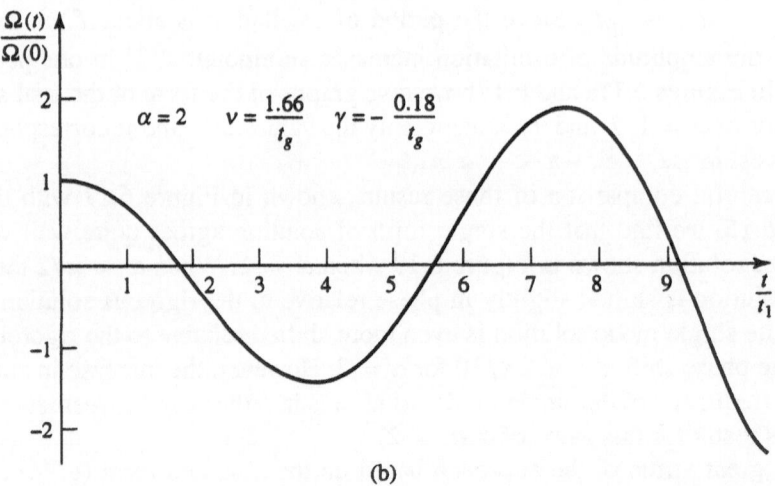

(b)

Figure 6.17. (b) $\Omega(t)$ as calculated using the form $e^{-\gamma t} \cos \nu t$ using only the lowest frequency mode.

484

the "answer" without giving us a feel for how the solution varies with different input parameters like α and t_1. This is often the case with results provided by computers.

6.3 Temperature Control Using Feedback

In Section 6.2 we discussed in detail a mechanical system under feedback control. In the present section we demonstrate that the control concepts which emerge in a mechanical case also appear in nonmechanical systems. We shall consider how feedback can be used to control the temperature of some system: it may be a home, or a thermostated laboratory water or oil bath, or it may be the body of a mouse or a man.

The case of temperature control permits us to go beyond "proportional feedback" which we discussed in the case of the steam engine. The thermal control system is also an interesting example because we are able to determine easily the response of the system to changes in the input or output variables over the entire range of these variables. This differs from the steam engine, where we characterized the system response only in the immediate vicinity of the normal operating point. Finally, the present example is very useful, practically, since temperature control systems are frequently needed in experimental physics and biology. The present section provides the principles and analysis by means of which one can design, construct, and operate such temperature control devices.

6.3.A. Temperature of a Heated (or Cooled) System without Feedback (Open Loop Operation)

(i) Equation for Temperature of a System as a Function of Time

In Figure 6.18 we show the basic means by which a house or temperature bath is brought to a temperature other than that of the surroundings. In the figure the heat input device is illustrated as an electrical heater. It could equally as well be an oil or coal furnace or the heat produced by the katabolism of food inside the body of a man.

The heat power into the house is denoted by $(dQ/dt)_{\text{in}}$. The magnitude of the heat input $(dQ/dt)_{\text{in}}$ is determined by setting the resistor that controls the current to the heater. Part of the heat that is put into the system (house) leaves it by conduction through the walls or by convection. We define this heat loss rate as $(dQ/dt)_{\text{out}}$. The difference between the heat inflow per second and the heat loss

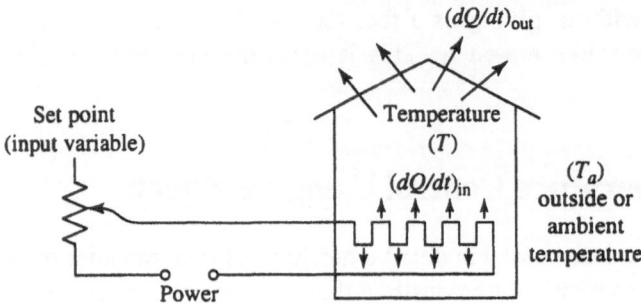

Figure 6.18. Schematic design of the heating of a house or other system.

from the system goes into raising the temperature of the system. If C is the heat capacity of the house in cal/°K, then the rate at which the temperature (T) rises is given by

$$\left(\frac{dQ}{dt}\right)_{\text{in}} - \left(\frac{dQ}{dt}\right)_{\text{out}} = C\frac{dT}{dt}. \tag{6-87}$$

The heat input $(dQ/dt)_{\text{in}}$ is presumed to be known and is set by the input variable. The rate at which heat leaks from the system will, in fact, depend upon the difference in temperature between the inside of the system, T and the ambient temperature outside, T_a. Let us presume that the heat loss rate is determined entirely by the difference between T and T_a in accordance with the equation

$$\left(\frac{dQ}{dt}\right)_{\text{out}} = \lambda(T - T_a). \tag{6-88}$$

Here λ is proportional to the "thermal conductivity" of the walls of the container. A well-insulated container has a small thermal conductivity and only a very little heat can leak from the container, even though there may be a large temperature difference between the system and the outside temperature. On using (6-88) in (6-87), we find that the temperature of the house or container changes with the time t in accordance with the equation

$$C\left(\frac{dT}{dt}\right) = \left(\frac{dQ}{dt}\right)_{\text{in}} - \lambda(T - T_a). \tag{6-89}$$

If we regard the temperature T of the system as the output variable and the input current to the heater as the input variable, then we can make a block diagram of the operation of this system as shown in Figure 6.19.

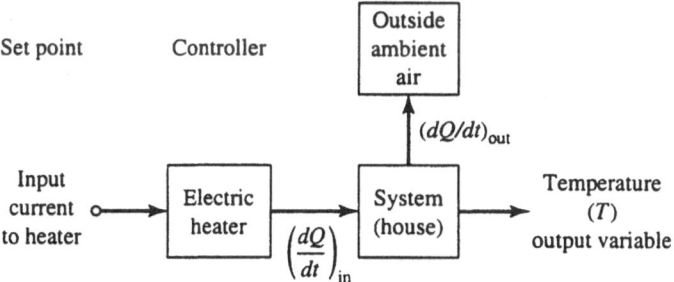

Figure 6.19. Block diagram of various components of an uncontrolled heating system. The diagram describes an "open loop" system without feedback.

The output of the system, i.e., the temperature, can be computed as a function of the heater power and the ambient temperature T_a as follows. Dividing both sides of (6-89) by the heat capacity C, we find

$$\frac{dT}{dt} = \frac{1}{C}\left(\frac{dQ}{dt}\right)_{in} - \left(\frac{\lambda}{C}\right)(T - T_a). \tag{6-90}$$

We may now define a time constant t_0 as

$$t_0 = \left(\frac{C}{\lambda}\right).$$

Furthermore, we can express $(dQ/dt)_{in}$ in terms of the final temperature that the system reaches. Calling this final temperature T_f, we observe that when the system is in equilibrium $(dT/dt) = 0$ at this time, $T = T_f$, and T_f is related to (dQ/dt) by setting $(dT/dt = 0)$ in (6-90). This gives

$$\frac{1}{C}\left(\frac{dQ}{dt}\right)_{in} = \left(\frac{T_f - T_a}{t_0}\right). \tag{6-91}$$

Thus, the final temperature of the system, without control, is related to the heat input and the ambient temperature by

$$T_f = T_a + \frac{t_0}{C}\left(\frac{dQ}{dt}\right)_{in}.$$ (6-92)

We may therefore finally write the equation for the time dependence of the output variable T (the temperature), using (6-91). This gives

$$\left(\frac{dT}{dt}\right) = \left(\frac{T_f - T}{t_0}\right).$$ (6-93)

This equation is perfectly analogous to (6-21) which we obtained in Section 6.2 for the steam engine, i.e.,

$$\frac{d\omega}{dt} = \frac{(\omega_f - \omega)}{t_0}.$$

In the present case we have the final temperature T_f and the system temperature T, while in the steam engine we have in place of these, the final engine speed ω_f and the present engine speed ω. Here t_0 can be called the "thermal response time," just as previously t_0 was the engine response time.

(ii)   Steady State Operating Temperature

We may now briefly examine the operation of this open loop system by considering the meaning of (6-92) and (6-93). Using the definition of $t_0 = \lambda/C$, we find from (6-92) that the final temperature of the system can be kept at any desired amount above (or below) the ambient temperature simply by controlling the input heating (or cooling) rate in accordance with the equation

$$T_f = T_a + \frac{1}{\lambda}\left(\frac{dQ}{dt}\right)_{in}.$$ (6-94)

The input power needed to establish any desired T_f is

$$\left(\frac{dQ}{dt}\right)_{in} = \lambda(T_f - T_a).$$ (6-95)

(In a system for which $T_f > T_a$ we use a heater and $(dQ/dt)_{in}$ is positive. To cool the system, $T_f < T_a$ and (dQ/dt) is a negative number: heat is removed from the container.) Clearly a well-insulated system is desirable, for then λ is small and the corresponding heater (or cooler) power needed to establish the desired difference $(T_f - T_a)$, being proportional to λ, is small. (Note that the final temperature is independent of the heat capacity C.)

The great disadvantage of the uncontrolled open loop system is that if the ambient temperature changes, the final temperature of the system changes by exactly the same amount. (This is analogous to the steam engine in the sense that changes in engine load in the uncontrolled engine produced marked changes in engine speed.) Clearly, it is a nuisance to have the house temperature fall when the outside temperature falls. It would be very desirable to have some feedback scheme that would sense the changes in ambient temperature and use these changes to control the heater in such a way as to maintain the temperature at the desired value.

(iii)   Transient Response to Changes in Ambient Temperature or Input Power

We may complete our discussion of the open loop heating (or cooling) system by examining the time-dependence of the temperature. Let us suppose that the system has temperature T_f at $t = 0$, and that at that time either the ambient temperature changes or that the heater power changes so that the new equilibrium temperature is T_f'. The time response of the house (or container) then obeys the equation

$$\left(\frac{dT}{dt}\right) = \frac{(T_f' - T)}{t_0},\tag{6-96}$$

with the initial condition that

$$T = T_f \qquad \text{at} \quad t = 0.\tag{6-97}$$

Our previous analysis of this simple equation shows that the temperature T approaches the new final temperature T_f' in accordance with the equation

$$T = T_f + (T_f' - T_f)\left(1 - e^{-t/t_0}\right).\tag{6-98}$$

This is drawn in Figure 6.20.

The temperature approaches the new final temperature exponentially with a time constant t_0 that depends on the heat capacity of the system. In the case of a house, t_0 can be quite long—of the order of many hours, or even a day for a big, well-insulated house. Thus a house with a long time constant will not be able to follow fluctuations in temperature of the ambient air provided that these fluctua-

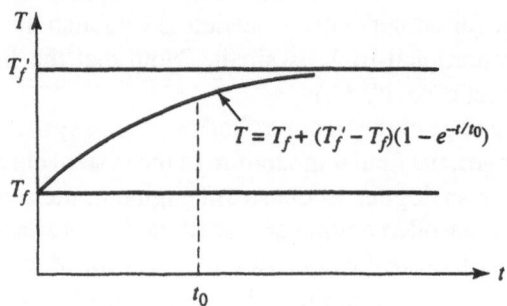

Figure 6.20. Temperature change with time when the final temperature takes on a new value.

tions take place in a time short compared to the thermal response time t_0. A house with a large value of $t_0 = (C/\lambda)$ can be said to have a large "thermal inertia," just as a flywheel with a large moment of inertia has a high mechanical inertia to changes in engine speed.

In order to reduce the effect of long-term changes in ambient temperature or input power, it is necessary to include feedback. We consider various feedback schemes in the next section.

6.3.B.   Temperature of a Heated (or Cooled) System with Feedback

(i)   On–Off Control System

There is a wide class of heaters (or coolers) which add heat at a single, fixed rate to the system they serve. For example, the oil burner in a home produces heat at some constant rate P_0 determined by the rate at which oil is pumped into the burner. Such constant power heaters can be controlled only by turning them on or off. We consider first the properties of a control system that employs such a power source.

To keep the temperature of a house or water bath controlled using such a heat source, we need a device which senses the actual temperature T and in effect measures the difference $T_s - T$ between the desired system "set temperature" T_s and the actual temperature (T). The device must then be connected to some "effector" which turns the heater *on* when the system gets so cold that $T_s - T$ becomes greater than some value ΔT, and then turns it off when the temperature T exceeds T_s by the amount ΔT. The effector must have the property that it remains off as the temperature falls until $T_s - T = \Delta T$ once again.

A bimetallic strip can serve to measure both, by its curvature, $T_s - T$, and also serve as an effector which turns the heater on or off. The on–off control system exhibits a "dead-band" in the sense that the system temperature is in fact only controlled within the limits $T_s - \Delta T < T < T_s + \Delta T$.

In Figure 6.21 we show a block diagram of such an on-off control system. The thermostat is denoted by the symbol $\otimes$. The feedback is evident because of the presence of the loop that represents the sensing of the output temperatures at the input thermostat. The feedback is negative in the sense that the thermostat, in effect, measures the set temperature (T_s) *minus* the output temperature (T).

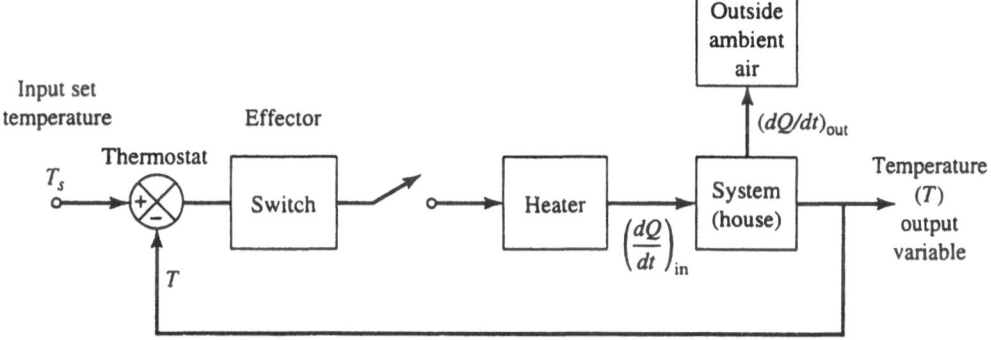

Figure 6.21. Block diagram of an on–off control system for a temperature control showing a closed loop between input (set temperature T_s) and output (system temperature T).

We may now analyze in detail how such an on–off system actually maintains the temperature within the limits $T = T_s \pm \Delta T$.

a) Equations for Temperature Changes in On–Off Control Systems.   If the ambient temperature were T_a, then we saw that to keep the house at temperature T_s with a constant heat input this input would have to be $(dQ/dt)_{in} = \lambda(T_s - T_a)$ (see (6-95)). Since in the present control system the furnace would have to be on all the time to reach the set temperature, we must put out more heat than $\lambda(T_s - T_a)$. Let us express the heat input from the furnace as

$$\left(\frac{dQ}{dt}\right)_{in} = \frac{1}{f}\lambda(T_s - T_a).$$ (6-99)

In this equation f is a pure number between 1 and 0. When $f = 1$, the furnace would just barely be able to heat the house for a fixed ambient temperature. For values of f less than unity the input power is greater than this threshold. We shall consider later a desirable value for this quantity (f). At this point we regard it as some fixed number which determines the "strength" of the furnace.

When the furnace is on, the temperature of the house varies in accordance with

$$\frac{dT}{dt} = \frac{1}{C}\left(\frac{dQ}{dt}\right)_{in} - \frac{\lambda}{C}(T - T_a)$$

or, using (6-99) and the definition $t_0 = C/\lambda$:

$$\left(\frac{dT}{dt}\right) = \frac{1}{f}\frac{(T_s - T_a)}{t_0} - \frac{(T - T_a)}{t_0}.$$ (6-100)

On the other hand, when the furnace is off the temperature T varies as

$$\left(\frac{dT}{dt}\right) = -\frac{(T - T_a)}{t_0}.$$ (6-101)

To examine how the house temperature varies with time, we shall begin by assuming that at $t = 0$ the heater is turned on. We then solve (6-100) with the initial condition $T = (T_s - \Delta T)$ until that time at which T becomes equal to $(T_s + \Delta T)$, at that point we then turn to (6-101), for the controller turns off then. This is solved until the temperature is reached for which $T = T_s - \Delta T$, and then the temperature cycle repeats.

The solutions of (6-100) and (6-101) can be found very simply when ΔT is small compared to the interval $T_s - T_a$. This will be the case when, for example, ΔT is say 2 °C and $(T_s - T_a)$ is 20 or 30 °C. Under these conditions we may regard the right-hand sides of (6-69) and (6-70) as being constant and equal to the value at which $T = T_s$. We can then integrate immediately. Using these assumptions, (6-100) and (6-101) become

(furnace on) $$\left(\frac{dT}{dt}\right) \cong \frac{(T_s - T_a)}{t_0}\left(\frac{1 - f}{f}\right)$$ (6-102) and

$$(\text{furnace off}) \qquad \left(\frac{dT}{dt}\right) \cong -\frac{(T_s - T_a)}{t_0}. \qquad (6\text{-}103)$$

b) "On-Times" and "Off-Times."   The heater stays on for a time (t_{on}) until the temperature rises an amount $2\Delta T$. That is, $(dT/dt)_{on} = 2\Delta T/t_{on}$. Similarly $(dT/dt)_{off} = -2\Delta T/t_{off}$. Using these in (6-102) and (6-103) gives us the "on-times" and the "off-times" for the heater

$$t_{on} = \frac{2\Delta T}{(T_s - T_a)} t_0 \left(\frac{f}{1-f}\right), \qquad (6\text{-}104)$$

$$t_{off} = \frac{2\Delta T}{(T_s - ta)} t_0. \qquad (6\text{-}105)$$

We plot, in Figure 6.22, the heater power and the system temperature as a function of time with this on–off control system.

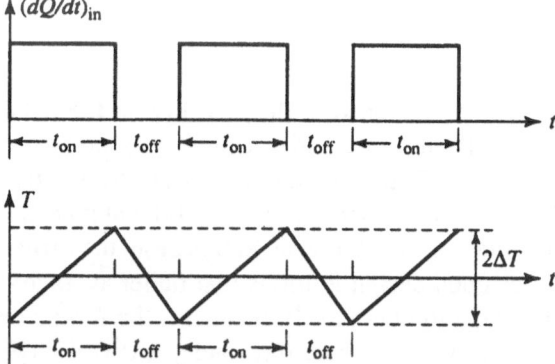

Figure 6.22. A graph of the input power and temperature of a system under on–off temperature control.

From the discussion above we see that the on–off system controls the temperature of the system by adjusting the length of the "on" interval. If the ambient temperature changes, the on- and off-times will automatically change in such a way as to satisfy (6-104) and (6-105) and maintain the house within the range $\pm\Delta T$ around the set temperature.

We observe from (6-104) and (6-105) that the characteristic time for response of the controlled system is much smaller than the response time t_0 of the uncontrolled system. In the case that $f = \frac{1}{2}$ the on- and off-times are equal and each is smaller than the uncontrolled response time t_0 in accordance with the equation

$$t_{controlled} = t_0 \left(\frac{2\Delta T}{(T_s - T_a)} \right). \tag{6-106}$$

c) **Duty Cycle.**   The quantity f is actually equal to the fractional amount of time that the heater is on. This fractional time is called the "duty cycle" and is defined as

$$\text{"Duty cycle"} = (t_{on}/(t_{on} + t_{off})). \tag{6-107}$$

Using (6-104) and (6-105), we find

$$\text{"Duty cycle"} = \frac{(f/(1-f))}{(f/(1-f)+1)} = \frac{f}{f+(1-f)} = f. \tag{6-108}$$

If the furnace is exactly matched to the heat loss to the outside, i.e., $f = 1$, then the furnace is on all the time, and it cannot summon on extra reserves to provide heat should the outside temperature fall markedly. On the other hand, if f is small compared to unity, the furnace is so powerful that it need be on only a very short fraction of the time to heat the house. In this case, the furnace is actually too powerful for its job. A good design figure is that under average conditions for T_s and T_a, f should be equal to one-half. In this case, the furnace will, on average, be *on* half the time on a normally cold day, and yet it will have enough reserve capacity to keep the house at the set temperature even should $(T_s - T_a)$ become twice as large as normal. This can occur on a very cold day.

In the present analysis we assumed instantaneous response of the thermostat: at the very instant that the temperature T reaches the limits $T_s \pm \Delta T$, the thermostat will turn the furnace on or off. Any real heating system will have a time delay (t_g). It can be shown (see the problems at the end of this chapter) that when the response time of the system (t_{off}) becomes comparable to t_g, the temperature "overshoot" of the system becomes as large as the control range ΔT itself, and the on–off system, in fact, fails to regulate.

(ii)   Proportional Control System

In the on–off control system the temperature constantly fluctuates up and down around the set temperature. It is often necessary to control temperature very steadily. For this purpose one must employ, first of all, a heater whose output power can be controlled more smoothly than by turning it on or off. With such a heater (or cooler) one can control temperature by a device that senses the difference between the set temperature (T_s), and the actual temperature (T), and then use this difference to control the power output of the heater. If the change in heater power is proportional to the difference ($T_s - T$), then the system is called a proportional control system. In the block diagram in Figure 6.23 we illustrate schematically how such a proportional feedback control of temperature operates.

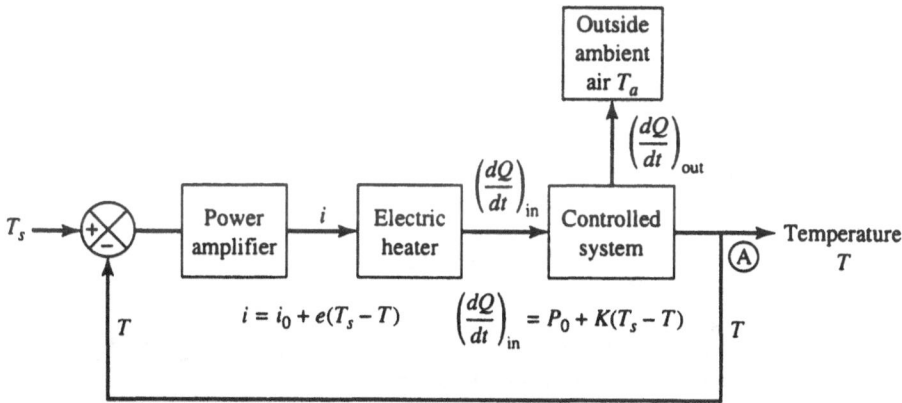

Figure 6.23. Block diagram of a proportional feedback control thermostat.

In this system we can imagine, for example, that the temperature sensor is a resistance thermometer whose electrical resistance is directly proportional to the system temperature (T). The electrical voltage across this thermometer is put into the ($-$) terminal of a differential amplifier. Into the other terminal one puts a voltage whose magnitude determines the set temperature. The power amplifier gives an output current proportional to the difference between the two input voltages, and hence proportional to the temperature difference $T_s - T$. The output variable (T), in effect, is reintroduced with a minus sign in this negative feedback system. The current out of the amplifier consists of two parts. A steady part i_0 plus a part proportional to ($T_s - T$). This current, when passed through the heater, produces a heat power given by

$$\left(\frac{dQ}{dt}\right)_{in} = P_0 + K(T_s - T). \tag{6-109}$$

The heater power consists of a steady part P_0, plus a part directly proportional to the temperature difference $(T_s - T)$. The constant of proportionality K depends on the precise details of the amplifier and temperature sensor. The units of K are those of power/°K as are the units of λ.

a) Equation for the Temperature of a Proportionally Controlled System as a Function of Time.   The temperature of the controlled system is obtained from the basic equation relating the net heat flow into the system to the rate of change of temperature. Again, using C as the heat capacity of the system, we may write

$$C\left(\frac{dT}{dt}\right) = \left(\frac{dQ}{dt}\right)_{in} - \left(\frac{dQ}{dt}\right)_{out}. \tag{6-110}$$

The heat flow out is, as before, given by

$$\left(\frac{dQ}{dt}\right)_{out} = \lambda(T - T_a), \tag{6-111}$$

where T_a is the ambient temperature. The input heating rate is given by (6-109). Putting (6-109) and (6-111) into (6-110), we find

$$C\left(\frac{dT}{dt}\right) = P_0 + K(T_s - T) - \lambda(T - T_a). \tag{6-112}$$

We shall arrange that P_0 be chosen in the following way: We will require in the steady state that for one representative value (T_{s_0}) of the set temperature, the system temperature is equal to the set temperature when T_a is equal to its mean value T_{a_0}. Under these conditions P_0 must be chosen to be equal to

$$P_0 = \lambda(T_{s_0} - T_{a_0}). \tag{6-113}$$

Let us, furthermore, express the set temperature T_s as

$$T_s = T_{s_0} + \Delta T_s \tag{6-114}$$

and we also express the actual ambient temperature as

$$T_a = T_{a_0} + \Delta T_a. \tag{6-115}$$

Using (6-113), (6-114), and (6-115), in (6-112) we find

$$\left(\frac{dT}{dt}\right) = \frac{K}{C}\Delta T_s + \frac{\lambda}{C}\Delta T_a - \frac{(\lambda + K)}{C}(T - T_{s_0}). \tag{6-116}$$

If we now recall that the time constant for the uncontrolled response of the system is

$$t_0 = \frac{C}{\lambda}, \tag{6-117}$$

and, further, if we define the dimensionless ration K/λ as

$$G = \left(\frac{K}{\lambda}\right) \tag{6-118}$$

we can express (6-116) as

$$\left(\frac{dT}{dt}\right) = \frac{G\Delta T_s}{t_0} + \frac{\Delta T_a}{t_0} - \frac{(1 + G)}{t_0}(T - T_{s_0}).$$

Or we may obtain a more recognizable form by rearranging this to obtain the expression

$$\left(\frac{dT}{dt}\right) = \{T_{f_1} - T\}\frac{(1 + G)}{t_0}, \tag{6-119}$$

where

$$T_{f_1} = T_s + \left(\frac{1}{1 + G}\right)(\Delta T_a - \Delta T_s). \tag{6-120}$$

If we define the time constant with feedback as

$$t_1 = \left(\frac{t_0}{1 + G}\right) = \frac{C/\lambda}{(1 + G)}, \tag{6-121}$$

we find, finally,

$$\left(\frac{dT}{dt}\right) = (T_{f_1} - T)\frac{1}{t_1}. \tag{6-122}$$

This equation is identical to (6-45) that describes the speed of the controlled steam engine, i.e.,

$$\frac{d\omega}{dt} = (\omega_{f_1} - \omega)\frac{1}{t_1}.$$

We observe at once that T_{f_1} is the final temperature of the controlled system. Also $t_1 = (C/\lambda)/(1 + G)$ is the time constant of the controlled temperature bath. The steam engine and the thermal bath have the same control equations!

b) Steady State Operating Temperatures of the Controlled Thermal Bath Under Changes in Set Temperature (Input) or Ambient Temperature (Load).
The final temperature of the thermal bath is T_{f_1}, for when $T = T_{f_1}$ in (6-122) the temperature no longer changes. T_{f_1} is related to ΔT_s and ΔT_a and T_s by (6-120), i.e.,

$$T_{f_1} = T_s + \frac{(\Delta T_a - \Delta T_s)}{(1 + G)}. \tag{6-123}$$

Let us examine in turn how the final system temperature responds to changes in the set temperature and the ambient temperature.

Suppose that we choose a set temperature T_s differing by the small amount ΔT_s from T_{s_0}, then the steady state final system temperature T_{f_1} will differ from the set temperature T_s by the small amount

$$(T_{f_1} - T_s) = -\frac{\Delta T_s}{(1 + G)}. \tag{6-124}$$

When the quantity G, which (as in the steam engine example) is called the open loop gain, is large compared to unity, then the final temperature of the system follows the set temperature very closely. In this way, we can use the controlled system to "sweep" the bath temperature by slowly sweeping the set point temperature. In the later discussion of the transient response of the system, we shall determine *how* slowly T_s must change so that the system temperature follows in accordance with (6-124).

If we define the quantity ΔT_{f_1} as

$$\Delta T_{f_1} \equiv (T_{f_1} - T_{s_0}).$$

Then since

$$\Delta T_s = (T_s - T_{s_0})$$

we may reexpress (6-124) in the form

$$\left(\frac{\Delta T_{f_1}}{\Delta T_s}\right) = \left(\frac{G}{1+G}\right). \tag{6-125}$$

The quantity $(G/(1+G))$ is called the "closed loop gain" of the system; it is equal to the ratio of the change in the input variable to the change in output variable when the control loop is closed. There is no amplification in this system when G is large, i.e., $\Delta T_{f_1} \approx \Delta T_s$, and the output change is nearly equal to the input change.

A great virtue of the closed loop feedback control system is that it reduces markedly the effect of changes in the ambient temperature. In the case of the uncontrolled heating system, if the ambient temperature changed by ΔT_a, the final temperature changed by the same amount. In the present controlled system, the difference between the final temperature and the set temperature (set at T_{s_0}), produced by a change ΔT_a in ambient temperature, is found from (6-94) to be

$$T_{f_1} - T_{s_0} = \frac{\Delta T_a}{(1+G)}. \tag{6-126}$$

We observe that in the proportional controller the change in ambient temperature produces an offset of the system temperature to a point $T_{s_0} + \Delta T_a/(1+G)$. The magnitude of this offset is smaller by a factor $1/(1+G)$ than that which occurs in the uncontrolled system.

We can understand physically the origin of this offset $\Delta T_a/(1+G)$ as follows. When the ambient temperature changes by an amount $-\Delta T_a$, the additional rate at which heat is lost is $\lambda \Delta T_a$. In the uncontrolled system this produces a change in system temperature of an amount ΔT_a. In the controlled system the increased power loss is made up primarily by the feedback and amplifier system as the system temperature changes from its initial value of T_{s_0}. To be exact, if the final temperature of the controlled system is $T_{s_0} - (\Delta T)_{\text{offset}}$, then the net additional heat input associated with this change is (see (6-112)):

$$K(\Delta T)_{\text{offset}} + \lambda(\Delta T)_{\text{offset}}.$$

The first term is provided by the feedback to the power amplifier, and the second term is the *reduction* in heat loss associated with the drop-in system temperature. This net additional heat input is equal to the loss $\lambda(\Delta T_a)$ produced by the change in ambient temperature. This determines $(\Delta T)_{\text{offset}}$ according to

$$(K + \lambda)\Delta T_{\text{offset}} = \lambda(\Delta T_a)$$

or

$$(\Delta T)_{\text{offset}} = \left(\frac{\lambda}{K + \lambda}\right)\Delta T_a = \left(\frac{1}{1 + G}\right)\Delta T_a.$$

Thus we see that the offset can be very small in the proportional control system because the effect of the small difference from the set point produces a big change in the correcting power of the heater.

In fact, one can show, just as in the case of the steam engine, that if we open the feedback loop at the point A in Figure 6.23 and insert a temperature change ΔT into the negative terminal of the power amplifier, the resultant change in system temperature will be $(K/\lambda)\Delta T$. Thus $G = (K/\lambda)$ is called the open loop gain.

We see then that the proportional control system can regulate the temperature quite close to the set temperature, but it always needs finite offset to control against changes in the ambient temperature.

c) Transient Response of the Proportionally Controlled Thermal Bath: Delay and Instability.   Equation (6-122) shows that if the final temperature T_{f_1} of the system is changed, by changing either the ambient temperature or the set point temperature suddenly, the system temperature will relax to the new final temperature exponentially with a characteristic time

$$t_1 = \left(\frac{t_0}{1 + G}\right),$$

which is a factor of $1/(1 + G)$ shorter than that for the uncontrolled system. The feedback can markedly decrease the response time of the system.

We may now examine how slowly the set temperature should be varied to permit the system temperature to follow it. Since the characteristic time for the response of the system is t_1, any finite desired change in temperature (ΔT_s) should be carried out in a long time interval compared to t_1, if we want the system temperature T to follow the set temperature closely. Alternatively, the rate $R = (dT_s/dt)$, at which the set temperature is swept, must be small enough so that $R \ll (\Delta T_s/t_1)$.

In considering the transient response of the system, we may also consider briefly the effect of a time lag in the response of the temperature sensor and heater. Any real system will take some finite time t_g to respond to changes in the temperature. As the gain in the feedback loop is increased, the response time of the controlled system $t_1 = t_0/(1 + G)$ can become comparable to the delay time t_g. When this occurs, the control system can become unstable. We can see the reason for this: Heat is put into the system at the rate $(dQ/dt) \sim K(\Delta T_a/(1 + G))$ by the power amplifiers when the ambient temperature changes by ΔT_a. We gave the reason for this in Section 6.3.B. If this continues for a time t_1, enough heat is put in to change the system temperature by the required offset amount $\Delta T_a/(1 + G)$. If, however, because of delay, the heat input is not turned off at this time, more heat is put in than is needed by a factor of $(t_1 + t_g)/t_1$. When $t_1 \sim t_g$, as much heat is put in during the delay time as was put in during the response time, and the system temperature overshoots the stable offset temperature difference by an amount approximately equal to the offset itself. At this point the control clearly fails.

We can obtain the equation for the thermal response of the system, including delay, by altering the input power equation to the form

$$\left(\frac{dQ}{dt}\right)_{\text{in}} = P_0 + K(T_s - T(t - t_g)).$$

If this is put into the equation

$$C\left(\frac{dT}{dt}\right) = \left(\frac{dQ}{dt}\right)_{\text{in}} - \left(\frac{dQ}{dt}\right)_{\text{out}}$$

we will obtain an equation identical in form to that which we obtained in the case of the steam engine, i.e., (6-60) or (6-61). Our analysis of the instabilities in that equation will apply exactly to the present case, and the temperature will change with time precisely as did the frequency of the steam engine as a function of the parameter $\alpha = (t_g/t_1)$. The onset of instability occurs when $\alpha = \pi/2$.

(iii)   Integral Control System

In the proportional control system there is always some offset from the set temperature if the ambient temperature changes. It is possible to eliminate this offset by arranging a different sort of heater response than that of proportional control. If the heater power is proportional, not to the temperature difference $T - T_s$, but to its time integral, we obtain a control system with very different properties from that of a proportional control system. Let us assume then that the heater power is controlled by the feedback in accordance with the formula

$$\left(\frac{dQ}{dt}\right)_{in} = P(t) = P_0 + K_I \int_0^t (T_s - T(t'))\, dt'. \tag{6-127}$$

Here t is the present time, t' is a dummy integration variable, and K_I measures the "strength" of the integral feedback control. K_I has the units of power/°K s. We see that the power input from the heater will tend to come to a steady value when the system temperature becomes exactly equal to the set temperature. If we now use (6-127) in our basic equation for the thermal bath, namely,

$$C\left(\frac{dT}{dt}\right) = \left(\frac{dQ}{dt}\right)_{in} - \left(\frac{dQ}{dt}\right)_{out},$$

we find

$$C\left(\frac{dT}{dt}\right) = P_0 + K_I \int_0^t (T_s - T(t'))\, dt' - \lambda(T - T_a). \tag{6-128}$$

This can be written in the form of a differential equation by differentiating both sides with respect to the time t. This gives

$$\frac{d^2T}{dt^2} + \left(\frac{\lambda}{C}\right)\left(\frac{dT}{dt}\right) - \frac{K_I}{C}(T_s - T) = 0. \tag{6-129}$$

Let us now define the temperature difference ΔT as

$$\Delta T = (T - T_s). \tag{6-130}$$

Using this in (6-129), we see that any temperature difference between the actual system temperature and the set temperature varies in time in accordance with the second-order differential equation

$$\frac{d^2}{dt^2}(\Delta T) + \frac{1}{t_0}\frac{d\Delta T}{dt} + \left(\frac{K_I}{C}\right)(\Delta T) = 0. \tag{6-131}$$

Here we used the definition

$$\frac{1}{t_0} = \frac{\lambda}{C}, \tag{6-132}$$

where t_0 is the response time of the uncontrolled system.

On examining (6-131), we observe that it has precisely the same form as that of the damped harmonic oscillator (discussed in Chapter 2). We see, once again, that widely different systems can, and often do, have identical equations to describe them. The damped harmonic oscillator equation occurs again and again in many fields. A knowledge of the characteristics of the motion of such an oscillator is useful far beyond the simple mechanical system where we first met it.

If we define the quantity K_I/C as

$$\frac{K_I}{C} \equiv \omega_0^2 \tag{6-133}$$

we may write (6-131) in the form

$$\frac{d^2}{dt^2}(\Delta T) + \frac{1}{t_0}\frac{d}{dt}(\Delta T) + \omega_0^2(\Delta T) = 0. \tag{6-134}$$

From our previous discussion of this equation in Chapter 2, Section 2.7, we may recall that the form of its solution depends in detail on whether $2\omega_0 t_0 > 1$ or $2\omega_0 t_0 < 1$. In the former case, the solution consists of an oscillation with frequency $\omega = \omega_0\sqrt{1 - (1/2\omega_0 t_0)^2}$ whose amplitude decays to zero exponentially as $e^{-t/2t_0}$. Thus, even though the system temperature finally becomes equal to the set temperature, it does not do so smoothly, but oscillates around the set temperature with decreasing amplitude.

Notice that this oscillation occurs even though the temperature sensor has *no time delay*. The integral control system shows this oscillation even in the absence of a delay because its response at the present time depends (through the integral in (6-98)) on the temperatures at earlier times. In essence, the integral controller has a "memory" of what happened at earlier times even though there is no delay in its sensor.

On the other hand, the system will control smoothly, without final offset, provided that $2\omega_0 t_0 < 1$. In the case of critical damping $2\omega_0 t_0 = 1$, the temperature relaxes *smoothly* from ΔT_0 to zero as $e^{-t/2t_0}$. The delay time is $2t_0$, about twice the characteristic time of the uncontrolled system. Thus, by choosing the feedback strength K_I in such a way that

$$\frac{K_I}{C}\left(\frac{2C}{\lambda}\right)^2 \equiv \omega_0^2 (2t_0)^2 = 1, \tag{6-135}$$

the integral control system will bring temperature deviations from T_s down to zero in a characteristic time of $\sim 2t_0$.

We then see that the integral controller eliminates the offset completely. It is said to have an infinite open loop gain. However, it does this at a price. The price is a very long response time: a response time twice as long as that of the uncontrolled system. The integral controller can oscillate when K_I is too large, and it can relax back to zero, even more slowly than $2t_0$ when K_I is smaller than that given in (6-135).

In actual practice a better control system than this one can be achieved by using a controller (heater) whose output depends on both the error $T_s - T$, and also on the integral of the error. The proportional part of the response is rapid, and the integral part of the response arranges the system temperature to become exactly equal to the set temperature.

6.4   Control of Body Temperature

6.4.A.   Body Temperature in Illness and in Health

A rise in body temperature is perhaps the most familiar herald of infectious disease, such as pneumonia, typhoid, or streptococcal infection.

Exposure to extremes of heat or cold can have very serious and even fatal effects on healthy persons. Heat exhaustion, brought on by exposure to high temperatures, is produced by marked dilatation of the peripheral blood vessels. If the subject engages in vigorous exercise the blood flow to the brain may be reduced to the point that the person may collapse and lose consciousness. Hypothermia, a lowering of body temperature that can occur by immersion in cold water, or by hiking in wet clothes on a cold rainy day, can be fatal.

The central body temperature of mass is normally at $37 \pm 2\,°C$ ($98.6 \pm 3.6\,°F$). If the core temperature rises to $41\,°C$ ($106\,°F$), the central nervous system begins to deteriorate and convulsions occur. A further rise in temperature to $45\,°C$ produces protein denaturation and death. On the other hand, if the central core temperature of the body falls below $33\,°C$ ($91\,°F$) nervous functions are depressed to the point that consciousness is lost. Below $30\,°C$ ($86\,°F$) the temperature regulating system fails and at $28\,°C$ ($82\,°F$), cardiac fibrillation and death can occur. These figures, combined with the recognition that man does in fact live and work in the outdoors in temperatures ranging from $40\,°C$ ($104\,°F$) to $-30\,°C$ ($-22\,°F$), show that the combination of proper clothing and the thermoregulatory activity of the body is capable of accurately maintaining body temperature within a quite narrow range, despite dramatic alterations in both ambient temperature and work output of the body.

6.4.B.   Thermal Properties of the Human Body

We can begin to understand the thermoregulatory system by recognizing the major heat flows into and out of the body. The net or total heating rate $(dQ/dt)_{tot}$ of the body can be thought of as consisting of three parts: The heat production rate associated with the average metabolic activity plus physical activity $(dQ/dt)_{met}$, the heat production (or removal) associated with the control system of the body $(dQ/dt)_{control}$ and, finally, the heat loss associated with passive modes of heat loss from the skin; convection and radiation $(dQ/dt)_{out}$. We may express the heat balance as

$$\left(\frac{dQ}{dt}\right)_{tot} = \left(\frac{dQ}{dt}\right)_{met} + \left(\frac{dQ}{dt}\right)_{control} - \left(\frac{dQ}{dt}\right)_{out}. \qquad (6\text{-}136)$$

We will now discuss each of these terms.

If for the moment we regard the entire body as being at a single temperature, i.e., that of the body core, then the rate of change of this temperature is related to (dQ/dt) by

$$\left(\frac{dQ}{dt}\right)_{tot} = C\frac{dT}{dt}, \qquad (6\text{-}137)$$

where C is the heat capacity of the body. The *specific* heat (heat capacity/unit mass) is $0.83\ kcal/°C\ kg$. We will assume that henceforth we deal with an average subject of $70\ kg$ mass. Thus, for this mass subject, we have $C = 58\ kcal/°C$ or

$$\left(\frac{dQ}{dt}\right)_{\text{total}} = 58 \text{ kcal/°C} \left(\frac{dT}{dt}\right). \qquad (6\text{-}138)$$

Consider next the term $(dQ/dt)_{\text{met}}$. We saw in Chapter 5 (Table 5.4) that the basal metabolic rate for a man of this size is

$$\left(\frac{dQ}{dt}\right)_{\text{basal}} = 70 \text{ kcal/h} \equiv P_0. \qquad (6\text{-}139)$$

Defining this basal rate as P_0, we can express the heat input associated with physical activity in terms of the basal rate as

$$\left(\frac{dQ}{dt}\right)_{\text{met}} = m P_0. \qquad (6\text{-}140)$$

In heavy activity $(dQ/dt)_{\text{met}}$ can be as large as 20 times the basal rate. Thus, in general, m is a pure number ranging from unity to 20.

If there were no heat flow out of the body, and no operation of the thermocontrol system, then the body temperature would change in accordance with the equation

$$C \left(\frac{dT}{dt}\right) = m P_0 \qquad (6\text{-}141)$$

or, using (6-139) and (6-138), we have

$$\left(\frac{dT}{dt}\right) = m 70 \text{ kcal/h} \times \frac{1}{58} \text{°C/kcal}$$

or

$$\left(\frac{dT}{dt}\right) = 1.2 m \text{°C/hr}.$$

Since m can be as large as 20, we see that without temperature control or heat loss the body temperature could rise as rapidly as 24 °C/h in heavy exercise.

Having seen the necessity of heat loss from the body, let us consider next the magnitude of the term $(dQ/dt)_{\text{out}}$. We include in this term only passive sources of heat loss from the body, such as convection and radiation. The heat loss associated with sweating (sudmotor activity) is produced by the control system.

The loss of heat from the skin by convection is proportional to $(T_{skin} - T_a)$, where T_a is the ambient temperature and T_{skin} is the skin temperature. We may write

$$\left(\frac{dQ}{dt}\right)_{\text{out convection}} = \lambda_1 (T_{skin} - T_a), \qquad (6\text{-}142)$$

λ_1 is a marked function of wind speed [12], and the clothing of the body. For a nude body of area $1.5m^2$, exposed to a wind speed of ~ 9 cm/s, the value of λ_1 is

$$\lambda_1 \approx 13 \text{ kcal/h}\,^\circ\text{C}. \qquad (6\text{-}143)$$

The body also loses heat by radiation. Analysis of this [12] shows that this source of heat loss has about the same magnitude as that from convection, i.e.,

$$\left(\frac{dQ}{dt}\right)_{\text{out radiation}} \cong 12 \text{ kcal/h}\,^\circ\text{C}(T_{skin} - T_a). \qquad (6\text{-}144)$$

Thus we can express the passive sources of heat loss as

$$\left(\frac{dQ}{dt}\right)_{\text{out}} = \lambda (T_{skin} - T_a), \qquad (6\text{-}145)$$

where $\lambda \approx 25$ kcal/h $^\circ$C for a nude subject in a light breeze. Clearly the quantity λ will vary widely depending upon the amount and type of clothing worn and upon wind speed. For a nude person λ can increase by a factor of 5 if the wind speed increases to 5 m/s($\sim$ 12 mph).

It is interesting to consider the properties of the human body with heat input from metabolism and heat output by convention and radiation, but without control. The discussion of Section 6.3.A, the open loop heating system, applies in the present case. We may estimate the characteristic time constant

$$t_0 = \frac{C}{\lambda}$$

first. Using $C = 58$ kcal/$^\circ$C and $\lambda = 25$ kcal/h for the nude subject above, we find

$$t_0 = \frac{58}{25} \sim 2.3 \text{ h} \qquad (6\text{-}146)$$

as the constant for the uncontrolled human body under these circumstances.

We can also examine how λ must change with metabolic activity if the body were to maintain its correct temperature even during physical exertion. This is found as follows. Let us first assume again that $T_{\text{skin}} = T$, the temperature of the core of the body, and that this is to be maintained at $T_{\text{set}} = 37\,°C$. Then, in the steady state we have that

$$\left(\frac{dQ}{dt}\right)_{\text{in}} = \left(\frac{dQ}{dt}\right)_{\text{out}} \tag{6-147}$$

or

$$m\,P_0 = \lambda(T_{\text{set}} - T_{\text{a}}). \tag{6-148}$$

Thus for the person to be comfortable, that is, not to require either heat production or heat removal by the thermoregulatory system, we have the following relation between the degree of metabolic activity m and the difference between ambient and body temperature

$$m = \left(\frac{\lambda}{P_0}\right)(T_{\text{set}} - T_{\text{a}}). \tag{6-149}$$

Choosing the conditions described above for a nude subject, we have

$$\left(\frac{\lambda}{P_0}\right) = 25 \text{ kcal/h°C h/70 kcal} = 1/2.8\,°C,$$

i.e.,

$$m = \left(\frac{T_{\text{set}} - T_{\text{a}}}{2.8\,°C}\right).$$

Thus, for example, if one is to be comfortable without clothes in an ambient temperature of $21\,°C$ ($70\,°F$), m must be

$$m = \left(\frac{37 - 21}{2.8}\right) \simeq 6.$$

Thus, he must be working at a rate about six times greater than the basal metabolic rate. He must be engaged in "moderate" physical activity (see Table 5.4). Equation (6-124) shows clearly that if one is to maintain body temperature without the use of the control system of the body, the quantities m and λ must be adjusted always so that

$$(T_{\text{set}} - T_{\text{a}}) = \frac{m P_0}{\lambda}.$$

When it is not possible to meet this requirement by changing m, the level of metabolic activity, or λ, the loss rate from the body, through the use of clothes, one must then draw upon the thermoregulatory control heat sources or sinks: $(dQ/dt)_{\text{control}}$. Also, to decrease the long time constant for the response of the body, it is necessary to have the aid of the $(dQ/dt)_{\text{control}}$ term.

6.4.C. The Control Elements for the Regulation of Body Temperature

Much progress has been made in identifying the detectors and effectors that control body temperature [13], and in measuring the heat output or cooling rate that these effectors can produce [13]. Detailed analyses have been made of the quantitative operation of the control loops. [14], [15].

In Figure 6.24 we present a block diagram that illustrates in a rough way the primary elements in the control system. In the first approximation one may look upon the body as being characterized by two temperatures: a core temperature T_{c} and skin temperature T_{skin}. Neurons in the brain and deep within the body detect

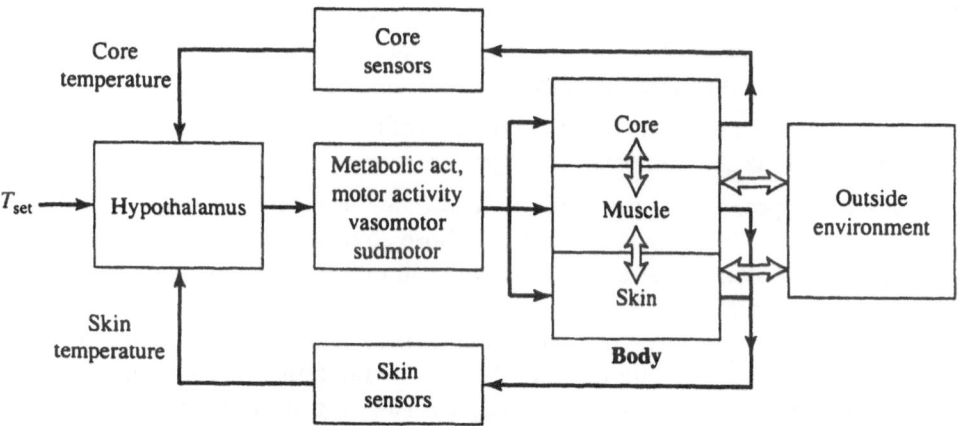

Figure 6.24. Elements in the human temperature control system.

the core temperature and transmit this information, in the form of nerve firing rates, to a control center in the hypothalamus. Here the core temperatures are compared with a set temperature T_s, which in the healthy person is about 37.1 °C. Experimentally, the temperature of the eardrum (tympanic membrane) can be taken as equal to the core temperature.

The skin temperature is sensed by cutaneous nerve endings of two types. One type detects cold, the other heat, and both can also respond in proportion to the rate of change of skin temperature. Signals are also set to the hypothalamus from these skin receptors.

The response of the body to changes in temperature depends in detail upon where these changes occur. The core temperature is under very stringent control, while the skin temperature is allowed to vary over a rather wide range. The body has the following mechanisms to provide heat to the core and to the skin:

(a) Increased metabolic activity and motor activity (shivering). This heat goes into the core and the muscles.

(b) Increased conduction of heat between the central core and the periphery, by vasomotor control of the diameter of blood vessels. Heat is conducted between parts of the body, mainly through the action of the flow of blood. (In the absence of blood flow the thermal conductivity of tissue is very low, being comparable to that of cork.) This mechanism can raise the skin temperature.

The body has the following mechanisms to cool itself:

(a) Sweating (sudmotor activity). Here heat is lost by the evaporation of sweat. The eccrine glands can be stimulated to sweat up to 1 l/h by nerve endings that secrete acetylcholine. Since the latent heat of evaporation of water is about 480 kcal/l, sweating can serve to remove heat at the rate of $\sim$ 580 kcal/h under favorable conditions of humidity, exposure, and air circulation.

(b) Vasomotor control. Here again conduction of the blood can be used to cool the *core* when the skin temperature is low.

It is not appropriate at this point to investigate in detail the control equations for heat flow into the core and the skin, whose solutions determine the response of the core and skin temperatures to changes in either metabolic activity $m P_0$ or ambient temperature.

We do wish, however, to demonstrate briefly the operation of the proportional control system used in limiting the *rise* in core temperature. Benzinger has shown [13] that the cooling of the body by sweating is produced as shown in Figure 6.25. Below a core temperature of 36.85° there is no sweating. Above 36.85 °C heat can be removed by sweating at a rate proportional to the differ-

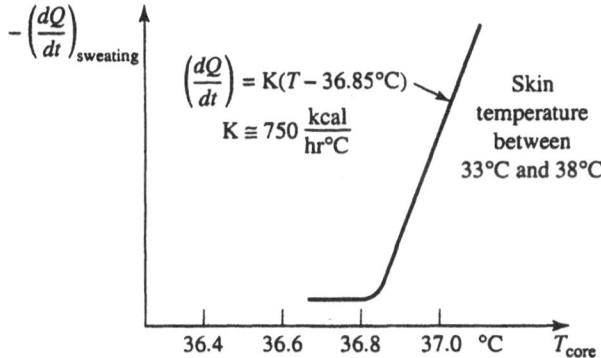

Figure 6.25. Rate of heat loss due to sweating as a function of core temperature.

ence ($T_{core} - 36.85\,°C$). This proportional control rate is independent of the skin temperature provided that the latter is in the range between $33\,°C$ and $38\,°C$. Using this proportional control device, let us examine how the system controls. We will make the crude approximation that as far as passive heat loss is concerned $T_{skin} - T_{ambient} = (T - T_{ambient})$, where T is the core temperature. Under these conditions we have the following heat balance equation

$$C\left(\frac{dT}{dt}\right) = P_0 + \left(\frac{dQ}{dt}\right)_{control} - \lambda(T - T_a). \tag{6-150}$$

Using

$$-\left(\frac{dQ}{dt}\right)_{control} = K(T - 36.85\,°C), \tag{6-151}$$

where

$$K \cong 750 \text{ kcal/h}\,°C, \tag{6-152}$$

we can compute the time constant for the controlled system and the open loop gain G. From our previous analysis in Section 6.3.A(ii) of the proportional control system, we see that the open loop gain for this core control system is

$$G = \frac{K}{\lambda} = 750 \text{ kcal/h} \,^{\circ}\text{C} \times \frac{1}{25}\text{h} \,^{\circ}\text{C/kcal} \cong 30, \qquad (6\text{-}153)$$

where we used the value of λ that applies to a nude figure in a light breeze.

The proportional control system that operates against a rise in core temperature thus has a gain of about 30. Changes in body temperature that would be produced in the uncontrolled system by changes in ambient temperature or physical activity are, in fact, reduced by a factor of about 30.

The response of the body under this control mechanism has a time constant t_1 that is smaller, by a factor $1/(1 + G) \sim 1/31$, than the time constant t_0 of the uncontrolled system. We estimated $t_0 \sim 2.3$ h previously. Thus the response time of such a system under proportional control will be $\sim 2.3/31$ h $= 4.5$ min.

We conclude this discussion of thermoregulation by briefly discussing fever. Fever results from a rise in the set point against which temperature is measured in the hypothalamus. Infections are associated with bacterial toxins called pyrogens that alter the set point in an as yet little understood manner. At the *onset* of fever the set point rises, but the body temperature is below the set temperature. The patient interprets this as feeling *cold*. The thermoregulatory centers constrict blood flow to the periphery to reduce heat loss. This results in skin that looks pale and feels "cold." The patient will cover himself with blankets. If these do not produce the required rise in temperature, the patient may start to shiver to produce greater internal heat. When the body temperature rises to the new set point, these symptoms disappear. After the infection subsides the set point returns to normal, and the patient feels too warm. He throws off the bed covers, and sweats until the body temperature falls to the normal set point.

Clinically, fever is regarded as being disadvantageous, and "antipyretic" drugs, such as aspirin, which block the action of the pyrogens, are used to reduce the set temperature to normal. Sometimes, in severe cases, physical methods are used to cool the body down.

6.5   Control of Blood Glucose Level

6.5.A.   Introduction: Glucose Tolerance Test

The carbohydrate, glucose, is a vitally important element in metabolism. The brain requires that the glucose concentration of the blood be maintained rather closely. A lack of an adequate glucose supply to the brain causes cellular damage, coma, and death within a few minutes. Glucose also provides energy for muscular contraction and glandular activity. The metabolism of glucose is important in the production of adenosine triphosphate (ATP) which, in turn, is vital for many metabolic processes in the cell.

To maintain the glucose concentration of the blood within close limits, despite variations in the intake of food and in metabolic activity, the body has developed an elaborate and effective control system. We can see some features of this control in the so-called glucose tolerance test. The normal concentration of glucose in blood is about 80 mg/100 ml, i.e., 100 ml of blood contains about 80 mg of glucose. In the glucose tolerance test the patient ingests 100 gm of glucose. The blood glucose concentration is monitored for several hours after this. The results of this monitoring is shown in Figure 6.26(a).

When insulin levels go up in the blood, the following consequences ensue:

(1) The insulin increase promotes the removal of glucose from the blood and its passage into certain cells, particularly muscle and fat tissue.

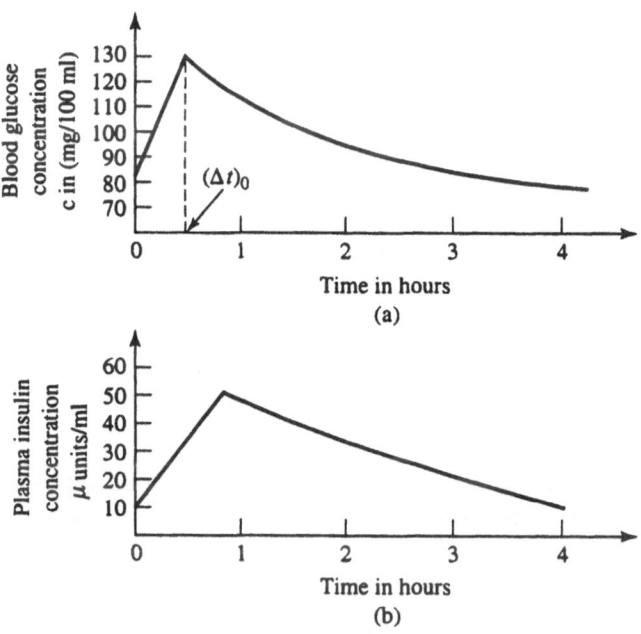

Figure 6.26. (a) Blood glucose concentration in a healthy subject versus time after the ingestion of 100 g of glucose. (b) Blood plasma insulin concentration in the same person versus time after ingestion of 100 g of glucose. (One micro unit of insulin is equal to $\frac{1}{24}$ ng.)

(2) Insulin stimulates the buildup of glycogen from glucose in the liver. Glycogen in essence is a polymeric form of glucose.

(3) Insulin inhibits the process of gluconeogenesis, i.e., the production of glucose in the liver from amino acids (proteins) and glycerol.

This sort of biochemical description of the action of insulin is really not very useful in our analysis of the control system because it does not specify the time-dependence of these processes. For our purposes, it is essential to know whether the processes above remove glucose from the blood at a rate proportional to glucose concentration, or at a rate proportional to the derivative or integral of the blood glucose concentration. As we develop the quantitative analysis of the blood glucose control system, we will describe the action of insulin and other hormones in terms of their contribution to control terms.

6.5.B.   Control Equation for Blood Glucose Concentration

Figure 6.26(b), showing the concomitant rise of insulin where glucose levels rise, suggests that proportional control is one element in the glucose control system. Also the long-term nature of the control system over periods longer than 8 h or so (not indicated in the figure) suggests the presence of slow integral control. Furthermore, the subsequent analysis of the glucose tolerance curve demonstrates that derivative control, as well, is present in the blood glucose control system. If Q is the total mass of glucose present in the blood, let us begin our analysis by assuming that the body's control system can supply glucose or remove glucose from the blood at the rate

$$\left(\frac{dQ}{dt}\right)_{\text{control system}} = K_D \frac{d(C_s - C)}{dt} + K(C_s - C) + R_{\text{ave}}. \qquad (6\text{-}154)$$

In this equation C is the actual concentration (in units of mass per unit volume) of glucose in the blood at time t; C_s is the set point against which changes in glucose concentration are compared. K_D is a proportionality constant for derivative control which has the units of volume. K is the constant of proportionality for proportional control, it has the units of volume per unit time. R_{ave} is the long-term average rate at which glucose is added or removed from the blood by the integral control elements.

The equation above contains a derivative control term. We have not previously described the effect of such a term in a control system. Its inclusion in the present case is essential.

Before proceeding further, let us descibe very briefly the physiological origin of each term in (6-129), keeping in mind that the organs or hormones acting are generally different depending on whether the concentration C is greater or less than the set point C_s.

The proportional control term represents the most rapid response. When the concentration exceeds C_s, this is sensed by the β islet cells of the pancreas, and insulin is released. The insulin rapidly facilitates the passage of glucose into muscle and fat cells, and the buildup of glycogen in the liver. When the glucose concentration C falls below C_s, the release of insulin does not occur. Instead glycogen stores in the liver are broken down rapidly into glucose and delivered to the blood. This process is called glycogenolysis. The liver can normally store about 100 g of glycogen for this short-time control function. Glycogen and glucose in muscle and other tissue can also be utilized to increase the blood glucose concentration.

Derivative control. When (dC/dt) is negative, i.e., when the glucose concentration is *decreasing* with time, this is sensed by a glucose receptor in the brain, probably in the hypothalamus. A signal is then sent via the sympathetic nervous system to the adrenal gland (the adrenal medulla) where epinephrine is released proportional to dC/dt. This epinephrine in turn produces glycolysis of the glycogen in the liver. When (dC/dt) is positive, we shall assume in our discussion that the coefficient K_D is the same as that when (dC/dt) is negative. The neuronal and hormonal elements associated with the $+dC/dt$ control are apparently not well established.

Finally, the slow $\sim$ 7 h integral control, when $C > C_s$ is associated with the process of fat storage—lipogenesis. On the other hand, when $C < C_s$, the process of gluconeogenesis, the production of glucose from proteins in the liver, is operative.

We now return to the quantitative features of the operation of the glucose control system.

Let $(dQ/dt)_{\text{total}}$ represent the total rate at which the amount (Q) of blood glucose is changing. Also let $(dQ/dt)_{\text{in}}$ be the rate at which blood glucose increases due to the ingestion of food, and let $(dQ/dt)_{\text{out}}$ be the consumption of glucose associated with the overall metabolism of the subject. The balance of flow of glucose into and out of the blood, including the effect of the control system, then leads to the equation

$$\left(\frac{dQ}{dt}\right)_{\text{total}} = \frac{d(CV)}{dt} = \left(\frac{dQ}{dt}\right)_{\text{in}} + \left(\frac{dQ}{dt}\right)_{\text{control}} - \left(\frac{dQ}{dt}\right)_{\text{out}}, \qquad (6\text{-}155)$$

where V is the total blood volume. Using (6-154), we find

$$\left(\frac{dC}{dt}\right) = \frac{1}{V}\left(\frac{dQ}{dt}\right)_{in} + \frac{K_D}{V}\frac{d}{dt}(C_s - C) + \frac{K}{V}(C_s - C) + \frac{1}{V}\left(R_{ave} - \left(\frac{dQ}{dt}\right)_{out}\right). \tag{6-156}$$

We wish to use this equation to explain the shape of the glucose tolerance test curves. Since the time scale over which this disturbance occurs is ~ 4 h, and this is short compared to the time scale of the integral control system (~ 7 h), we will assume that during the test, $(R_{ave} - (dQ/dt)_{out})$ remains equal to zero.

Second, we shall regard the dose ingested as being absorbed at some constant rate $(dQ/dt)_{in}$ until such time (Δt_0) that all the glucose ingested is absorbed into the blood, i.e., $(dQ/dt)_{in} = (\Delta Q/(\Delta t)_0)$. The time $(\Delta t)_0$ is presumed not to be affected by the glucose control system.

If ingestion occurs at $t = 0$, the equation that describes the glucose concentration as a function of time from $t = 0$ to $(\Delta t)_0$ is

$$\left(\frac{dC}{dt}\right) = \frac{1}{V}\left(\frac{dQ}{dt}\right)_{in} + \frac{K_D}{V}\frac{d}{dt}(C_s - C) + \frac{K}{V}(C_s - C), \tag{6-157}$$

$$0 < t < (\Delta t)_0,$$

while after $(\Delta t)_0$ the glucose concentration changes in accordance with

$$\left(\frac{dC}{dt}\right) = \frac{K_D}{V}\frac{d}{dt}(C_s - C) + \frac{K}{V}(C_s - C), \tag{6-158}$$

$$t > (\Delta t)_0.$$

Let us first solve (6-157) to see how the glucose concentration rises during absorption into the blood. For this purpose, it is convenient to define

$$C(t) = C_s + \delta C(t). \tag{6-159}$$

On placing this into (6-157) and rearranging a little, we find the equation

$$\frac{d}{dt}\delta C(t) = \left[\left(\frac{dQ}{dt}\right)_{in}\frac{1}{K} - \delta C(t)\right]\frac{K}{V} \times \frac{1}{(1 + K_D/V)}. \tag{6-160}$$

Now remember that the units of K are volume/time, and that the units of K_D are those of volume. We can therefore define, as the open loop gain associated with the derivative control system:

$$G_D = \left(\frac{K_D}{V}\right) \tag{6-161}$$

and we can also define, as the time constant of the proportional control system, the quantity

$$t_0 = \left(\frac{V}{K}\right). \tag{6-162}$$

We may therefore finally write the simple equation

$$\frac{d}{dt}\delta C(t) = \frac{[(\delta C)_f - \delta C(t)]}{t_0(1 + G_D)} \tag{6-163}$$

Here

$$(\delta C)_f \equiv \frac{(dQ/dt)_{in}}{K}. \tag{6-164}$$

The solution is

$$\delta C(t) = \delta C_f\left(1 - e^{-t/t_0(1+G)}\right). \tag{6-165}$$

At the time $(\Delta t)_0$ all the glucose has been absorbed. At that point

$$\delta C = \delta C_f\left(1 - e^{-(\Delta t)_0/t_0(1+G)}\right), \tag{6-166}$$

when $(\Delta t)_0 \ll t_0(1 + G)$, as appears to be the case, δC increases approximately linearly with time in accordance with the formula

$$\delta C(t) \cong \delta C_f\frac{t}{t_0(1 + G_D)}, \tag{6-167}$$

obtained from (6-166) on expanding the exponential and retaining only the term linear in t. When $t = (\Delta t)_0$, we can use the definitions for δC_f and t_0 to find

$$(\delta C)_{\max} = \delta C(\delta t_0) = \frac{\Delta Q}{(\Delta t)_0} \times \frac{1}{K} \frac{(\Delta t_0)}{(V/K)(1 + G_D)}, \tag{6-168}$$

i.e.,

$$(\delta C)_{\max} = \left(\frac{\Delta Q}{V}\right) \frac{1}{(1 + G_D)}. \tag{6-169}$$

We see that $(\Delta Q/V)$ is the level that the blood glucose would reach in the absence of any control whatever.

Following the time $(\Delta T)_0$ the concentration follows the (6-158) which is equivalent to (1-163) with δC_f set equal to zero. The solution for $\delta C(t)$ for $t > (\Delta T)_0$ is thus simply

$$\delta C(t) = (\delta C)_{\max} e^{(t - (\Delta t)_0)/t_0(1 + G_D)}. \tag{6-170}$$

In Figure 6.27 we plot this solution.

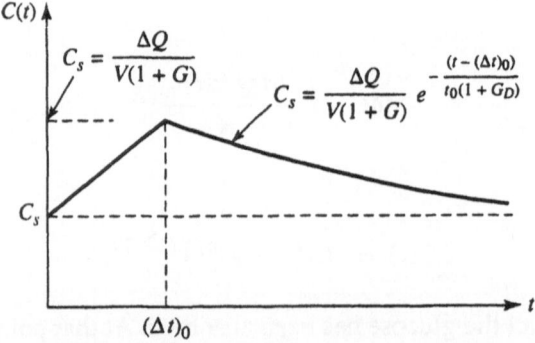

Figure 6.27. A graph representing the predicted shape of the glucose tolerance curves using the derivative plus proportional control.

6.5.C. Comparison Between Theory and the Experimentally Observed Glucose Tolerance Curve

We may use Figure 6.27 and the equations above to compare this model with the experimental data shown in Figure 6.26. Let us first observe that the total volume of the glucose compartment is equal to about 18 liters for a 70 kg man. This volume consists of both blood and extracellular fluid spaces in the body. Thus, on ingesting 100 gm of glucose, the maximum rise in glucose concentration (without the control loops) would be $(\Delta Q/V) = 100$ gm/18 liters $= 555$ mg/100 ml. The value of $(\Delta t)_0$ appears from Figure 6.26 to be about $\frac{1}{2}$ h. In this time all the glucose is absorbed. We see from the fact that the glucose concentration actually rises by only an amount

$$(\delta C)_{\text{max}} \equiv \frac{\Delta Q}{V}\left(\frac{1}{1 + G_D}\right) = 50 \text{ ml/100 ml}$$

that the gain G_D must be

$$1 + G_D = \frac{555}{50} = 11.3$$

or

$$G_D = 10.$$

Thus, the open loop gain G_D needed by the derivative control system, at least during the $+dC/dt$ phase, is about 10.

We now examine the decay period for the blood glucose, that is, the behavior for $t > (\Delta t_0)$. The characteristic time for this decay is about 2 h. Thus

$$t_0(1 + G_D) \simeq 120 \text{ min.}$$

This implies that the proportional control time constant t_0 is about 10 min or so. It is not unreasonable that the characteristic time for the release of insulin in proportional control is, in fact, of the order of 10 min.

Thus with the choices $G_D = 10$, $t_0 \sim 10$ min, this model of derivative and proportional control fits the experimental glucose tolerance test curves when the initial dose was $(\Delta Q/V) = 555$ mg/100 ml, and $(\Delta t)_0$ for that dose was 30 min.

It is easy to demonstrate that proportional control alone is incapable of explaining the shape of the glucose tolerance curve. In the case of proportional control, the equations for $\delta C(t)$ are (set $G_D = 0$):

$$\delta C(t) = \delta C_f\left(1 - e^{-t/t_0}\right), \qquad 0 < t < (\Delta t)_0,$$

and

$$\delta C(t) = \delta C_{\text{max}} e^{-(t-(\Delta t)_0)/t_0}, \qquad t > (\Delta t)_0.$$

Now to fit the tail of the glucose tolerance curve, we must choose $t_0 \simeq 120$ min. Since this is large compared to $(\Delta t)_0 = 30$ min, we find that

$$\delta C_{\max} \cong \delta C_f \left(\frac{(\Delta t)_0}{t_0} \right).$$

However, since $\delta C_f = (\Delta Q/(\Delta t)_0)/K$ and $t_0 = V/K$, it follows that

$$\delta C_{\max} = \left(\frac{\Delta Q}{V} \right).$$

But this maximum change in concentration is a factor of 10 larger than that observed experimentally. We, therefore, must conclude that proportional control is quite incapable of explaining the magnitude and shape of the glucose tolerance curves.

While the present model seems capable of quantitatively predicting the shape of the glucose tolerance curve, other evidence should be employed to test its correctness. Also, if the data on the shape of these tolerance curves were more precise, it could prove possible to determine whether such a coefficient as K_D is the same when (dC/dt) is positive or negative.

These considerations point the very great value of the *quantitative* analysis of physiologic feedback and control systems. This analysis provides a conceptual framework which quantitatively predicts response. It also stimulates specific questions as to the precise cellular, hormonal, and neural mechanisms that establish set points, detect differences from set points, and mediate corrections that reestablish the "necessary fixity of the internal environment."

6.6 References and Supplementary Reading

[1] *Automatic Control—A Scientific American* Book. Simon and Schuster, New York (1955).

[2] "Automatic Control—A State of the Art Report," by R. Oldenburger. *Mechanical Engineering*, pp. 39–45, (April 1965).

[3] *The Origins of Feedback Control* by O. Mayr. English translation MIT Press, Cambridge, MA (1970).

[4] "The Origins of Feedback Control," by O. Mayr. *Scientific American*, pp. 110–118 (October 1970).

[5] "On Governors," by J. C. Maxwell. *Proceedings of the Royal Society*, **16**, 270–283 (1868).

[6] *Selected Papers on Mathematical Trends in Control Theory*, edited by R. Bellman and R. Kalaba. Dover, New York (1964).

[7] *Die Regelung de Kraftmaschinen* by M. Tolle. Springer, Berlin (1921).

[8] "Feedback—The History of an Idea," by H. W. Bode. *Proceedings of the Symposium on Active Networks and Feedback Systems*. Polytechnic Press, Polytechnic Institute of Brooklyn (1960). (SEE ALSO PAPER #6, [6].)

[9] "Regeneration Theory," by H. Nyquist. *Bell System Technical Journal*, **11**, 126–147 (1932).

[10] *Cybernetics: or Control and Communication in the Animal and the Machine* by N. Wiener. MIT Press, Cambridge, MA (1948 and 1961).

[11] *The Wisdom of the Body* by W. B. Cannon. Norton, New York (1932).

[12] *Physiology and Biophysics*, edited by T. C. Ruch and H. D. Patton. W. B. Saunders, Philadelphia, 19th ed., pp. 1053 (1965).

[13] "Heat Regulation: Homeostasis of Central Temperature in Man," by T. H. Benzinger. *Physiological Reviews* **49**, 671 (1969).

[14] *Systems Physiology* by S. A. Talbot and U. Gessner. Wiley, New York (1973), Chap. 19.

[15] "Fine Control in the Human Temperature Regulation System," by R. W. Cornew, J. C. Houk, and L. Stark. *Journal of Theoretical Biology* **16**, 406 (1967).

6.7 Problems

1. On–off control of a heater. A bimetallic strip will bend, and its endpoint will be deflected by an amount proportional to the difference between the actual and set temperatures: $x = c(T - T_s)$. Use this strip to design a device that will turn the heater off if $T - T_s$ exceeds ΔT, and then *not turn it on again*, until T has first fallen below $T_s - \Delta T$. Also, equip your device with a means to adjust the value of T_s.

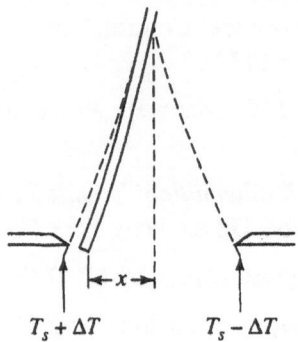

$$T_s + \Delta T \qquad\qquad T_s - \Delta T$$

2. Off–on control with a time lag. Suppose that in the off–on control system described in Section 6.3.A, there is a time lag δt between the time the "switch points" $\pm\Delta T$ are reached and the time the heater is actually shut off or turned on.

   (a) Show that in this case the amplitude of the temperature variation of the system is increased from $2\Delta T$ to a value $2\Delta T + 2(T_s - T_a)(\delta t/f t_0)$.

   (b) As one controls the temperature more closely, the on–off times get smaller and smaller. Show that the limits of temperature control are determined by the fact that the on–off times must be at least equal to δt.

   (c) Calculate the minimum temperature amplitude ΔT that can be achieved when a time lag δt is present, for $f = \frac{1}{2}$.

3. Effect of the daily outside temperature variation on the room temperature of a heated house. Let $(dQ/dt)_{in}$ be the *constant* heat input of the heating system of a house, C its heat capacity, and $\lambda(T - T_a)$ the heat loss to the outside. The room temperature T then satisfies the equation

$$\frac{dT}{dt} = \frac{1}{C}\left(\frac{dQ}{dt}\right)_{in} - \frac{\lambda}{C}(T - T_a).$$

Suppose that during a cold, sunny period, T_a has a daily fluctuation, approximately described by

$$T_a = \overline{T}_a + \Delta T_a \sin(\omega t), \qquad \omega = 2\pi/\text{day}.$$

   (a) Determine the value of $(dQ/dt)_{in}$ that will assure an *average* room temperature T_s.

(b) T will of course have a daily fluctuation too: $T = T_s + \delta T(t)$. Show that δT obeys the equation

$$\frac{d\delta T}{dt} + \frac{1}{t_0}\delta T = \frac{\Delta T_a}{t_0}\sin(\omega t), \qquad \frac{1}{t_0} = \frac{\lambda}{C}.$$

(c) This equation has a solution of the form

$$\delta T = a\sin(\omega t) - b\cos(\omega t).$$

Determine a and b.

(d) You get a more transparent expression for δT by writing

$$a = \sqrt{a^2 + b^2}\cos\phi_0,$$
$$b = \sqrt{a^2 + b^2}\sin\phi_0.$$

Show that δT may then be written as

$$\delta T(t) = \frac{\Delta T_a}{\sqrt{1 + (\omega t_0)^2}}\sin(\omega t - \phi_0).$$

Plot this curve alongside the curve $\Delta T_a \sin\omega t$. Give an intuitive interpretation of the reduction of the room temperature fluctuation amplitude for sufficiently large values of t_0.

(e) Show that the *phase shift* ϕ_0 is given by

$$\tan\phi_0 = \omega t_0.$$

(f) If the house is so well insulated that $t_0 = 12$ h, what is the amplitude of the room temperature fluctuations compared to ΔT_a?

(g) Assuming the maximum T_a occurs at 2 P.M., at what time will the rooms be warmest? ($t_0 = 12$ h.)

4. Suppose that in the case described in Problem 3, a proportional temperature control system is introduced, so that (6-112) applies, with $P_0 \equiv (dQ/dt)_{in}$.

   (a) Show that the oscillating part δT of the room temperature now satisfies the equation

$$\frac{d\delta T}{dt} + \frac{1}{t_1}\delta T = \frac{1}{t_0}\Delta T_a \sin(\omega T), \qquad \omega = 2\pi/\text{day},$$

where $t_1 = t_0/(1+G)$ and G is the open loop gain.

(b) By observing the similarity between the equation and that in Problem 3, part (b), show that δT is given by

$$\delta T = \frac{\Delta T_a}{1+G}\frac{1}{\sqrt{1+(\omega t_1)^2}}\sin(\omega t - \phi_1),$$

$$\tan\phi_1 = \omega t_1.$$

Discuss the effect of the gain on both the room temperature amplitude and on the phase shift ϕ_1.

5. **Effect of a change of load torque on a steam engine's angular speed.** Suppose that the load torque in a *controlled* steam engine is changed by a fixed small amount $\Delta\tau_\ell$, as illustrated in Figure 6.7. Show that the resultant change in the angular frequency of the flywheel is

$$\Delta\omega_1' = (\Delta\omega_0')/(1+G),$$

where G is the open loop gain, and $\Delta\omega_0'$ is the change of ω for the uncontrolled engine

$$\Delta\omega_0' = -\frac{\Delta\tau_\ell}{(d\tau_L/d\omega)_{\omega_0}}.$$

6. **Integral control.** The case of a rotating steam engine offers a natural example of integral control: Suppose we count the actual *number* of revolutions performed by the engine shaft from an initial time t_i to the present, t, and compare this with the number that would correspond to a steady angular velocity ω_s.

   (a) Satisfy yourself that the actual number of revolutions is

   $$n(t) = \frac{1}{2\pi}\int_{t_i}^{t} dt'\omega(t')$$

   and the set value is

   $$n_s(t) = \frac{1}{2\pi}\omega_s(t - t_i) = \frac{1}{2\pi}\int_{t_i}^{t} dt'\omega_s.$$

(b) The discrepancy $(n(t) - n_s(t))$ can be used as a control input, for instance, by generating a corrective torque

$$-(n(t) - n_s(T))\tau_c = -\frac{\tau_c}{2\pi} \int_{t_i}^{t} dt' (\omega(t') - \omega_s)$$

so that the equation for the angular velocity of the engine reads

$$I\frac{d\omega}{dt} = \tau_D - \tau_\ell(\omega) - \frac{\tau_c}{2\pi} \int_{t_i}^{t} dt' (\omega(t') - \omega_s).$$

Show that by taking the time derivative of this equation, you find a damped oscillator equation (provided τ_ℓ increases with ω), and that the steady state value of ω is ω_s, irrespective of load.

(c) Show that the characteristic damping time t_0 of the oscillation is

$$t_0 = \frac{2I}{(d\tau_\ell/d\omega)_{\omega_s}}.$$

(d) What should be the value of the control torque τ_c such that the oscillation produced by the control is critically damped?

Index

Continued from page ii

Earl W. Prohofsky, Department of Physics, Purdue University, West Lafayette, Indiana

Andrew Rubin, Department of Biophysics, Moscow State University, Moscow, Russia

Michael Seibert, National Renewable Energy Laboratory, Golden, Colorado

David D. Thomas, Department of Biochemistry, University of Minnesota Medical School, Minneapolis, Minnesota

Samuel J. Williamson, Department of Physics, New York University, New York, New York

PUBLISHED VOLUMES:

Russell K. Hobbie: Intermediate Physics for Medicine and Biology, *Third Edition*

George J. Hademenos and Tarik F. Massoud: The Physics of Cerebrovascular Diseases

M. Zamir: The Physics of Pulsatile Flow

George B. Benedek and Felix M.H. Villars: Physics with Illustrative Examples from Medicine and Biology, Mechanics, *Second Edition*

George B. Benedek and Felix M.H. Villars: Physics with Illustrative Examples from Medicine and Biology, Statistical Physics, *Second Edition*

George B. Benedek and Felix M.H. Villars: Physics with Illustrative Examples from Medicine and Biology, Electricity and Magnetism, *Second Edition*